Peptide pharmaceuticals

Peptide pharmaceuticals
Approaches to the design of novel drugs

Edited by David J. Ward

Open University Press
Milton Keynes

Open University Press
Celtic Court
22 Ballmoor
Buckingham
MK18 1XW

First Published 1991

British Library Cataloguing in Publication Data

Peptide pharmaceuticals: approaches to the design of novel
   drugs.
   1. Man. Antigens: Synthetic peptides
   I. Ward, David J.
   616.07'92

ISBN 0 335 09403 1

The processing and graphics for the colour images in this book was carried out on a Silicon Graphics Power Series graphics supercomputer and their help in the production of this book is gratefully acknowledged.

Typeset by Vision Typesetting, Manchester
Printed in Great Britain by Biddles Ltd, Guildford and King's Lynn

Contents

List of contributors   ix

Preface   xi

List of abbreviations   xiii

1 Introduction to peptide pharmaceuticals   1
D. J. Ward

1. Introduction   1
2. Peptide and protein structure, conformation, and dynamics   3
3. Drug discovery and development   8
4. Theoretical aspects of peptide drug design   10
5. Experimental aspects of peptide drug design   12
References   14

2 Conformational determination by NMR spectroscopy   18
H. Kessler and S. Steuernagel

1. Conformation in solution   18
2. Peptides and proteins – different molecules, different problems   20
3. Structural information from NMR spectra   22
4. Transformation of spectral information into three-dimensional
   structures   35
5. Conclusion   42
References   42

3 Crystallography in drug design   47
J. Murray-Rust

1. Introduction   47
2. Theory   50
3. Experimental methods   52
4. Results   61
5. Discussion   68
References   73

4 Theoretical approaches to peptide drug design   83
D. J. Ward, A. M. Brass, J. L i, E. Platt, Y. Chen and B. Robson

1. Introduction   83
2. Theory and methods   86
3. Development of descriptions for phase space   96
4. Development and testing of protocols for peptide modelling   105
5. Summary   127
References   129

5 Synthetic chemistry and the design of peptide-based drugs   135
V. J. Hruby, W. Kazmierski, A. M. Kawasaki, and T. O. Matsunaga

1. Introduction   135
2. Theory   139
3. Practical methodology   155
4. Results of approaches   160
5. Discussion – use of conformationally constrained ligands   162
References   177

6 Genetic approaches to peptide and polypeptide synthesis and design   185
J. Rosamond

1. Introduction   185
2. Isolation and expression of cloned DNA   188
3. Characterization of cloned DNA   196
4. Modifications of cloned DNA   200
5. Designs for the future   205
References   208

7 Physiological and pharmacological evaluation of peptide analogues   210
R. J. Knapp, T. P. Davis, T. F. Burks, and H. I. Yamamura

1. Introduction   210
2. Analysis of peptide-receptor interactions I: bioassay   212
3. Analysis of peptide-receptor interactions II: radioligand binding   217
4. The development of peptides for the μ and δ opioid receptors   228
5. Discussion   234
References   236

8 Expectations for peptide pharmaceuticals: a computational perspective 243
A. M. Brass, D. J. Ward, and J. Li

1. Introduction 243
2. The speeding of molecular dynamics structural refinement 244
3. The influences of hardware development 248
4. Drug–receptor interactions I: free-energy calculations 252
5. Drug–receptor interactions II: simulation procedures 254
References 256

Index 258

List of contributors

A. M. Brass, Department of Biochemistry and Molecular Biology, University of Manchester Medical School, Manchester M13 9PT, UK.

T. F. Burks, Department of Pharmacology, University of Arizona College of Medicine, Tucson, AZ 85724, USA.

Y. Chen, Department of Biochemistry and Molecular Biology, University of Manchester Medical School, Manchester M13 9PT, UK.

T. P. Davis, Department of Pharmacology, University of Arizona College of Medicine, Tucson, AZ 85724, USA.

V. J. Hruby, Department of Chemistry, University of Arizona, Tucson, AZ 85721, USA.

W. Kazmierski, Department of Chemistry, University of Arizona, Tucson, AZ 85721, USA.

A. M. Kawasaki, Department of Chemistry, University of Arizona, Tucson, AZ 85721, USA.

H. Kessler, Organisch-Chemisches Institut, Technische Universität München, Lichtenbergstr. 4, D-8046 Garching, FRG.

R. J. Knapp, Department of Pharmacology, University of Arizona College of Medicine, Tucson, AZ 85724, USA.

J. Li, Department of Biochemistry and Molecular Biology, University of Manchester Medical School, Manchester M13 9PT, UK.

T. O. Matsunaga, Department of Chemistry, University of Arizona, Tucson, AZ 85721, USA.

J. Murray-Rust, Department of Crystallography, Birkbeck College, University of London, Malet St, London WC1E 7HX, UK.

E. Platt, Department of Biochemistry and Molecular Biology, University of Manchester Medical School, Manchester M13 9PT, UK.

B. Robson, Department of Biochemistry and Molecular Biology, University of Manchester Medical School, Manchester M13 9PT, UK.

J. Rosamond, Department of Biochemistry and Molecular Biology, University of Manchester Medical School, Manchester M13 9PT, UK.

S. Steuernagel, Organisch-Chemisches Institut, Technische Universität München, Lichtenbergstr. 4, D-8046 Garching, FRG.

D. J. Ward, Department of Neuroscience, Institute of Psychiatry, University of London, Denmark Hill, London SE5 8AF.

H. I. Yamamura, Department of Pharmacology, University of Arizona College of Medicine, Tucson, AZ 85724, USA.

Preface

Peptide drug design is a multi-disciplinary persuit and this book has been organized and written so that people can gauge concurrent research in areas outside their own. No single book can simultaneously provide a complete introduction to a range of highly technical subjects, detail present and anticipated future areas of interest, and give some insight into the decision processes involved. However, that has been my aim.

Rather than describe them *de novo*, comparisons are drawn between peptide and protein structural analysis and between Nuclear Magnetic Resonance (NMR) spectroscopy and X-ray crystallography, in Chapters 2 and 3 respectively. There is increasing use of molecular dynamics simulation techniques for structural refinement, and the method is described in Chapter 4, together with other theoretical approaches and studies. Approaches to peptide synthesis, both chemical and genetic, are described in Chapters 5 and 6. Methods of testing, physiologically and pharmacologically, the peptides produced, are described in Chapter 7. Each chapter references major work and includes titles of papers. The structure of each chapter conforms, more or less, to the following style: Introduction, Theory/Methods, Practical Methodology, Studies and Results, Conclusion and Summary and most chapters describe studies on enkephalin, providing further continuity.

Computers are demonstrably playing an increasing role in all the subject areas described. Their application includes gene and protein sequence data bases, peptide and protein structural data bases, sequence analysis and molecular graphics, as well as for statistical analysis of data. Most of these are at least alluded to and referenced. Advances in the most computationally-intensive areas of computer modelling, for example quantum mechanics, are described in

Chapter 8, and provide a computational perspective of likely future advances in peptide drug design.

One take-home message from this book is that peptide drug design is not a linear process of structural analysis, synthesis and testing, but involves much feedback at each stage. Another is that the *rational* restriction, by synthetic means, of the conformational flexibility of a peptide generally makes it a better candidate for structural analysis, increases its stability to enzymes and hopefully leads to a selective drug.

This book does not include consideration of enzyme inhibitors as potential drugs; this is covered in *Design of Enzyme Inhibitors as Drugs* (Eds., M. Sandler and H.J. Smith) Oxford University Press (1989).

I am grateful to all the contributors to this book and all the staff at Open University Press for their considerable efforts.

List of abbreviations

ACE	angiotensin converting enzyme
APP	avian pancreatic polypeptide
BSA	bovine serum albumin
CAR	conformation–activity relationships
CNS	central nervous system
COSY	correlated spectroscopy
DBI	diazepam-binding inhibitor
GPI	guinea pig ileum
i.c.v.	intracerebroventricular
i.p.	intraperitoneal
i.t.	intrathecal
i.v.	intravenous
LHRH	luteinizing hormone-releasing hormone
MAD	multiple wavelength anomalous diffraction
MD	molecular dynamics
MIR	multiple isomorphous replacement
MIRAS	multiple isomorphous replacement and anomalous scattering
MPF	mitosis/maturation promoting factor
MVD	mouse vas deferens
NOE	nuclear Overhauser effects
PEG	polyethylene glycol
PTI	pancreatic trypsin inhibitor
QSAR	quantitative structure–activity relationship
RF	replicative form
s.c.	subcutaneous
SAR	structure–activity relationship
SAS	single anomalous scattering

SIR	single isomorphous replacement
SIRAS	single isomorphous replacement and anomalous scattering
SSR	sequence–structure relationships
TPA	tissue plasminogen activator
TRH	thyrotropin releasing hormone
TRNOE	transferred nuclear Overhauser effect

—— *Chapter 1* ————————————————————————

Introduction to peptide pharmaceuticals

D. J. Ward

1. Introduction

This book is about the development of pharmaceuticals based on the conformational behaviour and biological activity of the peptide hormones, neuropeptides, and proteins found in the body. It is only recently that peptides have been found in neurons with classical neurotransmitters,[1] but it has revealed that they can have a variety of biological actions.

Pharmaceuticals are drugs. The context in which the word 'drug' is frequently used in the media, with regard to *heroin* and *crack* for example, has lent it a frightening image, with overtones of illegality. The alternative choice by the media is often the 'wonder drug' reporting of preliminary laboratory results. The fact that many millions of people have substantially benefited from drug treatment, and some are kept alive by it, is often ignored or taken for granted.

The drug companies can be targets for public displeasure, especially when they are accused of charging a high cost for essential drugs at a time of extensive human tragedy, for example AIDS, and then reporting enormous profits. What is often not realized is that in terms of research and development costs it takes, on average, 10 years and £100 M to get a drug onto the market. For a 20-year patent, with perhaps half the time spent in development and clinical trials, there is not much time to recoup the outlay. *Generic* drugs, direct copies of the original by another company, can legally be produced once the patent has expired. However, some countries do not recognize the patent law, so they immediately copy the drug and sell it without suffering the huge development costs.

Another classification is the *orphan* drug, backed by legislation from 1983 to encourage development of drugs for 'minority' diseases, by which a manufacturer has exclusive rights for treatment of a particular disease for seven years.

There is no doubt that pharmaceuticals have a big effect on the financial markets. In the space of eight days in August 1989, two examples illustrated this. In the first, the share price of *Reckitt and Colman* rose, adding more than £100 M to the value of the company, when a paper was published in *Science* on a drug of theirs, marketed as a painkiller in the USA by *Proctor and Gamble*, which reduced cocaine addiction in monkeys.

The second case is even more dramatic. American research which showed that *Retrovir*, the anti-AIDS drug of *Wellcome*, can delay the onset of AIDS, added £1,300 M in one day, to the market value of *Wellcome*, and this rise contributed two-thirds of the *total* rise in the share index that day.

Another source of public disquiet is the perceived reluctance to admit when a drug already on the market is having serious side-effects. This is a particularly complex area, especially when the drug is being used on life-threatening illness, as there may be no way of knowing if the same result would have occurred without treatment. The problems outlined above are really questions of medical *ethics*, as is the treatment of animals in experimentation. In a book dealing with drugs, these issues have to be stated, partly to define the context in which the subject will be discussed.

Although drugs have been used in various forms for centuries, and drug design is not new,[2] it is only recently that peptides have become serious candidates as potential drugs.[3] We are now able to isolate specific genes, and mass-produce their polypeptide product,[4-6] for example insulin, but we have to bypass the digestive system and inject it directly into the blood. Chemical modification[7] other than replacement of one amino acid by another,[8] which is achieved in proteins by site-directed mutagenesis,[9] has to be 'manually' built in, so such techniques are restricted to peptides. Obviously, the chemist wants to synthesize the smallest possible sequence, and the systematic shortening of natural peptides has revealed that only parts of the molecule are required for particular effects.[10]

This book deals with development of ideas and approaches for the design of peptide-based drugs and a number of aspects of drug discovery are excluded, for example, drug delivery. Although there are immense problems in using peptides and proteins as drugs, a number are already available, for example insulin and growth hormone, supplied in their natural forms, as well as several luteinizing hormone-releasing hormone (LHRH) analogues.[11]

Peptides are rapidly broken down by enzymes, and for this reason the oral route is avoided for insulin and other unaltered forms of natural peptides. Particular parts of the molecule are recognized by the enzymes and these can be chemically modified to lengthen the half-life of activity.[12]

In summary, this book will examine ways of determining the three-dimensional shapes of peptides, in solution[13,14] and at the receptor.[15-17] It will describe the production of peptides, in modified[18] and unaltered forms for use as possible drugs, and ways of testing their biological activity.[19] In practice, drug discovery is not the simple linear process outlined above, and drug delivery,[20,21] drug targeting,[22,23] and clinical trials[24] are other procedures to be undertaken before the drug might be sold. Delivery and targeting are only considered by implication, in that our aims in this book include stabilizing the chemical structure,[12,18,25] thereby permitting oral intake,[26] and secondly, conformationally biasing in favour of the *active* form,[27] thereby attempting to prevent *nonspecific* interactions which is a form of *toxicity*. The progress towards *biopharmaceuticals*[28,29] has led to the formation of many new companies,[24] to exploit the increasing number of peptides being discovered with important physiological roles.[30]

2. Peptide and protein structure, conformation, and dynamics

Peptides and proteins are biological molecules whose 'building blocks' are *amino acids*. When amino acids are linked together they become *residues*. There is no defined point at which a peptide becomes a protein, and *polypeptide* can be used for the range of about 15 to 50 residues. Peptides and proteins are both composed of amino acids, so they have many features and properties in common, but also many differences. In this book, particularly in the chapters dealing with conformational analysis, some attention is given to comparative analysis of peptides and proteins, to provide insight into the properties of peptides, rather than have to describe them *de novo*.

Proteins appear in many roles being for example, receptors, enzymes, and structural components. By contrast, peptides can be hormones and neuropeptides, or substrates, and since these frequently interact with proteins, it has been proposed that they derive from receptor protein precursors,[31] and 'remember' how to interact and fold up. The mutual interaction, or 'folding', is part of the signal *transduction* process, which is the biochemical or physiological effect(s) of the binding.

A section of the amino acid sequence of a peptide or protein, containing the residues proline, alanine, glycine and tyrosine, could be represented as ... Pro–Ala–Gly–Tyr ... by giving the residue a three-letter code, or P–A–G–Y in the accepted one-letter representation[32,33] (see Table 1). The description can be expanded to the constituent atoms, as for alanine in Fig. 1, where the dots indicate the extent of the *residue* and correspond to the *peptide bond*. In the general case, the side-chain is denoted by R, and for alanine it is a methyl group

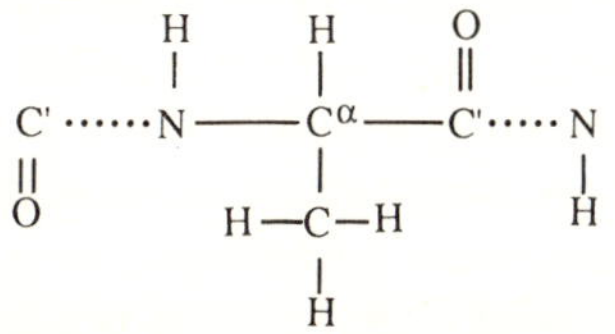

Figure 1 Structural formula for alanine. The dots indicate the extent of the residue and represent the peptide bond to residues before and after in the peptide/protein sequence.

Table 1 The naturally occurring amino acids, with their three-letter and one-letter codes

Amino acid	Three-letter code	One-letter code
Alanine	Ala	A
Isoleucine	Ile	I
Leucine	Leu	L
Phenylalanine	Phe	F
Proline	Pro	P
Methionine	Met	M
Tryptophan	Trp	W
Asparagine	Asn	N
Cysteine	Cys	C
Glutamine	Gln	Q
Glycine	Gly	G
Serine	Ser	S
Threonine	Thr	T
Tyrosine	Tyr	Y
Aspartate	Asp	D
Glutamate	Glu	E
Arginine	Arg	R
Histidine	His	H
Lysine	Lys	K

(CH_3). This is a two-dimensional description, and provides no information about how the atoms are arranged in space and how their three-dimensional disposition can vary, according to their environment. The environment might be in the hydrophobic core of a protein, adjacent to a lipid membrane,[34] or in aqueous solution, perhaps in the presence of ions.

The *chemical formula* can be defined as the atomic content, whereas the *structural formula* is the relative arrangement of these atoms in a molecule and can be described using two-dimensional descriptors. Configuration is specifically the relative arrangement of the atoms around a *chiral* centre. Chirality arises when a carbon atom has four different groups attached, as is the case for C^α atoms in amino acids (except glycine), and it is possible to have an optical *isomer*, in this case a mirror image. The different types are referred to as L and D amino acids, and all naturally occur in the L form (see Fig. 2). *Conformation* is the three-dimensional orientation of the atoms in the molecule. These terms are largely interchangeable, and 'three-dimensional structure' or sometimes just 'structure' is used to correspond to 'conformation'.

The conformation or three-dimensional (tertiary) structure (see Fig. 3 for a plot of the two conformers of deamino-oxytocin in the crystal[35] taken from the Cambridge Structural Database,[36]) is distinguished from the primary structure, which is the *sequence*, and the secondary structure which we will define according to the torsion (dihedral) angles of the backbone, see Fig. 4. Residue i, in this case alanine, is preceded by residue $i-1$, and followed by residue $i+1$.

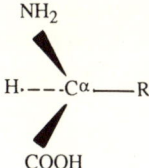

Figure 2 Definition of an L amino acid. COOH and NH_2 are out of the plane of the page, H is behind the plane, and R is in the plane.

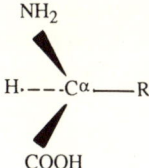

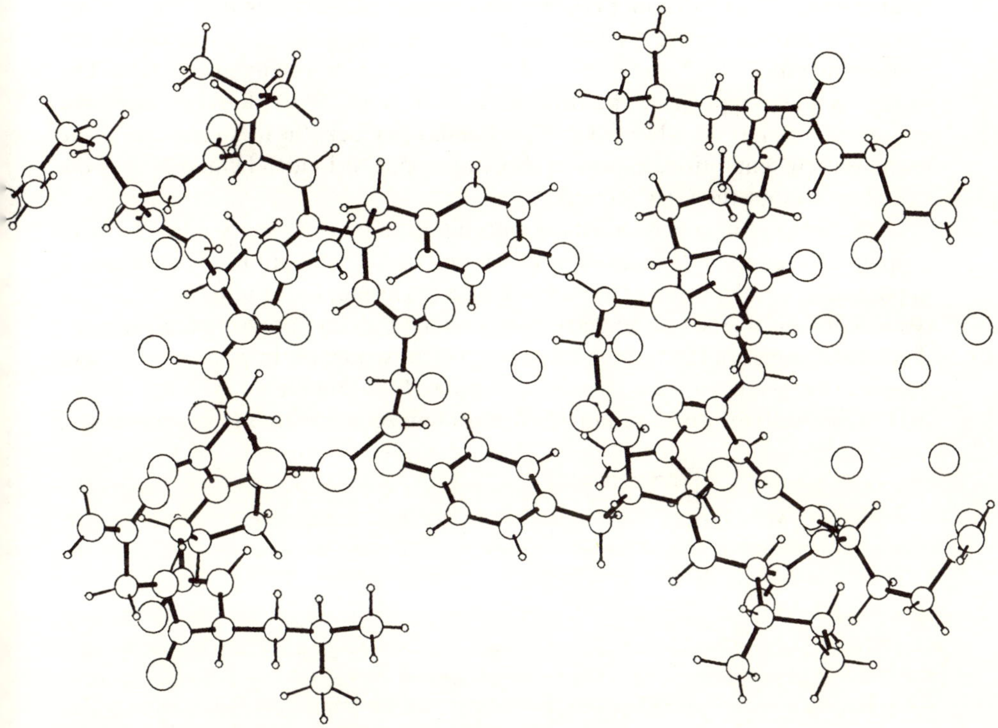

Figure 3 Crystal structure of deamino-oxytocin,[35] taken from the Cambridge Structural Database[36] and plotted using software supplied with the database.

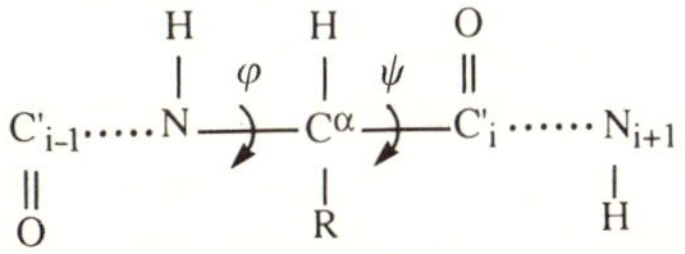

$$C'_{i-1} \cdots N \xrightarrow{\varphi} C\alpha \xrightarrow{\psi} C'_i \cdots N_{i+1}$$

Figure 4 As for Fig. 1, except that R is the generalized side chain, $i-1$ and $i+1$ represent the residues before and after, respectively, in the sequence. φ and ψ are the backbone torsion angles.

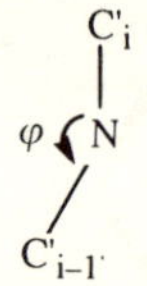

$$\varphi \big\langle N \overset{\displaystyle C'_i}{\underset{\displaystyle C'_{i-1}}{\big|}}$$

Figure 5 Definition of φ. Looking along the bond N–C$^\alpha$, in Fig. 4, C$^\alpha$ will be behind N. The angle made by the bonds C$^\alpha$–C$'_i$ and C$'_{i-1}$–N, considered as the 'hands of a clock', gives the angle φ.

Bond length is the distance between any two bonded atoms, and for the above example the N to C$^\alpha$ distance is generally around 1.47 Å (1 Å is 10^{-10} m) for any residue in any sequence. An example of a *valence angle* in the above case is the angle made at C$^\alpha$ by N–C$^\alpha$ and C$'_i$–C$^\alpha$ bonds, and is normally about 110°. The *torsion angle* φ represents the rotation around bond N–C$^\alpha$, looking along the bond and, taking C$'_{i-1}$–N and C$^\alpha$–C$'_i$ as hands of a clock; the torsion angle for the bond is their separation, in degrees. So, in Fig. 5, C is hidden behind N. ψ is the rotation around the C$^\alpha$–C$'_i$ bond.

A plot of φ against ψ is called a Ramachandran plot or *map* (see Fig. 6), with populated regions of the map illustrating preferred combinations of the angles. If a particular combination of angles is always preferred, for example $-55°, -45°$ (*helix*) or $-120°, 150°$ (extended), then we could always predict accurately the backbone angles of the residue in a new protein sequence. In practice, adjacent and nearby residues influence each other, but fully automatic computer programs have been written, which require only a protein sequence and give, on average,

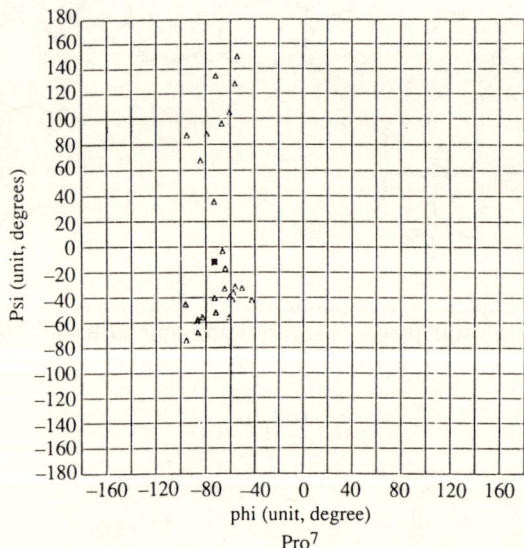

Pro^7

Figure 6 An example of a Ramachandran map, plotting φ against ψ for a range of stable and metastable conformers, here predictions for oxytocin, gives populated regions which may be characteristic for a residue, as for proline in this case.

(a)   (b)

Figure 7 Diagrams representing (a) *trans* and (b) *cis* configuration of the peptide bond. All the atoms are in the same plane.

about 65% agreement between predictions and secondary structure for proteins of known conformation.[37]

If we consider the backbone of a *dipeptide* unit, which is two amino acids linked together, the third angle after φ and ψ is ω, which is the rotation around the *peptide bond* and its value is defined by the four atoms C^{α}–C'–N–C^{α}. Whereas the N–C^{α} bond length is about 1.47 Å, the C'–N bond is about 1.32 Å. Electron *delocalization* in the CO–NH group favours a planar arrangement of the peptide bond, and in all naturally occurring amino acids, except proline, the *trans* form is very strongly favoured over the *cis* form, see Fig. 7 (a) and (b).

The sequence of a protein is determined by the arrangement of DNA bases in the gene. The message is converted from DNA to RNA in *transcription*, and the protein sequence is synthesized in the process of *translation*. *Post-translational* modification is the process by which residues like pyroglutamate and hydroxy-proline are formed, or sugars are added. Pyroglutamate (Glp,Z) shown in Fig. 8, must be at the N-terminus, because its amino group is *blocked*.

Although the conformation of proteins in crystals (see Fig. 9 for the C^{α}-plot of a structure of the enzyme lysozyme) appears to be very similar to that in solution, for peptides this is not the case. Although crystals are often highly solvated, one interpretation is that the higher ratio of internal to intermolecular interactions stabilizes the structure. An important factor is whether the structure of the molecule is characteristic of a protein *domain*, which has its own integral structure which cannot be subdivided without that structure falling apart, and the smallest domains in proteins are about 40–50 residues.

For example, avian pancreatic polypeptide (APP) has 36 residues, but the integral structure in solution is a dimer. Each monomeric part has a hydrophobic surface, and rather resembles a domain or small protein 'cut in half'. Structures smaller than APP tends to be a very flexible, but a *hairpin turn* structure tends to

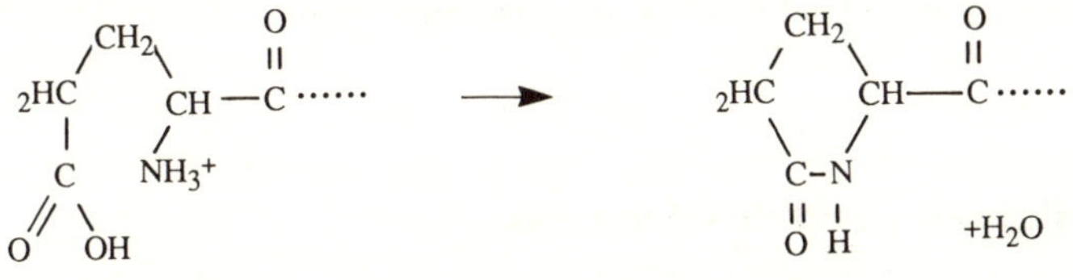

Figure 8 Formation of the residue *pyroglutamate* with ring closure giving a *cis* 'peptide bond'. This residue is found only at the N-terminus of peptides, because it is *blocked*.

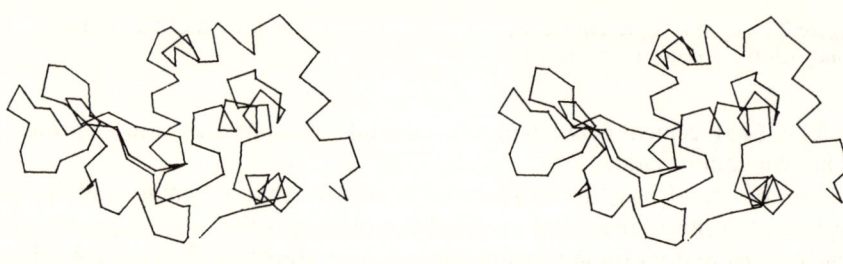

Figure 9 C^z stereoplot of the enzyme lysozyme.

be a dominant feature of the most abundant conformers in the population. Similar turn-like conformers are found above five residues, whereas at less than four residues, a turn is probably impossible by most criteria or definitions, and the structure tends to be roughly extended.

A problem of describing preferred conformations is that they might represent a minority of the conformations in solution, perhaps only 5%, with the remaining conformations being of a huge variety of types each of which is less than 5% abundance. Thus, whilst the crystal structure may be the favoured form in solution, the molecule might spend 95% of its time in a variety of other forms. Since only a few kcal/mol separate such conformers, the conformation of a peptide hormone or drug at a receptor can be as different from that in solution as may be the hydrated crystal form. It is vital, then, that any approach to peptide design recognizes that the interaction with the receptor could stabilize a conformation not found in solution. Interactions in the crystal or at the receptor are just special cases of perturbations, and changing solvent can have dramatic effects on the spectrum of conformational types of a smaller peptide. For example, use of dimethyl sulphoxide or any other non-aqueous solvent for the study of smaller peptides by NMR may produce results of very little biological interest.

3. Drug discovery and development

Historically, extracts of plants were used as cures for illness, and as chemical techniques became more sophisticated, the active ingredients were isolated. These have become the *leader substances* on which much of the drug development this century has been based. The development of peptide analogues has tended to be based on systematic modification of the leader substance,[8] with data from receptor binding[38] and *biological assay* being used to classify features of activity. The molecule isolated which has activity is the leader substance, and if it has a number of different actions then some *functional groups* may be unnecessary for the biological effect of interest.

The spatial orientation of the *pharmacophore*,[17,39,40] or essential functional groups, can serve as a *template*,[14] and this arrangement may be maintained in the binding process[41] by a *scaffold* of the other residues which are otherwise unnecessary for activity, and which could be replaced by other structural features. Designing a novel structure *rationally* is taken here to mean design on the basis of studying the conformational characteristics and biological activity of the *endogenous* (parent or native) substance, and closely related structures. Though the endogenous substance still represents the leader substance, in this case the resulting drug may not be obviously derived directly from the chemical structure of the original.[26]

Peptides are natural substances which have well-defined mechanisms for breakdown, and in the case of peptide hormones, form part of the feed-forward and feedback cycles which control the amount of molecule in circulation. This is an example of *homeostasis*, which is regulation of physiological conditions within a certain range, as exemplified by blood. Therefore, one possibility is that injecting a sample of natural hormone will cause the body to *down-regulate* its own production, so the end effect will be no change. The more usual case is that the injection causes several different actions. Where there is a variety of actions, for example the endocrine and neurohormone actions of a peptide hormone, we would wish to bias the conformation towards one or other of the *active* forms, and avoid the toxicity due to multiple actions.

Historically, models for peptide–protein interaction have centred on the idea of a *lock and key* of enzyme–substrate interaction, and whilst this is a useful description of the importance of complementary three-dimensional structures, or conformations, on the interaction of biological molecules,[41] it implies a rather local change. It was demonstrated by Anfinsen[42] that the three-dimensional structure of proteins is determined by its amino acid sequence, and the fact that proteins can become denatured, losing functionality and any similarity to the equilibrium conformation, under conditions like high temperature, suggests that activity is associated with well-defined conformations. These are disrupted by fast-moving water molecules as temperature rises. Structural modifications can also affect the conformational choices and preferences of a molecule. In addition, the solvent, ions, pH, and presence of other biomolecules will also have influence.

The multiple actions of peptide hormones and neuropeptides[1,11] could arise by different conformers being selected or enhanced in different receptor environments. Restrictions in conformational freedom, by chemical modifications, can lead to selected actions, which encourages this view. Restriction of conformational freedom increases the likelihood that theoretical or experimental conformational analysis will identify the active and solution conformers.[13] It also suggests that there is increased probability, over the native molecule, of the solution and active conformers being closely related.

Deductions regarding the active/binding site can be investigated by site-directed mutagenesis, providing *structure–activity relationship* (SAR) data for proteins.[9] Such SAR data has historically only been available for peptides, following chemical synthesis.[8] Thus, although peptide and protein engineering frequently use different procedures, they both aim to optimize structural and conformational features. It is possible to continue the above analogy to a *lock-and-key* action between *ligand* and receptor, where the ligand is any 'small' molecule interacting with a larger molecule. The ligand will have solution conformers, perhaps differing from the receptor-bound conformation, and again from the *active* form. These stages are not the same in all types of interactions. For example, a substance stimulating biological effect is an *agonist*, whilst an *antagonist* only binds; but by stopping an agonist binding, the antagonist is also having an effect (see Chapter 7).

4. Theoretical aspects of peptide drug design

Computer-aided drug design[40,43] is of recent origin and, especially in relation to peptide and polypeptide drugs, has exploited earlier developments in techniques for energy minimization[44] and molecular dynamics,[45] as well as molecular graphics, especially for the study of ligand–receptor interaction.[46,47] *Quantitative structure–activity relationships* (QSAR) have been applied to peptides,[48] and modelling can be considered a QSAR technique.[49] The receptor is usually protein and, even in the absence of crystal structures,[47] there is likely to be sequence homology[50] to a protein of known structure for use in modelling.[16,46]

At the most chemically detailed level, quantum mechanical techniques have been applied to small drug molecules,[51] and the move now is towards a quantum mechanical level of detail at the binding/catalytic region[52] and a molecular mechanical interpretation of events further away. Recent excitement has been caused by the development of free-energy perturbation methods,[53,54] with the potential for predicting changes in binding affinity of residue modification. In addition, advances in parallel processing[55,56] have prompted re-developments in artificial intelligence,[57] though in the guise of *neural networks* with application in protein structure prediction problems.[58,59]

A number of established companies sell molecular modelling packages, for example Chemical Design (founded in 1983) and Polygen and Biosym which were founded in 1984. The earliest companies were Molecular Design (founded in 1978) and Tripos (founded in 1979) which is now owned by Evans and Sutherland. Whilst Tripos tends to concentrate on QSAR techniques, Chemical Design offers a package, including transputers, for quantum mechanical calculations. Biosym, Polygen, and Biomos packages are mainly orientated towards peptide and protein calculations using molecular mechanics and dynamics, reflecting the background of their founders.[13,14,60]

To aid in the application and development of molecular dynamics techniques, we are interested in the *phase space* within which the molecular simulation is taking place. If there are N particles (usually atoms) in the *system* (which is everything being simulated) then the phase space has $6N$ dimensions, with each particle having descriptors x, y, and z describing its position, and p_x, p_y, p_z describing the momenta conjugate to x, y, and z.

The state of any system corresponds to a point in phase space, and its history corresponds to movement of that point through phase space, sweeping out a continuous trajectory. In any simulation, however, the actual trajectory is not continuous, but represents a number of discrete points. Formally this means that there is a hidden dimension to the problem and the information in the simulation, as such, represents a Poincaré section through the space of true dimensionality. This has important implications for simulations; for example, there is non-conservation of total energy of the system. The purest possible treatment of an isolated system, in which the total energy (kinetic plus potential) is conserved, is *Hamiltonian*, although the most realistic simulations would consider quantum mechanical effects.

One reason why these points are worth noting is because moves have been made to introduce quantum mechanical phenomena into simulations,[52] and to address the problem of correct treatment of the electrostatic interactions in a complex medium (see Chapter 8). For present purposes, events are considered as taking place in a simple well-behaved region of phase space known as the *manifold*.

In the $6N$-dimensional phase space there is an object (generally set of objects) in $6N-1$ dimensions called the *manifold*. It is the (hyperdimensional) *surface* of these objects to which the allowable trajectory is confined. The volume, shape, and topology of the manifold govern the properties of the system and the tractability of the problem which the simulation seeks to solve (see Chapter 4). Introducing interactions or changing the temperature affects these properties by changes of volume, shape, and topology, but in ways which are qualitatively predictable. Whatever changes are introduced by whatever means, the result is a particular manifold representing an *attractor*.

The attractor may represent more than one potential minimum in the energy surface, which is to say, for example, that a globular protein is rarely if ever confined to a simple harmonic state, and we can replace the *multiple minimum* problem by the *multiple attractor* problem. The shape of the final manifold is of crucial importance in considering to what degree free energy, and other thermodynamic properties, are being correctly calculated (see Chapters 4 and 8).

Given that the system under consideration is Hamiltonian, there are powerful constraints on which parts of the phase space the trajectory of the simulation can reach. We may need to modify the shape and topology of the manifold, to allow the simulation to reach the answer at point B from an initial point A, if A and B do not lie on the same continuous manifold, if on the same *iso-energy* level at all. This may be effected by changing the definition of the potential energy component to include a biasing or *target* function (see Chapter 4) based on external data.

Therefore, when selecting the 'best possible' starting conformation, it is not 'closeness' to the global minimum in terms of conformational similarity which matters, but whether it can be reached by the *algorithm*.

Only molecular dynamics represents a true use of the phase space by including the momentum component, though molecular dynamics with constant cooling has many similarities to energy minimization. Cooling a simulation to close to absolute zero can be used to characterize local minima on the trajectory, subsequently subjecting that more-or-less single conformation to analysis. The cooling process may leave the conformation *trapped* in a high-energy minimum, though the risk can be reduced by gradual cooling. Some of the examples in Chapter 4 involve molecular dynamics at various temperatures, which can be shown to cause regions of the iso-energy manifold to grow and fuse, so allowing transit from a point A in one region to a point B in another. In general, the manifold of a system at a higher temperature can be shown to enclose that at a lower temperature; by contrast simply running for longer periods at a lower temperature; may never allow one to reach A from B.

A major advantage of molecular dynamics over traditional energy-minimization methods is the ability to explore the potential surface, and thereby escape from a *false-solution* minimum that unresolved characterization places the molecular conformation (see Chapters 2 and 3). Escape from even a deep minimum can be achieved by raising the temperature, and/or by introducing distance constraints.

In addition to these ideas, and for peptide design, we can make use of a form of information theory to predict the favoured conformation of residues, for use in starting conformations, and thereby emphasize those interactions which are between residues close in the amino sequence (see Chapter 4). Physically, this corresponds to a condition known as the *theta condition*, analogous to the van der Waals point of a gas. The omission of the longer-range interactions produces a manifold of higher dimensionality, but lower genus. By fading the longer range interactions in gently, and combining this with a *target function* based on experimental data, we hope to reach the correct region of phase space.

Conformational preferences are determined by their calculated energy, dependent on the potential functions used, which are analytical representations of the energy of interaction between atomic centres.[60,61] The accuracy of the assignment of properties to each atom will therefore determine the quality of the result.

5. Experimental aspects of peptide drug design

Developments in nuclear magnetic resonance (NMR) spectroscopy[62] and X-ray crystallography techniques (Chapters 2 and 3, respectively) permit detailed conformations to be described for constrained peptides[13] and for proteins.[47] Experimental conformational analysis has identified that particular environments induce well-defined conformations of flexible peptides, for example, amphipathic helices are favoured in a lipid environment.[34,63] The study of molecular recognition processes involved in drug–receptor interaction,[64,65] as well as the development of new delivery systems,[66] requires a concerted experimental and theoretical approach.

Although peptides can often interact with a variety of receptors, the conformational flexibility which probably allows this can be reduced by chemical modification. As part of a design scenario this requires details of the biological activity and site of action of each analogue produced and, preferably, some knowledge of their preferred conformations.

The *opiate* morphine has been used for several centuries for the relief of pain, and it binds preferentially to the μ opioid receptor type, whilst the endogenous enkephalin *opioids* bind mainly to the δ type. A number of conformationally-constrained opioid ligands have been synthesized (see Chapter 5) and biologically tested (Chapter 7). For example, [D-Pen^2,D-Pen^5] enkephalin (DPDPE) is much more selective than the enkephalins for the δ receptors.[27] The reduction in the conformational freedom of the analogues makes them better candidates than the native peptide for experimental and theoretical conformational analysis,[13] and permits the investigation of *conformation*–activity relations.[27,67] Although morphine was identified before the enkephalins, it is the type of relatively rigid non-peptide analogue that peptide drug design aims at.[26]

Although there has been a consensus that enkephalins, and other opioids, are the natural equivalent of the opiates, there have recently been suggestions that opiates are either part of the diet, and possibly essential like vitamins, or are endogenous.[68] Similar suggestions have been made after benzodiazepine-like molecules were identified in brains stored since before the first benzodiazepine was synthesized.[69]

Genetic approaches to peptide and polypeptide synthesis (see Chapter 6) allow us to produce large quantities of natural substance. However, the sale of drugs based exactly on endogenous peptide hormones, for example growth hormone, insulin, and tissue plasminogen activator (TPA), brings into question the ownership of natural substances.[70,71] This aside, there is an enormous market for these molecules in deficiency diseases (see Chapter 6).

The main emphasis of this book is development of research methods and protocols to provide candidates for peptide-based drugs, and in this search expertise from many disciplines is being combined. Improvements in techniques for determining three-dimensional structures in NMR spectroscopy and X-ray crystallography, and in computer software and hardware, means that peptides and proteins are now widely studied in the context of molecular design. Chemical synthesis and genetic and protein engineering can provide novel molecules, suggested either *de novo* or from conformational analysis, as candidates for biological testing, as well as for further experimental and theoretical conformational analysis.

Most widely prescribed synthetic drugs have been developed from conformationally rigid templates in which the two-dimensional structural formula will tend to correspond closely to its three-dimensional counterpart. When a molecule is flexible, because rotation is allowed around interatomic bonds, a large number of different conformations can be possible, as exemplified by peptides and proteins. For the rigid structure, chemical changes can be correlated with particular biological effect, as has been practised for over a century. Such analyses can provide indications as to a mechanism of action, as well as to further changes that would improve effect. Hence, a *design* scenario is born. This book is about developments that permit the above type of studies for conformationally-flexible molecules, including most of the endogenous biologically-active molecules and their receptors.[72]

Acknowledgements

The description of phase space is the work of Barry Robson. The author was supported in part by SERC grant GR/D/98754 and EEC Biotechnology Action Program (BAP) contract 0149 (UK).

References

1. 'Coexistence of peptides with classical neurotransmitters' T. Hökfelt, D. Millhorn, K. Seroogy, Y. Tsuruo, S. Ceccatelli, B. Lindh, B. Meister, T. Melander, M. Schalling, T. Bartfai and L. Terenius (1987) *Experientia*, **43**, 768–780.
2. 'Science and the discovery of drugs' M. Weatherall (1987) *The Pharmaceutical J*. (Feb. 14), 210–212.
3. 'Neuropeptides and their processing: Targets for drugs design' J. W. van Nispen and R. M. Pinder (1987) *Ann. Reports Med. Chem.*, **22**, 51–62.
4. 'Quality assurance of products manufactured by recombinant DNA technology: Introduction and elements of a philosophy' J. van Noordwijk (1988) *Arzneim.-Forsch./Drug Res.*, **38** (II), 7, 943–947.
5. 'Biotechnology in bulk drug production: objectives and strategies of process development' P. N. Hess (1987) *Arzneim.-Forsch./Drug Res.*, **37** (II), 10, 1210–1215.
6. 'Quality control of protein drugs' H. Bachmayer (1988) *Arzneim.-Forsch./Drug Res.*, **38** (I), 4, 590–591.
7. 'Drug discovery – today and tomorrow: The role of medicinal chemistry' K. R. Freter (1988) *Pharmaceutical Res.*, **5** (7), 397–400.
8. 'Structure–activity relationships of enkephalin-like peptides' J. S. Morley (1980) *Ann. Rev. Pharmacol. Toxicol.*, **20**, 81–110.
9. 'SARs: The receptor side of the coin' H. D. Hollenberg (1987) *Trends Pharmacol. Science*, **8** (6), 197–200.
10. 'α-Melanotropin: The minimal active sequence in the frog skin bioassay' V. J. Hruby, B. C. Wilkes, M. E. Hadley, F. Al-Obeidi, T. K. Sawyer, D. J. Staples, A. E. deVaux, O. Dym, A. L. de L. Castrucci, M. F. Hintz, J. P. Riehm and K. R. Rao (1987) *J. Med. Chem.*, **30**, 2126–2130.
11. 'Small peptides – new targets for drug research' A. S. Dutta (1989) *Chem. in Britain* (February), 159–162.
12. 'Solid phase synthesis and biological properties of $\psi[CH_2NH]$ pseudopeptide analogues of a highly potent somatostatin octapeptide' Y. Sasaki, W. A. Murphy, M. L. Heiman, V. A. Lance and D. H. Coy (1987) *J. Med. Chem.*, **30** (7), 1162–1166.
13. 'Calculating three-dimensional molecular structure from atom–atom distance in-

formation: cyclosporin A' J. Lautz, H. Kessler, J. M. Blaney, R. M. Scheek and W. F. van Gunsteren (1989) *Int. J. Peptide Protein Res*, **33**, 281–288.

14. 'Theoretical simulation of conformation, energetics, and dynamics of peptides' A. T. Hagler (1985). In: *The Peptides* (S. Udenfriend and J. Meienhofer, eds.), **7**, pp. 213–299. Academic Press, New York.

15. 'Distance geometry approach to rationalizing binding data' G. M. Crippen (1979) *J. Med. Chem.*, **22** (8), 988–997.

16. 'Drug design by the method of receptor fit' P. J. Goodford (1984) *J. Med. Chem.*, **27** (5), 557–564.

17. 'Pharmacophore identification and receptor mapping' C. Humblet and G. R. Marshall (1980) *Ann. Reports Med. Chem.*, **15**, 267–276.

18. 'Elements for the rational design of peptide drugs' J.-L. Fauchère (1986). In: *Advances in Drug Research* (B. Testa, ed.), **15**, pp. 29–69. Academic Press, London.

19. 'Clinical perspectives on neuropeptides' D. A. Lewis and F. E. Bloom (1987) *Ann. Rev. Medicine*, **38**, 143–148.

20. 'Computer-aided dosage form design. I. Methods for defining a long-acting first-order delivery system of maximum formulating flexibility' T.-Y. Lee and R. E. Notari (1987) *Pharmaceutical Res.*, **4** (4), 311–316.

21. 'Computer-aided dosage form design. II. Methods for defining a zero-order sustained-release delivery system of maximum formulating flexibility (1987) T.-Y. Lee and R. E. Notari (1987) *Pharmaceutical Res.*, **4** (5), 385–391.

22. 'Engineering targeted *in vivo* drug delivery. I. The physiological and physicochemical principles governing opportunities and limitations' C. A. Hunt, R. D. MacGregor and R. A. Siegal (1986) *Pharmaceutical Res.*, **3** (6), 333–344.

23. 'Gastrointestinal dynamics and pharmacology for the optimum design of controlled-release oral dosage forms' N. W. Read and K. Sugden (1987) *CRC Crit. Rev. in Ther. Drug Carrier Systems*, **4** (3), 221–263.

24. 'European biopharmaceutical culture' J. Hodgson (1990) *Bio/Technology*, **8**, 720–723.

25. 'The design of metabolically-stable peptide analogs' D. F. Veber and R. M. Freidinger (1985) *Trends Neurosci.*, **8** (9), 392–396.

26. 'Design of potent, orally effective, nonpeptidal antagonists of the peptide hormone cholecystokinin' B. E. Evans, M. G. Bock, K. E. Rittle, R. M. DiPardo, W. L. Whitter, D. F. Veber, P. S. Anderson and R. M. Freidinger (1986) *Proc. Natl. Acad. Sci. USA*, **83** (13), 4918–4922.

27. 'The conformational properties of the delta opioid peptide [D-Pen^2,D-Pen^5]enkephalin in aqueous solution determined by NMR and energy minimization calculations' V. J. Hruby, L.-F. Kao, B. M. Pettitt and M. Karplus (1988) *J. Am. Chem. Soc.*, **110** (11), 3351–3359.

28. 'Biopharmaceuticals: Drugs of the future' L. P. Gage (1986) *Am. J. Pharmaceutical Educ.*, **50** (4), 368–370.

29. 'Biobusiness in the pharmaceutical industry' R. G. Werner (1987) *Drug Res.*, **37** (9), 1086–1093.

30. 'Biotechnology and the pharmaceutical industry: new cardiovascular drugs' M. D. Dibner and P. B. M. W. M. Timmermans (1986) *Hypertension*, **8** (11), 965–970.

31. 'Is there a relationship between DNA sequences encoding peptide ligands and their receptors?' A. Goldstein and D. L. Brutlag (1989) *Proc. Nat. Acad. Sci. USA*, **86**, 42–45.

32. 'IUPAC–IUB commission on biochemical nomenclature' (1970) *J. Mol. Biol.*, **52**, 1.

33. *Polypeptide and Protein Structure* A. G. Walton (1981) Elsevier, New York.

34. 'Presence of an amphipathic helical segment and its relationship to biological potency of calcitonin analogs' R. M. Epand, R. F. Epand and R. C. Orlowski (1985) *Int. J. Peptide Protein Res.*, **25**, 105–111.

35. 'Crystal structure analysis of deamino-oxytocin: conformational flexibility and

receptor binding' S. P. Wood, I. J. Tickle, A. M. Treharne, J. E. Pitts, Y. Mascaren-has, J. Y. Li, J. Husain, S. Cooper, T. L. Blundell, V. J. Hruby, A. Buku, A. J. Fischman and H. R. Wyssbrod (1986) *Science*, **232**, 633–636.

36. *Cambridge Structural Database* Cambridge Crystallographic Data Centre, University Chemical Laboratory, Lensfield Road, Cambridge CB2 1EW, UK.

37. 'Secondary structure prediction: combination of three different methods' V. Biou, J. F. Gibrat, J. M. Levin, B. Robson and J. Garnier (1988) *Protein Eng.*, **2** (3), 185–191.

38. 'Drug discovery at the molecular level: A decade of radioligand binding in retrospect' M. Williams and D. C. U'Prichard (1984) *Ann. Reports Med. Chem.* (R. C. Allen, ed.), **19**, pp. 283–292 Academic Press, London.

39. 'Peptide conformation and biological activity' G. R. Marshall, F. A. Gorin and M. L. Moore (1978) *Ann. Reports Med. Chem.* (C. Walsh, ed.), **13**, pp. 227–238 Academic Press, London.

40. 'Computer-aided drug design' G. R. Marshall (1987) *Ann. Rev. Pharmacol. Toxicol.*, **27**, 193–213.

41. 'Binding of flexible ligands to macromolecules' A. S. V. Burgen, G. C. K. Roberts and J. Feeney (1975) *Nature*, **253**, 753–755.

42. 'Principles that govern the folding of protein chains' C. B. Anfinsen (1983) *Science*, **181**, 223–230.

43. 'Computer-assisted drug design' A. J. Hopfinger (1985) *J. Med. Chem.*, **28** (9), 1133–1139.

44. 'Protein folding by restrained energy minimization and molecular dynamics' M. Levitt (1983) *J. Mol. Biol.*, **170**, 723–764.

45. *Dynamics of Proteins and Nucleic Acids* J. A. McCammon and S. Harvey (1987) Cambridge University Press.

46. 'Structure and energetics of ligand binding to proteins: *Escherichia coli* dihydrofolate reductase–trimethoprim, a drug–receptor system' P. Dauber-Osguthorpe, V. A. Roberts, D. J. Osguthorpe, J. Wolff, M. Genest and A. T. Hagler (1988) *Proteins: Structure, Function and Genetics*, **4**, 31–47.

47. 'Atomic structure of thymidylate synthase; target for rational drug design' L. W. Hardy, J. S. Finer-Moore, W. R. Montfort, M. O. Jones, D. V. Santi and R. M. Stroud (1987) *Science*, **235**, 448–455.

48. 'Peptide quantitative structure–activity relationships, a multivariate approach' S. Hellberg, M. Sjöström, B. Skagerberg and S. Wold (1987) *J. Med. Chem.*, **30**, 1126–1135.

49. 'Introduction: A review of QSAR methodology' C. J. Blankley (1983) in *Quantitative Structure–Activity Relationships of Drugs* (J. G. Topliss, ed.), pp. 1–21. Academic Press, London.

50. 'Multiple sequence alignment' D. J. Bacon and W. F. Anderson (1986) *J. Mol. Biol.* **191**, 153–161.

51. *Quantum Pharmacology* W. G. Richards (1977) Butterworth, London.

52. 'Quantum simulation of ferrocytochrome c' C. Zheng, C. F. Wong, J. A. McCammon and P. G. Wolynes (1988) *Nature*, **334**, 726–728.

53. 'The role of computer simulation techniques in protein engineering' W. F. van Gunsteren (1988) *Protein Eng.*, **2** (1), 5–13.

54. 'Free energy calculations by computer simulation' P. A. Bash, U. C. Singh, R. Langridge and P. A. Kollman (1987) *Science*, **236**, 564–568.

55. 'The connection machine' W. D. Hillis (1987) *Scientific American*, **256** (6), 86–93.

56. 'Computing with neural circuits: A model' J. J. Hopfield and D. W. Tank (1986) *Science*, **233** (4764), 625–633.

57. 'Artificial intelligence and the neurosciences' S. Ullman (1986) *Trends in Neurosci.*, **9** (10), 530–533.

58. 'Protein secondary structure prediction with a neural network' L. H. Holley and M. Karplus (1989) *Proc. Natl. Acad. Sci. USA*, **86**, 152–156.

59. 'Prediction of β-turns in proteins using neural networks' M. J. McGregor, T. P. Flores and M. J. E. Sternberg (1989) *Protein Eng.*, **2** (7) 521–526.
60. 'CHARMM: A program for macromolecular energy, minimization, and dynamics calculations' B. R. Brooks, R. E. Bruccoleri, B. D. Olafson, D. J. States, S. Swaminathan and M. Karplus (1983) *J. Comp. Chem.*, **4** (2) 187–217.
61. 'The design of biologically active polypeptides' B. Robson (1983) *CRC Crit. Rev. Biochem.*, **14** (4) 273–296.
62. 'Modern NMR techniques for structure elucidation' G. A. Morris (1986) *Magnetic Reson. in Chem.*, **24**, 371–403.
63. 'Studies on peptide conformation in the design of peptide agonists' E. T. Kaiser (1987) *Biochem. Pharmacol.*, **36** (6) 783–788.
64. *Drug Action at the Molecular Level* (G. C. K. Roberts, ed.) (1977) Macmillan, London.
65. 'Drug–receptor relationships, selection of therapeutic goals, and adaptive control of pharmacokinetic systems' R. W. Jelliffe (1987) *Fed. Proc.*, **46** (8) 2494–2501.
66. 'Alternative delivery systems for peptides and proteins as drugs' D. A. Eppstein and J. P. Longenecker (1988) *CRC Crit. Rev. in Ther. Drug Carrier Systems*, **5** (2), 99–139.
67. 'Topological similarities between a cyclic enkephalin analogue and a potent opiate alkaloid: A computer-modeling approach' J. DiMaio, C. I. Bayly, G. Villeneuve and A. Michel (1986) *J. Med. Chem.*, **29**, 1658–1663.
68. 'Morphine and codeine from mammalian brain' C. J. Weitz, L. I. Lowney, K. F. Faull, G. Feistner and A. Goldstein (1986) *Proc. Natl. Acad. Sci. USA*, **83**, 9784–9788.
69. 'Demonstration and purification of an endogenous benzodiazepine from the mammalian brain with a monoclonal antibody to benzodiazepines' A. L. De Blas and L. Sangameswaran (1986) *Life Sci.*, **39** (21), 1927–1936.
70. 'FDA puts new heart drug on hold' (1987) *Science*, **237** (4810), 16–18.
71. 'Companies vie over new heart drug' (1987) *Science*, **237** (4811), 120–122.
72. 'Cell-membrane hormone receptors: some perspectives on their structure and function relationship' C. C. Yip (1988) *Biochem. Cell Biol.*, **66**, 549–556.

—— *Chapter 2*————————————————

Conformational determination by NMR spectroscopy

H. Kessler and S. Steuernagel

I. Conformation in solution

It is the basis of our scientific understanding of nature that the properties of a molecule are determined by its structure. The meaning of the term *structure* has changed historically as our knowledge has accumulated. We now include not only the spatial structure but also the dynamics of the molecule. Peptides and proteins provide a challenge in modern structure elucidation, because they are more or less flexible biopolymers of varying size and complexity. They are used in nature for a variety of different functions. The 20-letter code of the naturally occurring amino acids with chemically different side-chains offers an easy access to an almost infinite number of biopolymers with different functions such as transmitting or receiving messages (hormones and their receptors), catalysing a chemical reaction (enzymes) or stopping it (inhibitors), or carrying recognition information (antibodies), to mention only a few.

The three-dimensional structure of the backbone of peptides and proteins is determined by the angles ϕ (defined as the dihedral angle $C'-N-C^\alpha-C'$) and ψ ($N-C^\alpha-C'-N$) about single bonds for each amino acid (see Fig. 1).[1] The barriers of rotation around these bonds are small (less than about 5 kcal/mol), which provides a high flexibility of the peptide chain.

The substitution at the α-carbon of each amino acid by different side-chains provides functionality and additionally introduces some restrictions in the conformational space. However, in addition to the rotations about bonds in the backbone, the side-chain rotamers also have to be considered. In general, every peptide is an extremely flexible molecule, which can change its shape in different environments, such as in the solid state or in solution,[2] depending on the nature

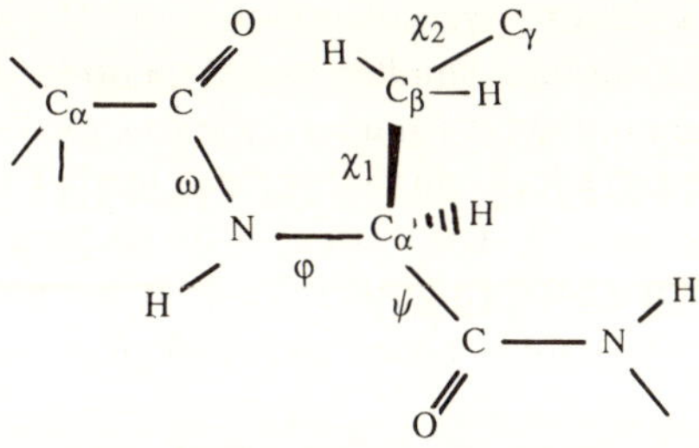

Figure 1 IUPAC definition of dihedral angles in peptides and proteins. The angle ω is almost exclusively near 180° except for peptide bonds involving N-alkylated amino acids, where angles of about 0° are also found.

of the solvent, but also when it binds to a biological receptor.[3] It is the aim of a pharmaceutical chemist to design a biologically active molecule with the knowledge of the 'bioactive conformation'. Any conclusion about structure–activity depends on the possibility of determining 'the' conformation under well-defined conditions. We must also keep in mind that the molecules in general often contain rigid domains and more or less flexible domains side by side. This is true for structures in solution as well as in the crystal, but in the latter normally only a small number of conformations is found due to the regular packing in the crystal.

Structure elucidation in solution starts with a search for the most stable conformation in a certain environment. Often, it is immediately apparent that there is no preferred conformation. It is then incorrect to derive a model of just one conformation, because NMR parameters (NOE, coupling constants, chemical shifts) do not linearly depend on structural changes.[4] Even when there might be a slightly preferred conformation, the relevance of this one to the 'bioactive conformation' remains doubtful. Without conformational restrictions[3,5] leading to a reduced number of spatial structures, no further work should be spent on such a project.

What kind of restrictions of the conformational space are possible? As mentioned above, steric hindrance by side-chains in the natural amino acids does not result in a drastic reduction of the conformational space. However, hydrogen bonds within the backbone and from side-chains form a secondary structure with increasing stability as the size of the peptide chain is increased. Hence, protein structures are normally well defined by β-sheets, α-helices, and other more or less rigid elements, but this is often not the case with the connecting loops between them. It is also a general rule that the structure in a core of a protein is much better defined than at the surface, where the side-chains of the amino acids often exhibit multiple conformations. This is true also for crystal structures, where multiple conformations became more evident as the structures are refined. Recently molecular dynamics calculations[6] have become an important method of detecting those equilibria.[7] However, crystallization has a strong tendency to freeze out a limited number of conformations due to the repeating arrangement of the molecules and the symmetry of the crystal.

In solution such restrictions are not present, and the number of populated conformations is much higher. Smaller, linear peptides in isotropic solution normally adopt a large number of rotamers, the so-called *random coil structure*. Mean NMR parameters which do not give information about the spatial arrangement are found, and the tendency to form single crystals is low. It often occurs that more than one conformation is found in a crystal of such a molecule and when different crystals are obtained the conformations found in them may differ considerably.[8]

To obtain a better understanding of the spatial requirement for biological activity, cyclizations may be performed to reduce the conformational space.[3,5] Nature uses disulphide bridges between cysteine side-chains for ribosomal synthesized proteins, but backbone to backbone, backbone to side-chain, or side-chain to side-chain cyclizations can be formed via enzymatic processes. Chemical synthesis provides a variety of ways to effect cyclization. If such a conformationally constrained molecule is still biologically active, all conformations which cannot be reached in the cyclic structure can be excluded as being the biologically active one. If this conformer is favoured in the cyclic molecule, the activity might well be much higher than in the parent compound. When conformational studies in different solvents show that the same conformation is always found, regardless of the environment, the observed structure is likely to be in a relatively deep energetic minimum (one strongly preferred conformation). If this molecule is biologically active then there is good reason to assume that this structure is close to the one when the molecule is bound to its receptor. Otherwise the change of the conformation would cost energy taken from the binding energy. This means that methods to determine the conformation in different environments are of utmost importance for drug design.

To summarize the problem of multiple conformations, we point out that for small peptides cyclization can be used to reduce the otherwise unacceptable large conformational space. In a few cases also, small peptides may prefer one conformation, especially in non-isotropic solvents (micelles or vesicles),[9,10] but in our experience this is the exception. For proteins the secondary and tertiary structure normally introduces enough stabilization to form a more or less rigid core, whereas the outer sphere still exhibits flexibility.

2. Peptides and proteins – different molecules, different problems

Peptides and proteins are both polymers formed by condensation of amino acids. There is no clear definition for distinguishing peptides and proteins. A polypeptide chain of 40 amino acids may well be called a large peptide, but it is also a small protein. However, with increasing molecular size, the properties and the methodology for investigating the structure changes.

Proteins are generally synthesized at the ribosome and therefore contain only the 20 natural amino acids (with the exception of the recently found additional amino acid selenocysteine, which is also coded by DNA).[11] Peptides are formed in nature either by cleavage of a longer peptide chain (of a protein precursor) or by enzymatic synthesis. The latter allows a larger variety of amino acids to be used, and hence more unconventional structures are found in smaller peptides. Also connectivities other than peptide bonds and disulphide bridges (such as ester bonds or homodetic cyclic structures) are common in peptide antibiotics but not in proteins. The same is true for chemical modifications. As proteins are in practice not available by classical chemical synthesis, but only by gene technology, the variability in the structure of the amino acids is restricted to the natural amino acids, whereas the fantasy of a chemist can create many structures differing from those found in nature.

In addition to the differences in chemical structure, peptides and proteins differ more or less in molecular size, in solubility, and last but not least in flexibility (see above). As far as molecular size is concerned, it is obvious that in principle a smaller molecule can be studied in greater detail than a large biopolymer. But this statement is only partially true. The unfavourable correlation time (mean time for the molecular tumbling in solution) of an oligopeptide containing 5–10 amino acids has prevented extensive quantitative NOE evaluations until recently, when this problem has been overcome by rotating frame experiments.[12,13] Also, as secondary structure elements are less fixed in peptides, conclusions about solution structures are less reliable unless the above-mentioned restriction elements are used.

However, the smaller size of peptides facilitates the application of heteronuclear NMR techniques on peptides that contain the magnetically active heteroatom isotope at natural abundance.[14] Recently, newly developed NMR techniques and improvements in spectrometer sensitivities have shifted this limit to increasing molecular size, such as medium-sized proteins.[15] In addition, the possibility of enriching the isotopes, such as ^{15}N, ^{13}C, or both, using gene technology allow the measurement of proteins of increasing size.[16] By these rather tedious procedures it seems possible to solve structures or at least partial structures of proteins up to a molecular weight of 100 kD. Conventionally the upper limit has been at about 10–20 kD,[15] depending on the structure, solubility, and amount of available material. It should also be mentioned that natural peptides and proteins often appear as mixtures of isomers which are difficult to separate (microscopic inhomogeneity) and this introduces enormous problems in structural analysis.

The solvent for proteins is almost exclusively water, but polar aprotic solvents such as DMSO are also used. Peptides, especially polar ones which contain a high amount of lipophilic amino acids, often dissolve better in solvents of low polarity, such as $CDCl_3$ or C_6D_6. The choice of the solvent depends on the problem to be tackled. As a peptidic drug finds its receptor mostly in the membrane, a solvent which mimics the membrane environment best would be ideal. However, measurements in membranes are difficult to perform. Often sodium dodecyl sulphate, which is commercially available, seems to be a good compromise. Before starting a conformational analysis a careful selection of the solvent should be done with respect to solubility, the best correlation time, conformational homogeneity of the peptide, and the lack of disturbing solvent signals in the interesting spectral range.

Yet another difference between peptides and proteins results from using molecular distances as the most important pieces of information about spatial structures. The relation between the outer surface to the volume increases with decreasing molecular size. As all protons which are near the surface exhibit a considerably smaller number of NOE effects than those in the inner core, the structures for small peptides are less accurately determined than those for larger proteins, when only NOEs are used. This is also true for groups at the surface of larger proteins, not forgetting their already higher flexibility due to less hindrance by neighbouring groups. In conclusion, the reliability of a peptide structure depends more on the number of data that can be experimentally provided, than is the case for a protein. As most restraint molecular dynamics calculations are performed *in vacuo* (at least at the beginning of the structure determination), peptide structures are more sensitive to structural artifacts due to the neglect of the solvent.

3. Structural information from NMR spectra

The extraction of structural information from NMR spectra using the advantages of multidimensional techniques such as 2D-NMR,[17] or their 1D analogues,[18] or more recently also 3D-NMR,[19,20] has been shown in a lot of applications. For both peptides and proteins, similar methodologies have been developed for this purpose.[21,22] The evaluation of structural relevant parameters from an NMR spectrum has to be achieved by following a simple scheme, which can be divided into two major steps:

- In the first step, qualitative information has to be collected. Dealing with the investigation of peptides and proteins, the particular spin systems of the amino acids have to be identified. Each spin system has then to be assigned to its specific sequential position within the polypeptide chain.
- Secondly, quantitative information has to be evaluated, which is more tedious.

When all signals in the NMR spectrum are assigned to specific atoms in the molecule, the extraction of the quantitative data can be started. Valuable quantities are either J-coupling constants to determine dihedrals or NOE *build-up rates* to calculate interproton distances, the latter being of major importance for investigation of proteins.

3.1. QUALITATIVE INFORMATION

Sequential resonance assignment includes the rather time-consuming steps of the assignment of each signal in the spectrum to a specific type of spin system (first-level assignment) *and* the successive assignment of these spin systems to their sequential positions (sequential assignment). Recently the so-called main-chain strategy has been used for the assignment of protein signals where the primary structure is known, as it is in almost every case. Only a few 'labels' are needed to assign fragments from which the sequence is derived, using only NH and H_α protons and a few characteristic resonances in the side-chain. To us, this procedure seems straightforward and does not differ principally from the conventional stepwise procedure, i.e. identification and subsequent sequencing. In practice both steps work hand in hand, more or less simultaneously.

First of all, since many amino acids exhibit a characteristic pattern of ^{1}H

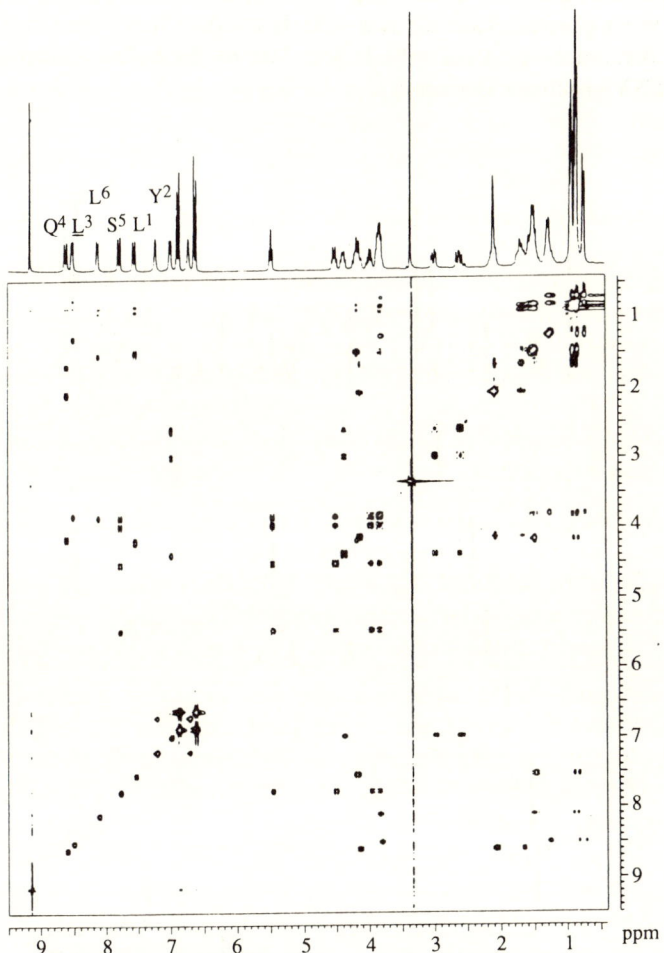

Figure 2 (a) 250 MHz TOCSY spectrum of cyclo[$-$Leu^1$-$Tyr^2$-$D-Leu^3$-$Gln^4$-$Ser^5$-$Leu^6$-$]. For amino acids ideally correlations of all signals of a spin system to the NH resonance are seen in the TOCSY.

signals, they can easily be detected in two-dimensional spectra making use of correlations via J-coupling. These scalar spin–spin couplings cause cross-peaks in COSY spectra with different frequencies as coordinates.[23] They indicate that two nuclei resonating at these frequencies are two or three bonds apart. These spectra contain significant information about particular structural elements. The TOCSY experiment[24] supports the assignments of COSY spectra in a useful manner. Here additional correlations are found for protons which are not directly coupled but have common coupling partners. Therefore it is possible to achieve assignments for protons with strongly overlapping signals. Indeed, the TOCSY experiment can substitute for the COSY experiment, since the duration of the spin-lock mixing sequence determines the efficiency of the transfer: Experiments with short mixing times reveal only correlations of directly coupled nuclei, while for long mixing times *all* signals of a scalarly coupled spin system are correlated (*TO*tal Correlation Spectroscop*Y*). In Fig. 2 the method of assignments by means of a TOCSY spectrum is shown.

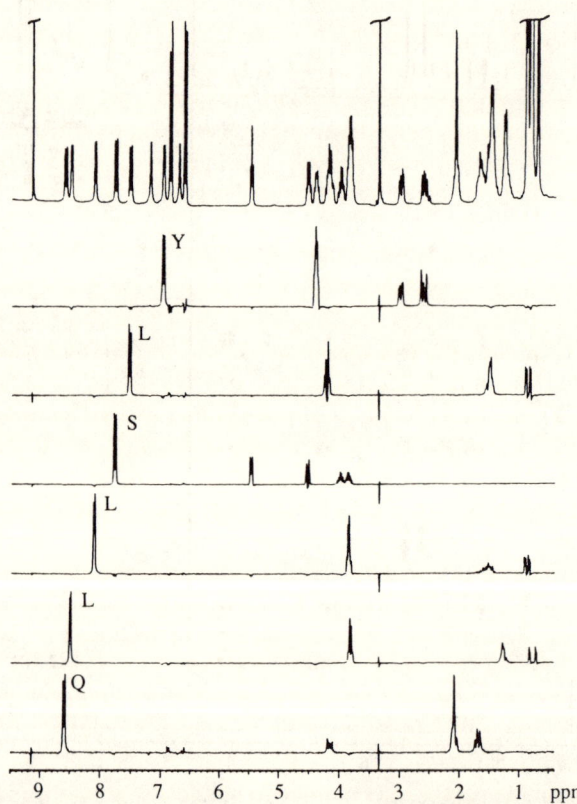

Figure 2 (b) Traces through the TOCSY spectrum of (a) at the resonances of the NHs. For all six amino acid residues all signals of the side-chains can be assigned.

For the investigation of proteins, the acquisition of TOCSY spectra with several mixing times is recommended to assign the signals of the individual spin systems. The occurrence of only in-phase signals yields a much better sensitivity than the antiphase patterns observed in COSY spectra. However, for the determination of coupling constants it is necessary to acquire COSY spectra[23,37] or directly use the E.COSY technique.[25]

In the following, we want to discuss some general features in connection with the assignment of NMR spectra of peptides and proteins. The spin systems of the naturally occurring amino acids can be divided into two groups. For the amino acids Ala, Arg, Gln, Glu, Gly, Ile, Leu, Lys, Met, Pro, Thr, Val belonging to the first group, characteristic spin systems arise that are typical only for each of these amino acids. The amino acids Asn, Asp, Cys, His, Phe, Ser, Trp, Tyr belong to the second group and exhibit very similar AMXY spin systems which have to be distinguished in each spectrum. The similarity of these spin systems arises due to the fact that the scalar coupling terminates at the position of the β-protons. Amino acids occurring more than once in the peptide sequence have to be assigned to their correct sequential position. However, in all these situations no additional information can be obtained from COSY or TOCSY spectra. For this purpose additional techniques have to be used. If the measurements are not restricted to homonuclear (^{1}H) techniques, heteronuclear correlations (especially ^{1}H–^{13}C) provide further information. However, owing to the low sensitivity of ^{13}C nuclei at natural abundance, the limitations are reached with a certain molecular size. More sensitive (inverse) techniques which have been developed recently, now enable the measurement of even small proteins.[15,31] Isotope-edited NMR spectra of specifically labelled compounds also enable the investigation of particular regions within a molecule (see below).[26]

Restriction to only homonuclear (^{1}H) measurements make the NOESY[27] or ROESY[12,13] experiments the techniques of choice to complete the sequential assignments of all signals. In these cases, dipole–dipole interactions of protons are interpreted yielding a large number of structurally interesting pieces of information. These are not only useful for signal assignments but also supply distances between protons in a molecule by using *build-up rates* quantitatively. This is discussed in the next section 3.2.2.

In Fig. 3 correlations of a backbone amide proton are shown. The broken arrows indicate NOEs within the same amino acid residue. They can be used to differentiate amino acids which exhibit similar spin systems. For example, the proton connected to R in Fig. 3 might not be scalarly coupled to the β-protons, as it is in the case of aromatic amino acids Phe, Trp, and Tyr. Here, the NOE between the NH and the aromatic proton (H)-R helps to assign the type of the amino acid involved. The solid arrows represent short-range NOEs to protons of the amino acid preceding in the peptide sequence. This kind of NOE provides sequential information. It is important whether or not independent knowledge about the peptide sequence is available. Two different situations are noteworthy. In the case where peptide or protein sequences are known, the strategy differs from the strategy where partially unknown compounds are investigated. The

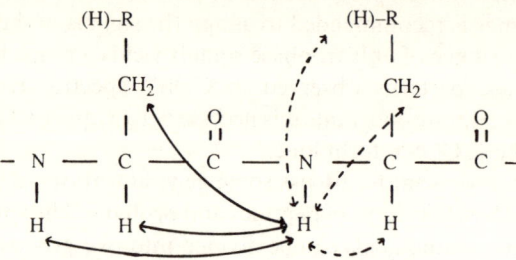

Figure 3 Part of a polypeptide sequence. Short-range NOEs are symbolized by broken arrows for correlations between protons of the same residue and solid arrows for correlations between protons of sequentially adjacent residues. R-(H) indicates protons which are present in the side-chain but not scalarly coupled to the β-protons.

known position of an individual spin system can be used to assign the sequential neighbourhood using the knowledge of the amino acid sequence. The same counts for synthetically known small peptides. Here, sequential resonance assignment might be used as an independent proof of the existence of a distinct peptide bond, such as the one formed during a cyclization.

In the case of investigating unknown compounds, a somewhat different strategy has to be used. Here, each spin system has to be assigned separately to its type of amino acid before any information about the amino acid sequence can be extracted. Considering the repeated occurrence of amino acids with similar spin systems, and the appearance of identical amino acid residues, it becomes obvious that a complete sequential assignment of a peptide or a protein by NMR measurements can only be achieved up to a molecular size of about 20 or 30 residues.

Some new techniques[28,29] have been presented which are based on the well-known relayed-NOESY experiment.[30] As schematically shown in Fig. 4, cross-peaks of different intensities on both sides of the diagonal occur in these spectra connecting the NH signals of two amino acids adjacent in the peptide sequence. The asymmetry originates from the subsequent performance of two different transfer steps. The first transfer between two protons only takes place if an ROE is observed, the second one is due to J-coupling. By this technique it is possible to connect two NH-protons of adjacent amino acids. As mentioned above, there is no observable J-coupling between a specific NH-proton and the H_α-proton of the amino acid preceding in the sequence. Since the ordering of the two transfers (J,ROE) is fixed for a single experiment, transfer is only possible in one direction and is prohibited in the other, thus producing asymmetric cross-peaks (see Fig. 4). The NH region of such a ROTO spectrum[28] is well suited for sequential assignments, as is shown in Fig. 5. However, if there exists a direct ROE between NH-protons of adjacent amino acids, cross-peaks symmetrical to the diagonal are also observed. Due to overlapping of ROTO-peaks with those symmetrical signals from direct ROEs, the interesting asymmetric signals can sometimes

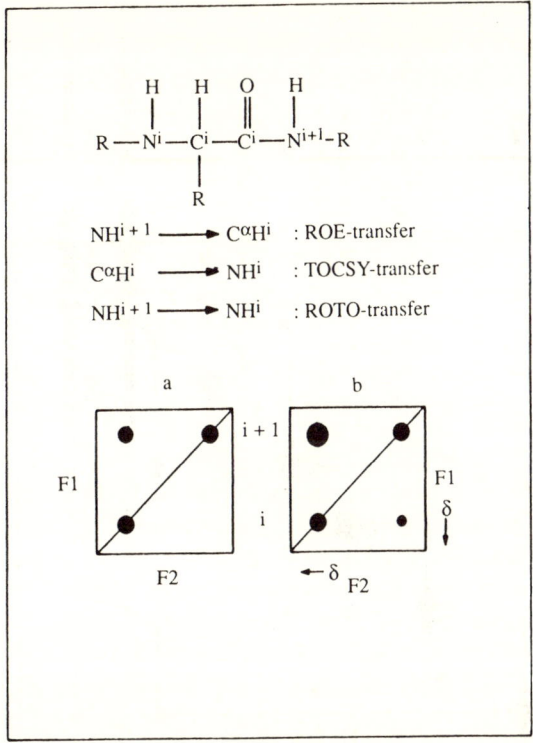

Figure 4 Schematic representation of $NH(i) - NH(i+1)$ cross-peaks in a ROTO spectrum. These peaks occur in asymmetric positions with respect to the diagonal (frame a). Their positions are used to determine the correct sequential ordering. In frame b it is shown that these peaks might be superimposed from direct $NH(i)-NH(i+1)$ cross-peaks.

hardly be interpreted. Processing procedures which at least minimize symmetrical contributions enable a better recognition of these peaks.[28,29] However, the knowledge of the ROESY spectrum for a correct interpretation of the ROTO spectrum seems indispensable in order to determine the individual magnetization transfers. For proteins the application of the analogous NOESY–TOCSY experiment seems to be favoured.[29] However, in these cases a lot of direct NH–NH NOEs can obscure the desired asymmetric cross-peaks.

Dependent on the molecular size, sophisticated multidimensional NMR experiments can be performed even with heteronuclei. Recent developments of inverse techniques yield a gain in sensitivity owing to the direct acquisition of proton magnetization.[15] Heteronuclear experiments are especially useful for providing additional information in the case of degenerate or overlapping ^{1}H signals. For proteins like tendamistat even correlation spectra via heteronuclear two- and three-bond couplings are available.[31] These spectra have been used to disentangle overlapping ^{13}C- and ^{1}H-signals which otherwise lead to ambiguities

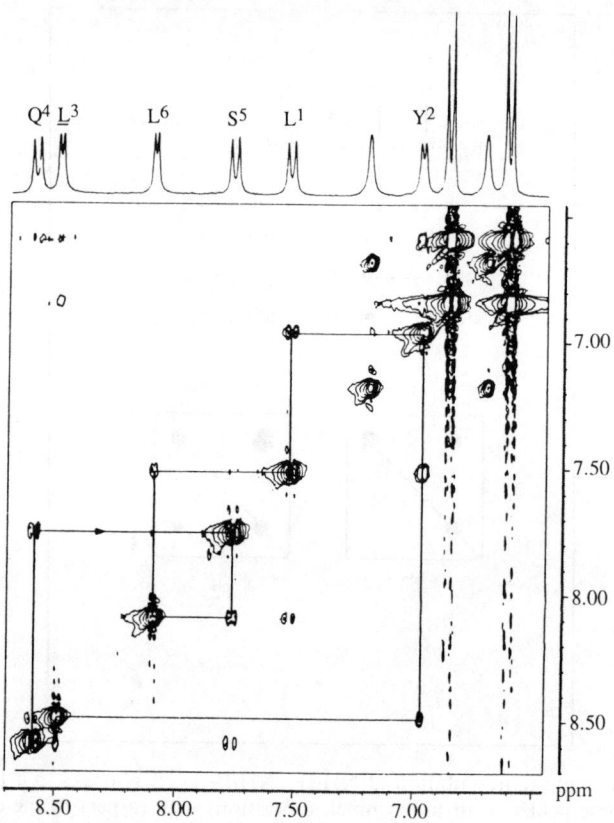

Figure 5 250 MHz ROTO spectrum of cyclo[–Leu^1–Tyr^2–D-Leu^3–Gln^4–Ser^5–Leu^6–]. By means of the asymmetric NH(i)–NH($i+1$) cross-peaks, a sequential assignment can easily be carried out. Some of the asymmetric cross-peaks are overlapped by direct NH–NH cross-peaks.

in assignments. The proton spectra of tendamistat have been carefully studied by Wüthrich *et al.*[32] and all signals have been assigned with the exception of some tentative assignments. Those could be proven to be correct by means of heteronuclear spectra. Furthermore, complete assignments of all carbon signals with the exception of the carbonyl-carbons are available. For example, for seven Ala residues the α-carbon signals and all α- and β-carbon signals of eight Thr residues could be assigned using the corresponding two- or three-bond couplings of the methyl protons (see Fig. 6).

Heteronuclear experiments of larger molecules are normally not very efficient when the nucleus of interest is in its natural abundance. However, isotopical enrichment increases the potential of NMR spectroscopy of proteins dramatically. The ^{15}N nucleus is synthetically easy to introduce, either in the form of a uniform labelling by growing the material in labelled ammonium salts or in

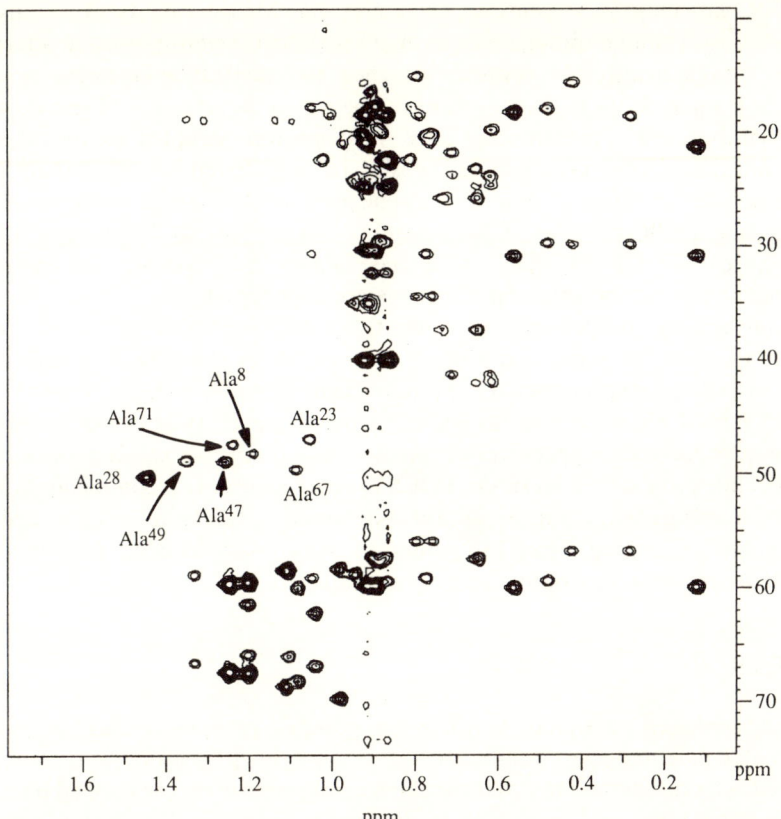

Figure 6 600 MHz inverse H,C-correlation via long-range couplings of tendamistat. The assignments are indicated for the correlations of all Ala methyl-protons to the α-carbon signals.

selective labelling by feeding distinct amino acids. In the first case the chemical shift dispersion of the nitrogen is used as additional disentangling by performing heteronuclear 3D-NMR techniques.[33,34] In specifically labelled molecules, hetero-filters are used to suppress unlabelled parts of the molecule or to suppress the labelled part.

Both approaches seem to become extremely useful and provide hope that the molecular size which can be studied by NMR will increase from about 100 amino acids, the upper size for conventional techniques, by almost an order of magnitude. A nice example of such a procedure is a recently studied pepsin–inhibitor complex.[34] Recently also, carbon labelling has been used for similar experiments but also to establish connectivity.[16]

In connection with investigations of proteins, isotope-edited NMR spectroscopy has to be mentioned.[35] [15]N labelled amino acids are extremely useful but ^{13}C labelling also seems to be interesting. Measurements can easily be achieved by a simple spin-echo experiment with 180° hetero pulses on alternate scans. Magnetization of protons attached to labelled heteronuclei is thus inverted, while all the other protons are unaffected. Subtraction of spectra with and without 180° pulse reveals only those protons coupled to the labelled heteronucleus. This principle can easily be transferred to almost all two-dimensional techniques. Also the application of X-filtering techniques[26] is interesting for the purpose of assignments of large proteins. If only one sort of amino acids is ^{15}N labelled, the corresponding spin systems can easily be assigned while all the other spin systems are suppressed.

Summarizing the previous considerations, the methodology for assigning NMR spectra of peptides and proteins does not differ strongly. For unlabelled proteins, when one is restricted to homonuclear ^{1}H measurements, TOCSY and ROESY/NOESY experiments, and a combination of these seem to be the techniques of choice. For spectra with signals in strongly overlapping regions, the ROTO (for proteins the NOESY–TOCSY) experiments are recommended, from which a sequential connectivity for the amino acid spin systems can be determined. Crowded regions in proteins can be analysed using homonuclear and heteronuclear three-dimensional techniques.

3.2. QUANTITATIVE INFORMATION

Having assigned all signals to the corresponding protons in their particular sequential positions, the conformational analysis of peptides and proteins can be started. The subsequent elucidation of the conformation can be divided into two major parts. One part refers to the backbone conformation, which plays a major role for the conformational description of proteins. Here, the representation of helices and sheets is done mainly by means of only α-C, α-N, and C′O. The analysis of the side-chain conformation requires extraction of homo- and heteronuclear J-coupling constants, but sometimes NOE effects can also be used. Often the latter procedure is not unambiguous. As molecular dynamics simulations are mostly preferred *in vacuo*, calculated side-chain conformations often have the tendency to form a globular structure in contrast to experimental bindings in solution. For smaller molecules it is therefore recommended to derive the χ_1 angle from $^3J_{HH}$ and $^3J_{CH}$ about the α–β bond.

3.2.1. *Determination of coupling constants*

From three-bond coupling constants, using the *Karplus* equations,[36] dihedral angles are derived from these data which are the interesting structural parameters. For the determination of homonuclear (^{1}H) coupling constants, E.COSY[25] as well as the DISCO[37] techniques might be of general interest. However, both techniques are limited to small molecules which do not exhibit too broad lines. Analogously, dihedral angles can be calculated from heteronuclear coupling constants.[38] However, for measurements of molecules consisting of heteronuclei at natural abundance (especially ^{13}C) no method has yet been found yielding reliable values within reasonable measuring time. For small proteins, on the other hand, information about dihedral angles is often provided by evaluation of interproton distances (see Section 3.2.2.).

3.2.1.1. *Determination of homonuclear coupling constants*

In the following we describe procedures for the extraction of coupling constants from ^{1}H-NMR spectra. With increasing molecular size the number of well-isolated multiplets decreases, thus preventing the extraction of coupling constants from one-dimensional spectra. In two-dimensional spectra the multiplicities of particular resonances are represented in the cross-peak fine structures. However, the procedures described below result from the phase behaviour of the multiplet components. As can be seen from the transfer functions of an α,β cross-peak, e.g. in a phenylalanine.^{17}

$$\sin(\pi J_{\alpha,\beta} t_1) \cos(\pi J_{\beta,\beta,} t_1) \sin(\pi J_{\alpha,\beta} t_2) \cos(\pi J_{\alpha,\beta,} t_2) \cos(\pi J_{\alpha,NH} t_2)$$

the resulting two-dimensional multiplet has in both dimensions antiphase splitting with respect to the α,β coupling ('active coupling', sin terms) and in-phase splitting with respect to all the other couplings ('passive couplings'; cos terms). Therefore, the individual lines of the multiplets have opposite signs leading to at least partial cancellation. The aim of both procedures described below is a reduction of the number of lines in the multiplets in order to avoid cancellation achieved by partial decoupling.

Using DISCO^{37} (*D*ifferences and *S*ums Within *CO*SY Spectra) traces through cross-peaks of *one* spin system are added (or sometimes subtracted) yielding multiplets of a reduced number of lines with splittings exhibiting sums (or differences) of coupling constants. In these multiplets cancellation is diminished owing to a reduced number of lines with an additional larger splitting.

Another technique to reduce the number of lines in the two-dimensional cross-peaks is provided by E.COSY.^{25} Only so-called connected transitions are observed and the multiplicity is drastically reduced. Several methods have been proposed to obtain cross-peak with so-called E.COSY fine structures. The most reliable way seems to be the linear combination of several multiple-quanta filtered COSY spectra. The general principle of how to obtain E.COSY-type cross-peaks is shown in Fig. 7.

3.2.1.2. *Determination of heteronuclear coupling constants*

As mentioned above, due to the low sensitivity of heteronuclei like ^{13}C and ^{15}N at natural abundance the use of information from these data is strongly limited. The fact that ^{1}H–^{13}C coupling constants are spread over a wide range make an extraction of these values difficult. Indeed, H,C correlations via the large one-bond coupling constants of 100–200 Hz can easily be done. However, the interesting information is provided by three-bond coupling constants which are dependent on the dihedral angles. The values of these constants are in a range of 0–10 Hz. Since cross-peak intensities in COLOC spectra^{39} strongly depend on the magnitude of coupling constants, a semiquantitative determination of

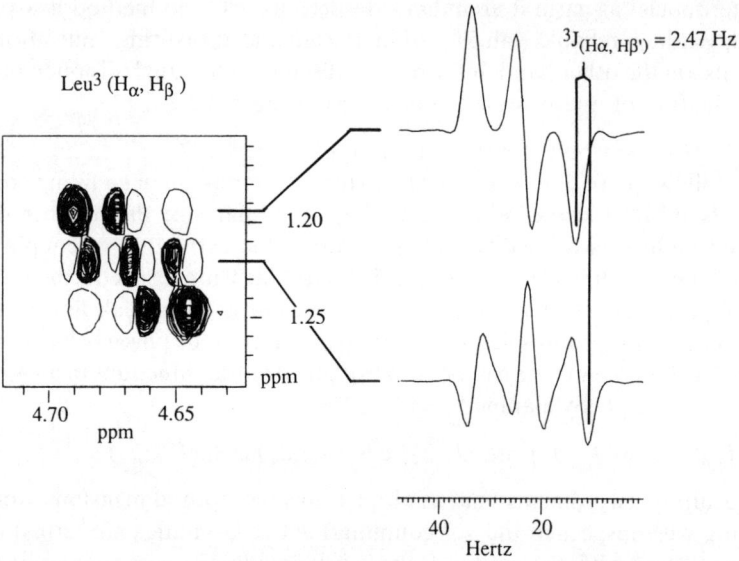

Leu^3 (H$_\alpha$, H$_\beta$)

^{3}J$_{(H\alpha,\ H\beta')}$ = 2.47 Hz

1.20

1.25

ppm

4.70   4.65
ppm

40   20
Hertz

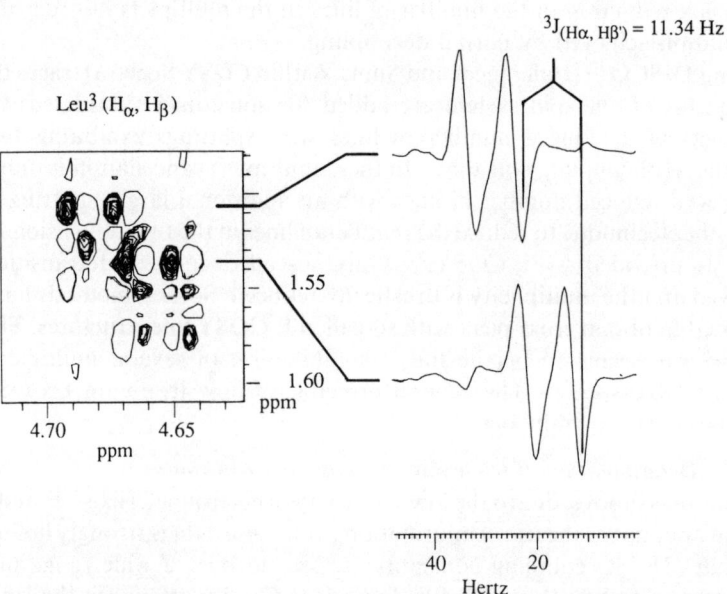

Leu^3 (H$_\alpha$, H$_\beta$)

^{3}J$_{(H\alpha,\ H\beta')}$ = 11.34 Hz

1.55

1.60

ppm

4.70   4.65
ppm

40   20
Hertz

Figure 7 Part of an E.COSY spectrum of didemnin A.^{44} The two cross-peaks shown are H$_\alpha$–H$_\beta$ and H$_\alpha$–H$_\beta$, cross-peaks from Leu^3. From each of the two cross-peaks the passive coupling constant can be determined most exactly due to a reduced number of lines. Note that for each cross-peak only the 'passive' coupling can be determined exactly due to its in-phase pattern, i.e. for the H$_\alpha$–H$_\beta$ cross-peak the ^{3}J$_{H\alpha H\beta'}$, coupling is the passive one and vice versa.

coupling constants is possible.[40] As will be shown, this is of major interest for the determination of the side-chain conformation. The occurrence of H^β–C'O cross-peaks, for example, can be used to distinguish between possible side-chain conformations (see Section 4).

3.2.2. Determination of NOE and ROE

The structures of polypeptides are only incompletely described by evaluation of homonuclear coupling constants. For example, no information can be provided for the values of the angles ω and ψ (see Fig. 1). Coupling constants, however, can only be used for the description of a single amino acid residue, especially for the angles ϕ and χ. For the description of the spatial arrangement of the particular spin systems, interproton distances contain indispensable information. There-fore, dipole–dipole interactions of ^1H are measured in NOESY and/or ROESY spectra.[12,13,27] Both experiments provide analogous results; it depends on the molecule under investigation which technique has to be applied.

For NOEs (nuclear Overhauser enhancements) as well as for ROEs (rotating frame Overhauser enhancements) relaxation effects realize the transfer of magnetization between spatially adjacent nuclei. These interactions will only occur if the corresponding nuclei are less than approximately 500 pm apart.[41] In contrast to the scalar coupling which is mediated through bonds, the *dipolar interactions* are mediated *through space*.

We now wish to describe the differences in the determination of the structural information from these spectra. The molecular size in principle determines which one of the two effects is used for the determination of intramolecular distances. The most important features of NOEs and ROEs can be extracted from Fig. 8. The magnitude of the measured effects are shown to be dependent on the molecular size expressed by the molecular correlation time τ_c at a given spectrometer frequency ω_0. It is very reasonable to use the NOESY technique for large molecules where strong negative effects can be observed, and for small molecules with positive effects. However, for medium-sized molecules only very small or even zero NOE effects are observed. For example, this is true for penta- and hexapeptides (with molecular weight of about 1000) at resonance frequencies of about 300 MHz. In these cases the measurement of ROEs overcomes these problems. Here, the observed effects are more or less independent of the molecular size.[12,13]

Cross-peaks appearing in NOESY spectra are based on the cross-relaxation of *longitudinal* magnetization. The intensities of these signals contain the desired structural information. It can be shown, that the cross-relaxation rate σ, which determines the NOE transfer, is obtained by[41]

$$\sigma = w_2 - w_0 = \frac{\gamma^4 \hbar^2 \tau_c}{10 r^6}\left(\frac{6}{1 + 4\omega^2\tau_c^2} - 1\right) \tag{1}$$

At a given spectrometer frequency ω_0 the size of σ is predominantly determined by the intramolecular correlation time τ_c and the interproton distance (r_{ij}).[6] For

$\omega_0\tau_c$ equals $\sqrt{(5)}/2$, the term in parentheses equals zero, and no NOE transfer takes place in NOESY spectra. In these cases ROESY spectra have to be measured.[42,43]

In ROESY spectra, however, cross-relaxation of *transversal* magnetization generates cross-peaks with intensities containing the interesting distance information. In these experiments the transversal components are locked by a weakly irradiated lock field. Now, the relaxation rates σ determining the ROE can be expressed by:[12,13]

$$\sigma = u_2 - u_0 = \frac{\gamma^4\hbar^2\tau_c}{10r^6}\left(\frac{3}{1+\omega^2\tau_c^2}+2\right) \tag{2}$$

In contrast to the NOE effects, ROE effects are always positive and never vanish. This can be seen from Fig. 8.

Since absolute values of correlation times are not available, it will not be possible to calculate interproton distances r_{ij} from equations (1) and (2). Assuming isotropic tumbling of the whole molecule the equations given above can be rewritten as

$$\sigma^{NOE} \sim \frac{1}{r^6}$$

and

$$\sigma^{ROE} \sim \frac{1}{r^6}$$

It becomes evident that the ratio of the cross-relaxation rates of two pairs of adjacent protons equals the ratio of the interproton distances. Therefore, if one of the two distances is well known, the other distance can easily be calculated. These procedures are described in detail in Section 4.

However, some remarks on the experimental methods for determining cross-relaxation rates have to be made. The duration of cross-relaxation transfer processes, which can be determined by the mixing time, also determines the intensities of the cross-peaks in NOESY or ROESY spectra. Monitoring the temporal development of the peak intensities shows that the mixing time must not be too long, since for large molecules spin diffusion processes occur. For very short mixing times on the other hand, for which (at least theoretically) the best values of relaxation rates should be obtained, the measured cross-peak volumes are too small, increasing the experimental error. Therefore, recording a series of spectra with several mixing times is recommended. From these the initial *build-up rates* represent the cross-relaxation rate and yield the most reliable results (see Section 4).

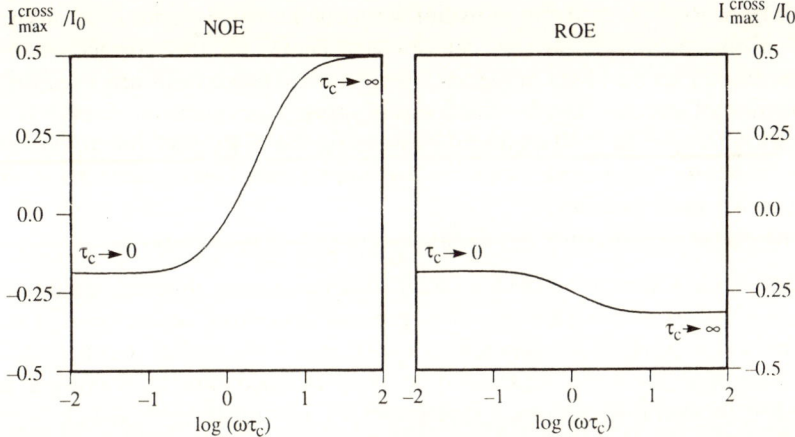

Figure 8 Maximum intensities I_{max}^{cross} of a cross-peak in 2D-NOESY and 2D-ROESY for a two-spin system, scaled on the intensities I_0 of the diagonal signal at mixing time $\tau_m = 0$. In NOESY spectra the cross-peak shows approximately 20% of the diagonal intensity for small molecules ($\tau_c \to 0$), vanishes at $\omega_0\tau_c = \sqrt{5}/2$ and reaches approximately 50% for large molecules ($\tau_c \to \infty$). In ROESY spectra the cross-peak intensity is always negative and is in the range of approximately 20% up to 34% of the diagonal.

4. Transformation of spectral information into three-dimensional structures

4.1. GEOMETRIC DATA FROM NMR PARAMETERS

As mentioned above, the following pieces of NMR information about the spatial structure can be used:

- *dipolar relaxation*, measured as NOE or ROE effects[41]
- *vicinal J coupling*. The size of a three-bond coupling constant depends on the orientation of the dihedral angle via the *Karplus relationships*[36]
- *chemical shift values*. They depend on the through-space orientation (anisotropy effects) and through bond effects of adjacent groups[21]
- *NH proton chemical shifts*. Signals of nuclei which are exposed to the solvent are shifted to higher field with increasing temperature.[45] They also change strongly when the solvent is changed. Line broadening by the addition of free radicals can be used to find solvent exposure of protons.[46]

4.2. THE CALCULATION OF INTERPROTON DISTANCES

The most important information for the spatial structure is obtained from the dipolar cross-relaxation rate σ, which can be measured by the initial build-up of NOE or ROE[47]. As $\sigma_i r_i^6$ is constant, with one known distance r the unknown interproton distances can be determined. These are transformed into the geometric structure of the molecule. Very similar procedures are used for the extraction of structural information out of cross-peak intensities from NOESY and ROESY spectra.[43] The quantitative evaluation of ROESY spectra has to be done carefully, as contributions by J-coupling and offset effects have to be suppressed or considered.[42,43]

The practical procedure for distance determination involves:

- Recording of several NOESY or ROESY spectra using different mixing times under well-controlled conditions (temperature control, suppression of undesired cross-peaks from coherent transfer processes, i.e. via J-coupling). All measurements are recommended to be performed directly following each other, to guarantee identical spectrometer conditions.
- The integrals of the cross-peaks in the base plane corrected spectra are determined and the peak volumes as a function of the mixing times are plotted obeying the equation

$$I(\tau_m) = 1 - \exp\{-\sigma \tau_m\}$$

The derivative of this equation yields the slope of this curve. For a mixing time equal to zero, the initial *build-up rates* are obtained which are directly proportional to the cross-relaxation rates:

$$dI(\tau_m = 0)/d\tau_m = \sigma$$

For proteins, however, due to cross-peak overlap or low signal to noise ratio, a simplified procedure is sometimes used. The volumes of the cross-peaks are estimated by counting well-defined contour lines. Then all values are classified in three groups of large, medium, and small effects. The distances thus obtained are also divided into three classes of long, medium, and short range, respectively.[21]

- For the accurate calculation of distances, *build-up rates* of known interproton distances have to be used.[48] Known distances which are provided by fixed structural elements are shown in Fig. 9. Usually the distances of diastereotopic geminal protons (178 pm), the well-separated aromatic protons of tyrosine (241 pm), or the NH proton of tryptophan and its peri-standing aromatic neighbour (282 pm) are used. However, calibrating *build-up rates* of ROEs has to be done cautiously, since for scalarly coupled nuclei the cross-peaks may contain contributions from net magnetization transfer via J-coupling as well which reduces their intensities.[42]

In smaller cyclic peptides, an iterative procedure can also be used, where the mean deviation of all distances in a molecular dynamics calculation is minimized.[49]

The method described here works under several assumptions: only a rigid molecule gives correct data, because different intramolecular mobility yields different effective correlation times τ_c and the equation given above no longer

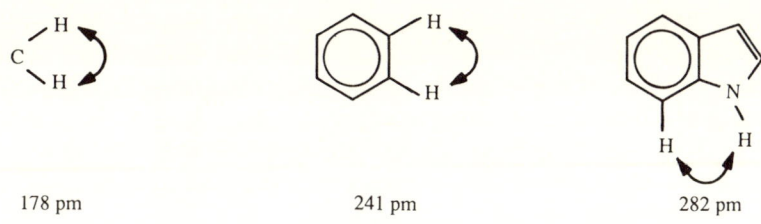

178 pm   241 pm   282 pm

Figure 9 Fixed structural elements with known interproton distances.

holds. Also, an isotropically tumbling molecule is assumed, but it seems that this approximation is more or less fulfilled for the molecules described here. In larger proteins, spin diffusion becomes more and more effective. To obtain reasonably high NOE values the mixing time has in practice to be at least 50 ms. During this time considerable spin diffusion occurs. Thus, the real initial *build-up rate*, which is proportional to the cross-relaxation rate, is difficult to obtain, and longer distances in particular have a tendency to be experimentally too short in comparison to real values. Recently procedures have been described to overcome these problems.[50,51] The interproton distances thus obtained are used for distance geometry calculations and restraint molecular dynamics calculations.

4.3. THE CONFORMATION OF THE BACKBONE

The transformation of intramolecular distances into a molecular geometry is not a 1:1 correlation, for several reasons. First of all, only short-range distances are known. Often signal overlap also prevents exact values being obtained even for well-expressed NOE cross-peaks. In addition there are errors in the measurement of each distance (normally about 10%), and the lack of data in regions where only a few protons are present leads generally to an underdetermined system. However, the constitution of the molecule provides additional information. For example, the structure of a benzene ring is not derived from NOE data between the aromatic protons but inserted as a rigid body in the starting model.

Analysing the conformation of small molecules, one can start directly from a ball-and-stick model and use experimental distances to obtain the spatial structure. With increasing molecular size, not only is such a procedure too tedious, but it may only lead to a local minimum as it is impossible to handle larger systems directly and to find the best agreement between many parameters. Efficient computer programs have been written for the transformation of distances in the molecular geometry.[52] These programs often yield a number of solutions which are compatible with all distances. As an example, different conformations of a nonadecapeptide[53] generated by distance geometry procedures are shown in Fig. 10.

The selection of the 'best structure' may be done by the minimum deviation of the distances in the structural model from the experimental values. However,

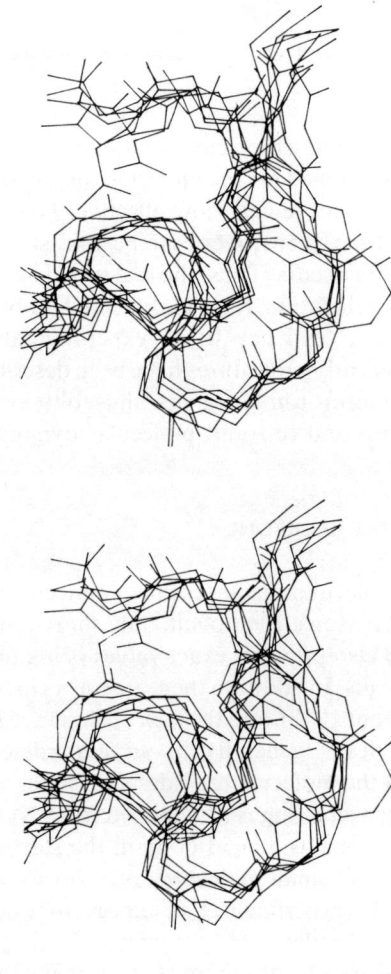

Figure 10 Stereoplots of DISGEO structures of the polycyclic nonadecapeptide Ro 09-0198.[53] The backbone atoms of $N-C^z-C-O$ of seven structures are superimposed. A set of 228 distance restraints from NOE *build-up rates* has been used. The mean deviation of all structures is less than 100 pm.

distance-geometry calculations do not include the energies of the structures and it is therefore necessary to perform a structure refinement by molecular dynamics calculations with constraints from the experimental distances.[54] As a result, often different families of distance-geometry structures then yield comparable solutions which satisfy the experimental data set. The energies of the thus obtained structures are taken as a criterion for the 'best' conformation. It may turn out that the 'best' structure is not identical with a distance-geometry structure which seemed to be the best when only deviations between experimentally observed and calculated distances were considered.

It is important to mention, that the result of a molecular dynamics calculation is a 'trajectory'; this means a large number of slightly different structures following the time evolution during the calculation is obtained (normally time steps of the order of femtoseconds are calculated). We can now sum up at least the rigid parts of the molecule and take the highest probability of the atoms in a certain configuration and call this mean structure 'the conformation'.

The reliability of a structure is determined by the number of experimental data available. Due to the lack of enough experimental data in smaller peptides it is necessary to check the reliability of the obtained structure by looking at other experimental parameters, mainly vicinal coupling constants. They can be calculated from the trajectory of the MD calculation by calculating the J value in each time-step conformation and forming the mean over the whole trajectory. It is not correct to take the dihedral angles of the 'mean conformation' directly to derive the J value.[4]

As restraint MD calculations in the beginning of a structure determination are normally performed *in vacuo*, electric charges will over-emphasize their influence in the force field. We therefore recommend reducing charges for solvent-exposed NH protons in the calculation to take into account solvation effects.[43] It is possible to determine the orientation of NH protons to the solvent by determining the temperature effects on their chemical shifts.[45] The backbone structures of peptides obtained by these procedures seem to fit the knowledge of structures obtained from X-ray analysis very well.

An example is given in Fig. 11. The cyclic hexapeptide was crystallized in two crystalline forms which contain three only slightly different conformations.[43] The NMR-derived conformation was obtained before the X-ray analysis was done. All structures exhibit a very close agreement in all dihedral angles of the backbone.[55] Differences are only observed in the orientation of the side-chains. One must keep in mind that we know from NMR data (see below) that there is more than one conformation populated, e.g. only 60% of the Phe⁶ residue is actually found in this conformation in the equilibrium in solution. One might prove the consistency of the structure (with side-chains fixed in the most stable conformation, see below) by performing a restraint MD calculation including the solvent. Unfortunately only water and carbon tetrachloride force fields are available so far. To reduce computer time normally only a short time range (5–20 ps) is used. Since the viscosity of the solvent prevents the molecule moving far

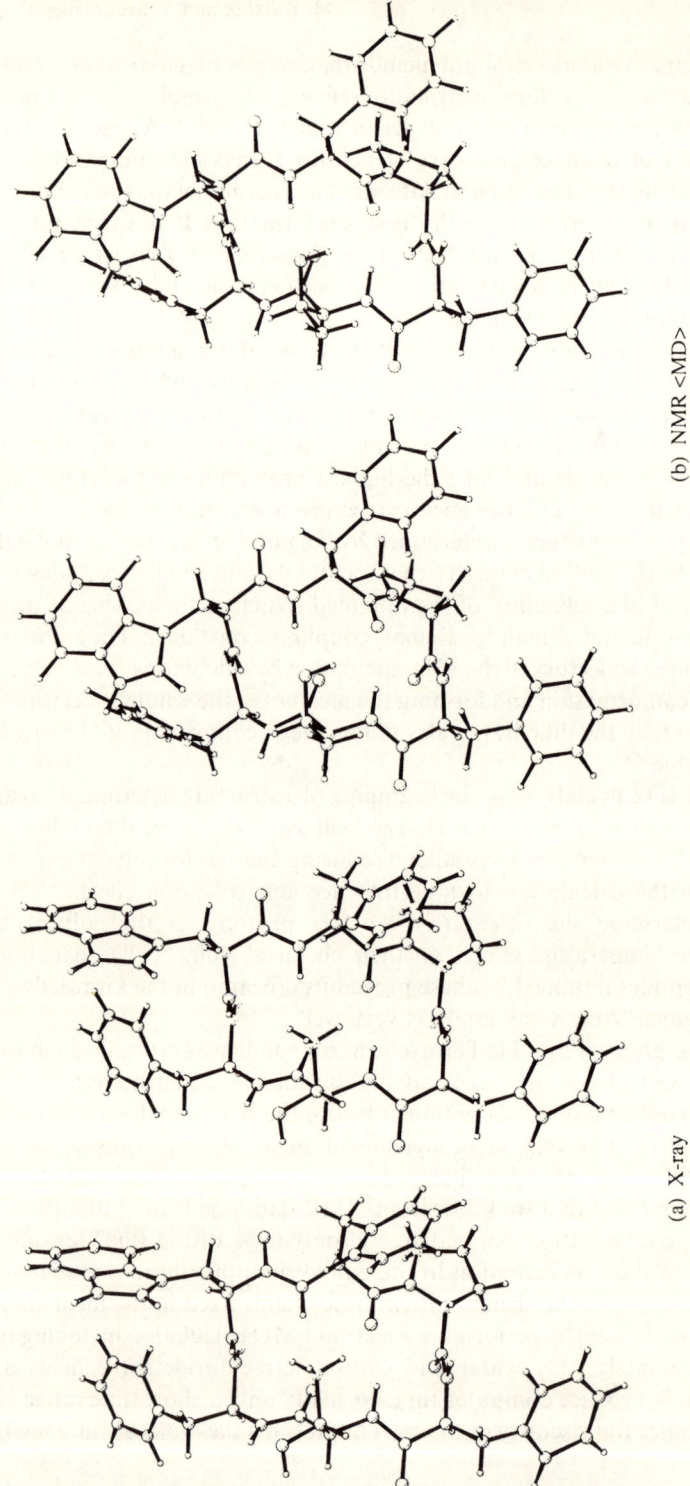

(a) X-ray

(b) NMR <MD>

Figure 11 Stereoplots of the structures of cyclo-[-D-Pro-Phe-Thr-Phe-Trp-Phe-]. (a) Conformation in the crystal. (b) Mean of a 40 ps trajectory obtained from restraint molecular dynamics. The side-chains are shown in the most populated conformation. In the crystal the hydroxyl-proton of the side-chain of Thr is involved in an *intermolecular* hydrogen bond, while in solution it forms an *intramolecular* hydrogen bond.

from the initial structure in such a short time space, a good starting structure is essential for these calculations.[49]

4.4. THE CONFORMATION OF THE SIDE-CHAINS

The orientation of side-chains of the amino acids in a protein are often obtained by restraint MD calculations using only NOE-derived distances. As we mentioned above, this is not true near the surface and for smaller peptides. As MD calculations are normally performed *in vacuo*, *van der Waals* interactions and/or charges induce the tendency to fold the side-chains back to the peptide backbone even when there is experimental evidence for the side-chains to stick out into the solvent.[43,56] We recommend determining the preferred conformation about the angle χ_1 by using coupling constants, and fixing this conformation during the MD run.[43]

Dihedral angles can be derived from three-bond coupling constants via empirically determined *Karplus*-type equations.[38] The *Karplus* function[36] does not give a single solution: up to four angles normally satisfy one experimental J value. However, when several different coupling constants which determine one dihedral angle can be determined, this ambiguity can be removed.

For example, the population of the rotamers with respect to χ_1 can be obtained from proton–proton coupling constants assuming only staggered conformations (Fig. 12) which contribute to the equilibrium.[57]

For the assignment of these rotamers one can use the fact that rotamers I and II contain one *gauche* and one *trans* coupling, whereas rotamer III possesses only two small *gauche* couplings. Therefore from proton coupling constants, only rotamer III can be differentiated from I and II. Assuming a 2.6 Hz *gauche* coupling constant and a 13.6 Hz *trans* coupling the populations of the rotamers I and II can be calculated (but without distinction!) while rotamer III is directly determined. To discriminate I from II one needs the assignment of both diastereotopic β-protons, which can be obtained from stereospecific synthesis or

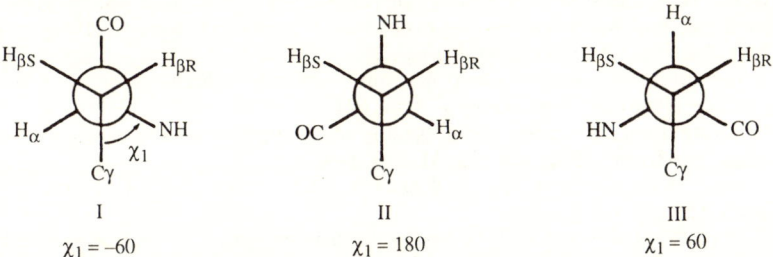

Figure 12 The three staggered conformations about the angle χ_1 of aromatic amino acids.

more easily using heteronuclear coupling constants to the carbonyl carbon.[40] In a few cases also NOE values of the β-protons can be used, especially in the core of a protein, when there are many NOEs in the neighbourhood.[58]

An extension further into the side-chain is difficult to achieve, but often those groups show increasing flexibility along the chain (monitored by carbon longitudinal relaxation times) and the presentation of a specific conformation should not be done.

5. Conclusion

NMR spectroscopy allows us to determine spatial structures of peptides and proteins in solution. Several structures have been obtained in recent years which can be compared to X-ray structures. These provide evidence of certain similarities in the structures but also often distinct differences between the conformation in solution and in the crystal. Often, the backbone conformation is quite similar in both environments, whereas the more flexible side-chains change their orientation. To summarize the results, it seems to us that one of the most often found differences results from *inter*molecular hydrogen bond formation between different molecules in the crystal, whereas these bonds are disrupted when the molecule is dissolved. Often, *intra*molecular hydrogen bonds are then formed, e.g. from hydroxyl groups of Thr[43] or substituted Thr residues.[2] These differences in the conformation have to be taken into account when the 'bioactive conformation' is discussed, and a careful consideration of all possible spatial structures is necessary. There is no doubt that modern NMR spectroscopy now provides an alternative way for structure determination, which can be used also for non-crystalline samples.

References

1. 'IUPAC–IUB commission on biochemical nomenclature' (1970) *J. Mol. Biol.* **52**, 1, *ibid.* (1970) *Biochemistry* **9**, 3471.
2. 'Assignment of the ^{1}H-, ^{13}C-, and ^{15}N-NMR spectra of cyclosporin A in $CDCl_3$ and C_6D_6 by a combination of homo- and heteronuclear two-dimensional techniques' H. Kessler, H. R. Loosli and H. Oschkinat (1985) *Helv. Chim. Acta*, **68**, 661–681.
3. 'Konformation und Wirkung von cyclischen Peptiden' H. Kessler (1982) *Angew. Chem.*, **94**, 509–520; *ibid. Int. Ed. Engl.* **21**, 512–523.
4. 'On the nature of molecular conformations inferred from high-resolution NMR' O. Jardetzky (1980) *Biochim. Biophys. Acta*, **621**, 227–232.
5. 'Conformational restrictions of biologically active peptides via amino acid side-chain groups' V. Hruby (1982) *Life Sci.* **31**, 189–199.
6. *Dynamics of Proteins and Nucleic Acids* J. A. McCammon and S. C. Harvey (1987) Cambridge University Press.
7. 'Effect of anisotropy and anharmonicity on protein crystallographic refinement. An evaluation by molecular dynamics' J. Kuriyan, G. A. Petsko, R. M. Levy and M. Karplus (1986) *J. Mol. Biol.* **190**, 227–254.
8. See, for example: 'Conformational analysis of enkephalin and conformation–activity

relationships' P. W. Schiller (1984). In: *The Peptides*, **6**: *Analysis, Synthesis, Biology* (S. Udenfriend and J. Meienhofer, Eds.) pp. 219–268. Academic Press, Orlando, Florida.

9. 'Evidence for a folded structure of Met-enkephalin in membrane mimetic systems: ^{1}H-NMR studies in sodium dodecylsulfate, lyso-phosphatidylcholine, and mixed lyso-phosphatidylcholine/sulfatide micelles' L. Zetta, A. De Marco and G. Zannoni (1986) *Biopolymers*, **25**, 2315–2323.

10. 'Sequential resonance assignments in protein ^{1}H nuclear magnetic resonance spectra. Glucagon bound to perdeuterated dodecylphosphocholine micelles' G. Wider, K. H. Lee and K. Wüthrich (1982) *J. Mol. Biol.*, **155**, 367–388.

11. 'Gene for a novel tRNA species that accepts L-serine and cotranslationally inserts selenocysteine' W. Leinfelder, E. Zehelein, M. A. Mandrand-Berthelot and A. Boeck (1988) *Nature*, **331**, 723–725.

12. 'Structure determination of a tetrasaccharide: transient nuclear Overhauser effects in the rotating frame' A. A. Bothner-By, R. L. Stevens, J. Lee, C. D. Warren and R. W. Jeanloz (1984) *J. Am. Chem. Soc.*, **106**, 811–813.

13. 'Practical aspects of two-dimensional transverse NOE spectroscopy' A. Bax and D. G. Davis (1985) *J. Magn. Reson.*, **63**, 207–213.

14. 'Modern nuclear magnetic resonance spectroscopy of peptides' H. Kessler, W. Bermel, A. Müller and K.-H. Pook (1986). In: *The Peptides*, **7**: *Analysis, Synthesis, Biology* (S. Udenfriend, J. Meienhofer and V. J. Hruby, Eds.) pp. 437–473. Academic Press, Orlando, Florida.

15. 'Toward the complete assignment of the carbon nuclear magnetic resonance spectrum of the basic pancreatic trypsin inhibitor' G. Wagner and D. Brühwiler (1986) *Biochemistry*, **25**, 5839–5843.

16. 'Correlation of carbon-13 and nitrogen-15 chemical shifts in selectivity and uniformly labeled proteins by heteronuclear two-dimensional NMR spectroscopy' W. M. Westler, B. J. Stockman and J. L. Markley (1988) *J. Am. Chem. Soc.*, **110**, 6256–6258.

17. 'Zweidimensionale NMR-Spektroskopie, Grundlagen und Übersicht über die Experimente' H. Kessler, M. Gehrke and C. Griesinger (1988) *Angew. Chem.*, **100**, 507–554; *ibid. Int. Ed. Engl.*, **27**, 490–536.

18. 'Transformation of homonuclear two-dimensional NMR techniques into one-dimensional techniques using Gaussian pulses' H. Kessler, H. Oschkinat, C. Griesinger and W. Bermel (1986) *J. Magn. Reson.*, **70**, 106–133.

19. 'Novel three-dimensional NMR techniques for studies of peptides and biological macromolecules' C. Griesinger, O. W. Sørensen and R. R. Ernst (1987) *J. Am. Chem. Soc.*, **109**, 7227–7228.

20. 'Three-dimensional J-resolved NMR spectroscopy' G. W. Vuister and R. Boelens (1987) *J. Magn. Reson.*, **73**, 328–333. 'Heteronuclear Three-Dimensional NMR Spectroscopy. A Strategy for the Simplification of Homonuclear Two-Dimensional NMR Spectra' S. W. Fesik and E. R. P. Zuiderweg (1988) *J. Magn. Reson.*, **78**, 588–593.

21. *NMR of Proteins and Nucleic Acids* K. Wüthrich (1986) John Wiley, New York.

22. 'Cytoprotective peptides – conformational analysis of a superpotent cyclic hexapeptide using NMR Spectroscopy and MD calculations' H. Kessler, G. Gemmecker, A. Haupt, J. Lautz and M. Will (1988). In: *NMR Spectroscopy in Drug Research* (J. W. Jaroszewski, K. Schaumburg, and H. Kofod, Eds.) pp. 138–155. Munksgaard, Copenhagen.

23. 'Two-dimensional spectroscopy, application to nuclear magnetic resonance' W. P. Aue, E. Bartholdi and R. R. Ernst (1976) *J. Chem. Phys.*, **64** (5), 2229–2246.

24. 'Coherence transfer by isotropic mixing: application to proton correlation spectroscopy' L. Braunschweiler and R. R. Ernst (1983) *J. Magn. Reson.*, **53**, 521–528. 'MLEV-17-based two-dimensional homonuclear magnetization transfer spectroscopy' A. Bax and D. G. Davis (1985) *J. Magn. Reson.*, **65**, 355–360.

25. 'Correlation of connected transitions by two-dimensional NMR spectroscopy' C. Griesinger, O. W. Sørensen and R. R. Ernst *J. Chem. Phys.*, **64**, 6837–6852.
'Practical aspects of the E.COSY technique. Measurement of scalar spin–spin coupling constants in peptides' C. Griesinger, O. W. Sørensen and R. R. Ernst (1987) *J. Magn. Reson.*, **75**, 474–492.

26. 'Simplification of two-dimensional ^{1}H-NMR spectra using an X-filter' E. Wörgötter, G. Wagner and K. Wüthrich (1986) *J. Am. Chem. Soc.*, **108**, 6162–6167.
'Editing of 2D ^{1}H-NMR spectra using X-half-filters. Combined use with residue-selective ^{15}N-labelling of proteins' G. Otting, H. Senn, G. Wagner and K. Wüthrich (1986) *J. Magn. Reson.*, **70**, 500–505.

27. 'Investigation of exchange processes by two-dimensional NMR spectroscopy' J. Jeener, B. H. Meier, P. Bachmann and R. R. Ernst (1979) *J. Chem. Phys.*, **71**, 4546–4553.

28. 'Relayed-NOE experiments in the rotating frame for sequence analysis of peptides' H. Kessler, G. Gemmecker and B. Haase (1988) *J. Magn. Reson.*, **77**, 401–408.
'Improvement of relayed-NOESY type experiments by implantation of spin-lock sequences' H. Kessler, G. Gemmecker, B. Haase and S. Steuernagel (1988) *Magn. Reson. Chem.*, **26**, 919–926.

29. 'NOESY–TOCSY, eine vorteilhafte 2D-NMR-Technik zur Analyse von Peptid-sequenzen' H. Kessler, G. Gemmecker and S. Steuernagel (1988) *Angew. Chem.*, **100**, 600–603; *ibid. Int. Ed. Engl.*, **27**, 564–566.
'Structural studies of a-bungarotoxin. 2. ^{1}H NMR assignments via an improved relayed coherence transfer nuclear Overhauser enhancement experiment' V. J. Basus and R. M. Scheek (1988) *Biochemistry*, **27**, 2772–2775.

30. 'Two-dimensional relayed coherence transfer-NOE spectroscopy' G. Wagner (1984) *J. Magn. Reson.*, **57**, 497–505.

31. 'Complete assignment of non-carbonylic carbon-13 resonances of tendamistat' H. Kessler, P. Schmieder and W. Bermel (1990). *Biopolymers*, in press.

32. 'Complete sequence-specific proton nuclear magnetic resonance assignment for the α-amylase polypeptide inhibitor tendamistat from *Streptomyces tendae*' A. D. Kline and K. Wüthrich (1986) *J. Mol. Biol.*, **192**, 869–890.
'Studies by ^{1}H nuclear magnetic resonance and distance geometry of the solution conformation of the α-amylase inhibitor tendamistat' A. D. Kline, W. Braun and K. Wüthrich (1986) *J. Mol. Biol.*, **189**, 377–382.

33. 'Three-dimensional NMR spectroscopy of a protein in solution' H. Oschkinat, C. Griesinger, P. J. Kraulis, O. W. Sørensen, R. R. Ernst, A. M. Gronenborn and G. M. Clore (1988) *Nature*, **332**, 374–376.

34. 'Isotope-edited NMR spectroscopy' S. W. Fesik (1988) *Nature*, **332**, 865–866.

35. 'Application of isotope-filtered 2D NOE experiments in the conformational analysis of atrial natriuretic factor(7–23)' S. W. Fesik, R. T. Gampe Jr. and T. W. Rockway (1987) *J. Magn. Reson.*, **74**, 366–371.

36. 'Contact electron-spin coupling of nuclear magnetic moments' M. Karplus (1959) *J. Chem. Phys.*, **30**, 11–15.
'Vicinal proton coupling in nuclear magnetic resonance' M. Karplus (1963) *J. Am. Chem. Soc.*, **85**, 2870–2871.

37. 'Differences and sums of traces within COSY spectra (DISCO) for the extraction of coupling constants: "Decoupling" after the measurement' H. Kessler, A. Müller and H. Oschkinat (1985) *Magn. Reson. Chem.*, **23**, 844–852.
'Spektrenvereinfachung zur Ermittlung von Kopplungskonstanten aus homo-nuklear-korrelierten 2D-NMR-Spektren' H. Kessler and H. Oschkinat (1985) *Angew. Chem.*, **97**, 689–690; *ibid. Int. Ed. Engl.*, **24**, 690–691.
'Fine structure in two-dimensional NMR correlation spectroscopy' H. Oschkinat and R. Freeman (1984) *J. Magn. Reson.*, **60**, 164–169.

38. 'NMR spectroscopy of large peptides and small proteins' V. F. Bystrov, A. S. Arseniev and Yu. D. Gavrilov (1978) *J. Magn. Reson.*, **30**, 151–184.
    'Spin–spin coupling and the conformational states of peptide systems' V. I. Bystrov (1976) *Prog. Nucl. Magn. Reson. Spectrosc.*, **10**, 41–81.

39. 'Assignment of carbonyl carbons and sequence analysis in peptides by heteronuclear shift correlation via small coupling constants with broadband decoupling in t_1 (COLOC)' H. Kessler, C. Griesinger, J. Zarbock and H. R. Loosli (1984) *J. Magn. Reson.*, **57**, 331–336.

40. 'Conformational analysis of amino acid side chains in peptides by quantitative evaluation of cross peaks in COLOC spectra' U. Anders, G. Gemmecker, H. Kessler and C. Griesinger (1987) *Fresenius Z. Anal. Chem.*, **327**, 72–73.
    'Recognition of NMR proton spin systems of cyclosporin A via heteronuclear proton-carbon long-range couplings' H. Kessler, W. Bermel and C. Griesinger (1985) *J. Am. Chem. Soc.*, **107**, 1083–1084.

41. *The Nuclear Overhauser Effect* J. H. Noggle and R. E. Schirmer (1971) Academic Press, New York.

42. 'Frequency offset effects and their elimination in NMR rotating frame cross-relaxation spectroscopy' C. Griesinger and R. R. Ernst (1987) *J. Magn. Reson.* **75**, 261–271.
    'Separation of cross-relaxation and J cross-peaks in 2D rotating-frame NMR spectroscopy' H. Kessler, C. Griesinger, R. Kerssebaum, K. Wagner and R. R. Ernst (1987) *J. Am. Chem. Soc.*, **109**, 607–609.

43. 'Peptide conformations. 46. Conformational analysis of a superpotent cytoprotective cyclic somatostatin analogue' H. Kessler, J. W. Bats, C. Griesinger, S. Koll, M. Will and K. Wagner (1988) *J. Am. Chem. Soc.*, **110**, 1033–1054.

44. 'Assignment of proton and carbon NMR signals of didemnin A in solution' H. Kessler, M. Will, G. M. Sheldrick and J. Antel (1988) *Magn. Reson. Chem.*, **26**, 501–506.

45. 'Konformation und Wirkung von cyclischen Peptiden' H. Kessler (1982) *Angew. Chem.*, **94**, 509–520; *ibid. Int. Ed. Engl.*, **21**, 512–523.

46. 'Conformations of cyclic octapeptides. 3. cyclo-(D-Ala-Gly-Pro-Phe)₂. Conformations in crystals and a $T_{1\rho}$ examination of internal mobility in solution' K. D. Kopple, K. K. Bhandary, G. Kartha, Y.-S. Wang and K. N. Parameswaran (1986) *J. Am. Chem. Soc.*, **108**, 4637–4642.

47. *Principles of Nuclear Magnetic Resonance in One and Two Dimensions* R. R. Ernst, G. Bodenhausen and A. Wokaun (1987) Clarendon Press, Oxford.

48. 'Build-up rates of the nuclear Overhauser effect measured by two-dimensional proton magnetic resonance spectroscopy: implications for studies of protein conformation' A. Kumar, G. Wagner, R. R. Ernst and K. Wüthrich (1981) *J. Am. Chem. Soc.*, **103**, 3654–3658.

49. 'Synthesis and conformational analysis of cyclic alanine-analogues of thymopoietin by NMR spectroscopy and molecular dynamics calculations *in vacuo* and in solution' H. Kessler, R. Kerssebaum, A. G. Klein, R. Obermeier and M. Will (1989) *Liebigs Ann. Chem.*, 269–294.

50. 'A theoretical study of distance determination from NMR. Two-dimensional nuclear Overhauser effect spectra' J. W. Keepers and T. L. James (1984) *J. Magn. Reson.*, **57**, 404–426.

51. 'Protein structures from NMR' R. Kaptein, R. Boelens, R. M. Scheek and W. F. van Gunsteren (1988) *Biochemistry*, **27**, 5389–5395.

52. *Theory and Application of Distance Geometry* L. M. Blumenthal (1970) Chelsea, New York.
    'The ellipsoid algorithm as a method for the determination of polypeptide conformations from experimental distance constraints and energy minimization' M. Billeter, T. F. Havel and K. Wüthrich (1987) *J. Comp. Chem.*, **8**, 132–141.

'An evaluation of the combined use of nuclear magnetic resonance and distance geometry for the determination of protein conformations in solution' T. F. Havel and K. Wüthrich (1985) *J. Mol. Biol.*, **182**, 281–294.

'Solution conformation of proteinase inhibitor IIA from bull seminal plasma by H-1 NMR and distance geometry' M. P. Williamson, T. F. Havel and K. Wüthrich (1985) *J. Mol. Biol.*, **182**, 295–315.

53. 'Complete sequence determination and localisation of one imino and three sulfide bridges of the nonadecapeptide Ro 09–0198 by homonuclear 2D-NMR spectroscopy. The DQF-RELAYED-NOESY-experiment' H. Kessler, S. Steuernagel, D. Gillessen and T. Kamiyama (1987) *Helv. Chim. Acta*, **70**, 726–741.

'The structure of the polycyclic nonadecapeptide Ro 09–0198' H. Kessler, S. Steuernagel, M. Will, G. Jung, R. Kellner, D. Gillessen and T. Kamiyama (1988) *Helv. Chim. Acta*, **71**, 1924–1929.

54. 'Local and collective motions in protein dynamics, mobility and function in proteins and nucleic acids' M. Karplus, S. Swaminathan, T. Ichiye and W. F. van Gunsteren (1983) *Ciba Found. Sympos.*, **93**, 271–290.

'Computer simulation as a tool for tracing the conformational differences between proteins in solution and in crystalline state' W. F. van Gunsteren and J. C. Berendsen (1984) *J. Mol. Biol.*, **176**, 559–564.

'Use of molecular dynamics computer simulations when determining protein structure by 2D-NMR' W. F. van Gunsteren, R. Kaptein and E. R. P. Zuiderweg (1984) *NATO/CECAM Workshop on Nucleic Acids Conformations and Dynamics* (Olsen, W. R., ed.), Orsay, pp. 79–82.

Groningen Molecular Simulation Package (GROMOS) Library Manual, GROMOS W. F. van Gunsteren and H. J. Berendsen (1987) Biomos B. V., Nijenborg 16, NL-9747 AG Groningen; 1–129.

55. 'Conformational analysis of cyclic peptides in solution' H. Kessler, J. W. Bats, K. Wagner and M. Will (1989) *Biopolymers*, **28**, 385–395.

56. 'Molecular dynamics simulation of cyclosporin A: the crystal structure and dynamics modelling of a structure in apolar solution based on NMR data' J. Lautz, H. Kessler, R. Kaptein and W. F. van Gunsteren (1987) *J. Comp. Aid. Mol. Design* **1**, 219–241.

'On the dependence of molecular conformation on the type of solvent environment: a molecular dynamics study of cyclosporin A' J. Lautz, H. Kessler, W. F. van Gunsteren, H.-P. Weber and R. M. Wenger (1990) *Biopolymers*.

57. 'Nuclear magnetic resonance study of some α-amino acids–I. Coupling constants in alkaline and acidic medium' K. G. R. Pachler (1963) *Spectrochimica Acta*, **19**, 2085–2092.

'Nuclear magnetic resonance study of some α-amino acids–II. Rotational isomerism' K. G. R. Pachler (1964) *Spectrochim. Acta*, **20**, 581–587.

'The proton magnetic resonance spectra of some substituted ethanes: the influence of substitution on CH–CH coupling constants' R. J. Abraham and K. G. R. Pachler (1963) *Mol. Phys.*, **6**, 165–182.

58. 'Protein structure determination in solution by nuclear magnetic resonance spectroscopy' K. Wüthrich (1989) *Science*, **243**, 45–50.

—— *Chapter 3* ————————————————————

Crystallography in drug design

J. Murray-Rust

1. Introduction

Since the publication of the X-ray crystal structure determinations of myoglobin[1] and haemoglobin,[2] crystallography has provided the complete description of the three-dimensional organization of several hundred proteins and several thousand organic molecules. Although the advent of 2D and 3D NMR[3,4] offers another technique, with the apparent advantage of being carried out in solution – a more 'biological' environment – crystallography is still the only method useful for larger proteins (greater than about 100 amino acids). Comparison of the two methods[5-8] shows that the overall chain-folds deduced are similar, but the N- and C-termini are often more poorly defined by NMR, which may indeed reflect real solution behaviour. Experimental requirements govern the choice of technique to some extent, in that crystallography needs crystals of about 0.1– 0.3 mm in size (a non-trivial exercise), and NMR a solution several millimolar in protein which is unaggregated. Combination of data from both methods, together with molecular dynamics simulations, is a rapidly expanding field which will provide even more powerful tools for investigations of protein conformations and structural mobility.[9]

1.1. HISTORY AND DEVELOPMENT

Early work on the crystallization of biologically active compounds centred around those which crystallized relatively easily; some small molecules and a few proteins. However, it made contributions at a very fundamental level to knowledge of the chemical formulae of important compounds such as cholesteryl iodide[10] and vitamin B_{12}.[11] The various levels of structural organization of polypeptides have been investigated from small chains such as enkephalin (see below) and oxytocin,[12] through intermediate sizes[13,14] to small globular proteins such as insulin[15] and on to very large assemblies such as rhinovirus[16] which has a molecular weight of about 8.5 million.

Analysis of peptides therefore spans the full range of crystallographic techniques from those used for 'small' molecules to those used for macro-molecules, and technical progress continues to be made in all areas. The information from these studies, combined with physicochemical and biochemical results, has shown

- how individual enzymes and families of enzymes operate
- combined with molecular biological techniques, the effects of mutations on their behaviour
- some general principles for protein folding, structure, and design

to arrive at a sophistication where the first 'designer proteins' have been produced.[17]

1.2. WHY CRYSTALLOGRAPHY?

Because of its speed and objectivity, crystallography is the method of choice for determination of small molecule structures (not peptides in particular), and many pharmaceutical companies regularly determine tens of structures per year in this way. Determination of chemical identity is straightforward, and so is relative (or even absolute) chirality, which is particularly important for molecules with several chiral centres, or where different isomers have different activities. Crystallography is still the only way of obtaining complete three-dimensional structural information for proteins much in excess of approximately 100 amino acids in size. The techniques of crystallography are all in a period of rapid advance which has been described as a new revolution,[18] and the result will be the solution of larger and more complicated problems, and some extension from essentially static to kinetic experiments.

Naturally-occurring polypeptides are most commonly linear arrangements of amino acids, and of the 20 commonest all except glycine are chiral and almost exclusively in the L configuration. Small peptides, of say 3–20 amino acids, tend to be able to adopt several low-energy conformations, depending on their environment, so that changes in the crystallization solvent can lead to different conformers being favoured (e.g. enkephalin, discussed below). As molecular size increases, recognizable secondary structure develops, so that a small protein such as insulin has helices and β-sheet, and assembles round a hydrophobic core. Although there is a conserved core, there is still considerable conformation variation in insulin crystallized in various conditions. Larger proteins have complex combinations of secondary structural motifs, such as the β-barrel or the

Rossman fold, and may have several well-defined domains.[19] The overall structure is then less sensitive to the environment, so that the influence of solvent is less marked, although mobility in loop regions, for example, is important for activity.

A drug designer might wish to ask several types of question about the molecules in the system under investigation, such as

- what are the low-energy conformations?
- are they different in solution and when the molecule(s) are bound?
- what is the detailed chemical mechanism of the reaction?
- how does a molecule bind to its receptor?
- which regions of a protein are responsible for receptor binding and which for biological activity?

The objective might then be to design a molecule which has greater specificity for a particular receptor, is a better agonist or antagonist, replaces a deficiency in some individuals, or disrupts the process of infection, e.g. by preventing viral entry into a cell. The scope and detail of the contribution crystallography can make here varies with the size of the molecules, as well as with how well particular systems are already characterized by other techniques. The crystal structures of individual small non-cyclic peptides indicate low-energy conformations, and might suggest modifications to make these more rigid, but say relatively little about the bound state(s). For this, crystallization of a complex between an enzyme, for example, and its peptide substrate is needed, and can directly show the conformation most favourable for binding. Structures of proteins can give insight into the working of individual proteins, or into families of proteins where there is homology between the members. Proteins can be modified, either chemically or by site-directed mutagenesis, to test hypotheses about binding or reaction, or to change aggregation or solubility behaviour. Interactions between proteins, or between proteins and other macromolecules such as DNA, are crucial to understanding biological systems, which is necessary if they are to be changed.

The 'information explosion' of small molecule structures several years ago has been followed by a similar dramatic increase in the number of published protein structures, so that use of databases is essential. The Cambridge Crystallographic Database[20] ($> 55\,000$ organic molecules) has facilities for rapid search for related structures and geometrical and statistical calculations. The Brookhaven Protein Data Bank[21] contains refined structures of about 300 proteins. The structural data, and the software to search and analyse them, are a major tool for molecular design. Combination with protein sequence information and alignment procedures allows modelling of homologous structures where one member of a family is already known, and since there are many more sequences than protein structures, one eventual goal is the prediction of protein folding from sequence data alone.

The aim of this chapter is to give some general pointers to the theory and practice of crystallography, although in no way being comprehensive, and to show where recent developments extend the size and scope of the systems which can be examined, and the speed with which this can be accomplished.

2. Theory

We need to use X-ray diffraction because, although electron microscopy is advancing rapidly, and electron crystallography has shown the structures of some two-dimensional arrays at around 20 Å resolution (e.g. cholera toxin[22] and tetanus toxin[23]) there is as yet no physical lens that allows direct imaging of molecules analogous to that of a powerful visible-light microscope. The result of the X-ray experiment – a 'solved' structure – gives the positions of the atoms, from which molecular geometry may be calculated, and thermal parameters which are related to the thermal movements of the atoms, and possibly to disorder. The wavelengths of X-rays (about 2 to 0.6 Å) are similar to the interatomic distances in the crystal, and the arrangement of the atoms is ordered and periodic in three dimensions, so that the crystal acts as a three-dimensional diffraction grating. The general theory of X-ray crystallography and structure solution is well laid out in several monographs[24-28] but brief descriptions of some terms might be useful. (There is a glossary in Glusker and Trueblood).[28]

A crystal can be thought of as a translationally repeating pattern in which the minimally repeating unit (*unit cell*) is arranged on a three-dimensional lattice. There are 14 available (Bravais) lattices, and combining these with *symmetry elements* (mirror planes, rotation axes, screw axes, glide planes) gives rise to the 230 *space groups*. The unit cell of any crystal is conventionally chosen to exhibit the full symmetry of the lattice. The space groups are listed, together with their component symmetry elements, in Volume A of the *International Tables for X-ray Crystallography*.[29] The *asymmetric unit* is the smallest part of a crystal structure to which application of the space group symmetry operations will generate the complete unit cell. It is not necessarily a chemically sensible entity; it may contain only a portion of the molecule, if the molecule lies on a symmetry element, or more than one molecule, unrelated by exact symmetry. The presence of several molecules in the asymmetric unit increases the number of parameters needed to define the structure, and may make solution more difficult; on the other hand multiple copies related by non-crystallographic symmetry (as in the icosohedral viruses) allow powerful averaging techniques to be employed to improve the electron density.

X-rays are scattered by the electrons in the crystal, and the atoms are therefore assumed to be spheres located at maxima in the electron density (except for a few very accurate studies of electron density, which use more sophisticated models).[30] The aim of the single crystal diffraction experiment is to obtain from the directions in which the X-rays are scattered information on the shape and size of the unit cell, and from their intensities their electron density and thus the spatial arrangement of the atoms in the unit cell. The *Bragg equation*

$$\lambda = 2d_{hkl} \sin \theta$$

shows the direction of the diffraction for a reflexion with Miller indices hkl, and the intensity (I) is proportional to the square of the structure factor (F). The structure factor equation for a particular reflexion (hkl) is

$$F_{(hkl)} = \sum_j f_j \exp 2\pi i \, (hx_j + ky_j + lz_j)$$

which is equivalent to $|F|\exp i\alpha(hkl)$. Thus every atom in the structure contributes to each reflexion.

If the Fs are known, the electron density at xyz is given by

$$p(xyz) = \frac{1}{V}\sum_{all}\sum_{hkl}\sum F(hkl) \exp -2\pi i \, (hx + ky + lz)$$

Unfortunately in the diffraction experiment only the magnitudes ($|F|$) and not the phases (α) are measured – this is the 'phase problem' which is basic to crystallography, and which the various methods of structure solution have been designed to overcome. The above equations assume atoms at rest; thermal vibration in fact causes a decrease in the scattered intensity which can be described by a factor $\exp(-B\sin^2\theta/\lambda^2)$ where the isotropic *temperature factor B* is related to the mean square displacement. Where more data are available the thermal motion may be represented by six anisotropic thermal parameters, to give the directions of the displacements. *Refinement* is then the process of improving the trial atomic positions and thermal parameters to obtain the best fit between observed and calculated structure factors, conventionally represented by the R-factor

$$R = \sum |(|F_o| - |F_c|)|/\sum |F_o|$$

The *intensity* (I) measured is proportional to $F^2 \times (\text{Lp}) \times (\text{Abs})$ where Lp is a geometric correction depending on the exact geometry of the equipment used and Abs is a correction for absorption of some X-ray energy by the crystal. The former is calculated and applied in *data reduction*; the latter may be measured experimentally, approximated from the crystal geometry, or even neglected.

Friedel's law states that $I_{hkl} = I_{\bar{h}\bar{k}\bar{l}}$, that is the diffraction pattern is centrosymmetric. However, near to an X-ray absorption edge for a given atom, 'anomalous scattering' causes this law to break down. The differences between Friedel pairs of reflexions arising from anomalous scattering (the anomalous differences) can be used to solve the phase problem (see, for example, Hendrickson[31]) and to help with the determination of chirality. The effect on the intensities is fairly small, but useful where there are metal ions in proteins (either native, such as in the iron-transport transferrins, or introduced deliberately to produce heavy atom derivatives). In small molecules, such as sugars, the anomalous scattering from oxygen atoms can be sufficient to determine the chirality of the molecule.

A term frequently used in protein crystallography is *resolution*, which is roughly the smallest distance which can be distinguished between two points. For a low resolution (>6 Å) structure, only a gross description of the molecular shape is possible, whereas at high resolution (< 1 Å) individual atoms can be resolved, and thermal motion described. Most small molecule structures are at high resolution, but proteins may fall anywhere on the high–low continuum, depending *inter alia* on the quality of the crystals and the stage the data collection/solution/refinement process has reached; it is quite usual to publish an initial solution at low resolution and then to further extend the phasing to obtain a more detailed picture.

3. Experimental methods

The essential stages of any single crystal structure determination are

- crystallization
- data collection
- structure solution
- refinement

The techniques for small and large molecules overlap considerably, but as the size of the problem increases some changes in approach are necessary. Once crystals have been obtained, it is often possible, for a small-molecule structure, for data collection, solution, and refinement to be carried out by automated diffractometers and program packages, with a minimum of manual intervention; this is not yet true of proteins. Each stage of the process has its own difficulties, but in each area noticeable progress is being made.

Much of the methodology is described in the texts referred to above, and in two volumes of *Methods in Enzymology*[32] devoted to protein crystallography, so what follows is only a brief summary of each area, mainly to point out recent developments.

3.1. CRYSTALLIZATION

For routine crystallographic measurements, crystals of side 0.1–0.5 mm are generally needed, though in practice long thin needles or extremely thin plates may be all that is available. In spite of attempts at systematization, there is still an element of art (or luck) in crystallization, and many time-consuming trials may be required to find the optimum conditions, so that the time scale varies from hours to months (or infinity).

In general small-molecule crystals, whether peptides or not, are grown from a single solvent, or possibly a combination of two solvents, by heating/cooling, evaporation, or vapour diffusion. Solvents and conditions are screened until large enough crystals are produced; in some cases, crystals can be obtained from several solvents, either in several crystal forms or only one. Proteins are normally crystallized from a buffered solution, because the effect of pH is much more marked, with the protein being least soluble around the isoelectric point. The solubility is also affected by other factors such as the ionic strength of the solution, temperature, presence of organic solvents, and counter-ions.[33] The importance of purity for successful crystallizations has also been stressed.[34] Many methods are used; those above plus dialysis, changes in pH and addition of precipitants such as methylpentanediol (MPD) and polyethyleneglycol (PEG), so that several parameters may have to be varied. Cryocrystallization[35] uses rapid freezing so that data can be collected at low temperatures, where thermal motion is reduced. The Biological Macromolecular Crystallization Database[36] is being compiled to form a record of crystallization conditions and crystal data for over 600 molecules including proteins, nucleic acids, and viruses, to assist in experiment design. Crystallization of macromolecules still tends to be very labour-intensive, although the recent developments in robotic methods[37-39] should enable it to become more efficient and reproducible. Crystallization in microgravity (i.e. on spacecraft) has been shown to produce good quality crystals, and new apparatus for this method is being developed[40] but cost is likely to prevent it being a routine procedure. Some beautiful examples of macro-molecular crystals are shown in reference 41.

Once crystals have been produced they are mounted, usually either by being glued to a glass fibre (for small molecules which are air-stable) or by being sealed in a thin-walled glass capillary tube containing a drop of the mother liquor (for proteins).

Preliminary photographs will show the resolution and quality of the diffrac-tion and the space group and cell dimensions can be determined. Figure 1 shows typical X-ray diffraction patterns of enkephalin and insulin. Measurement of the density, or its approximate calculation, then gives the number of molecules in the asymmetric unit, and the percentage of solvent in the unit cell.

At this point procedures for 'small' and 'large' molecules diverge somewhat, in that usually only crystals of the compound itself are required for small molecules, whereas for proteins several heavy-atom derivatives are needed for the multiple isomorphous replacement method. For this, crystals are soaked in solutions containing heavy-atom compounds, and then photographed; the aim is to find compounds which will bind to a small number of sites in the protein, without disrupting the essential structure. The photograph will then show very similar cell dimensions to the native crystal (< 1%), but small changes in the intensities of the diffracted spots. Again, this is time-consuming; often the crystal is destroyed, or no changes are observed, or there is a large change in the cell dimensions. A database of heavy-atom derivatives which have successfully been used in solved structures is being developed,[42] and may offer guidelines for this process. Very high-molecular-weight structures, such as the nucleosome core particle, need heavy-atom cluster compounds to make useful derivatives.[43] Another approach becoming more feasible is to modify or introduce specific metal binding sites by site-directed mutagenesis, as for example the introduction of additional cysteines into phage T4 lysozyme[44] and colicin A.[45] Of course in some cases a similar structure may already have been solved, so that only a native protein dataset is collected, and the structure is solved by molecular replacement methods.

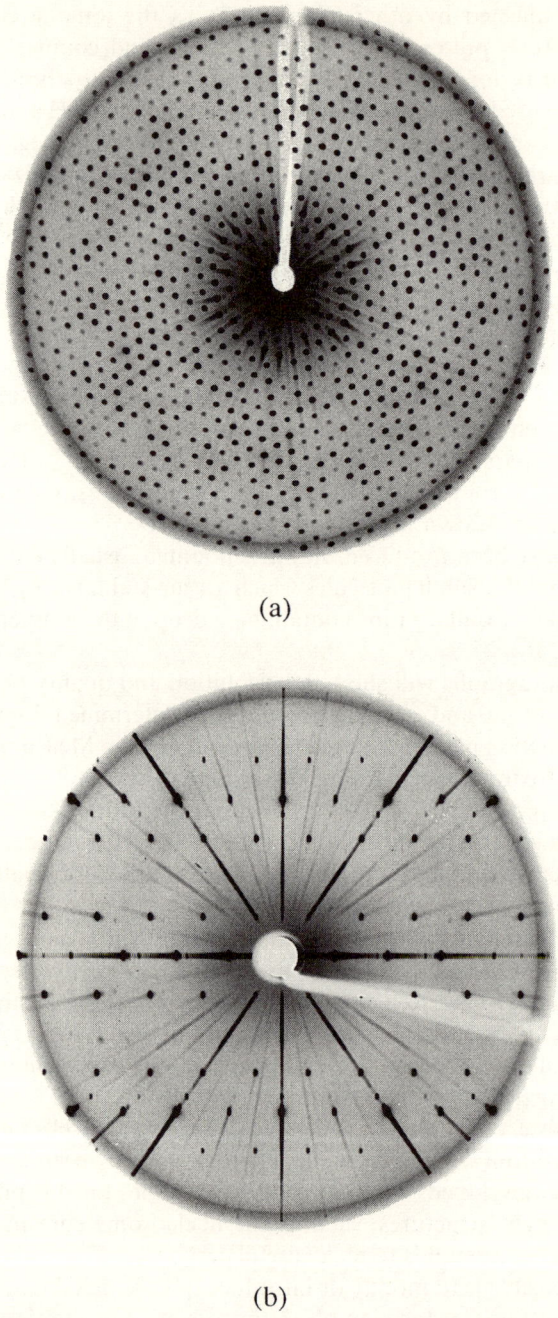

(a)

(b)

Figure 1 X-ray diffraction pictures of (a) insulin and (b) enkephalin. The inverse relationship between cell dimensions and spot separation is very clear.

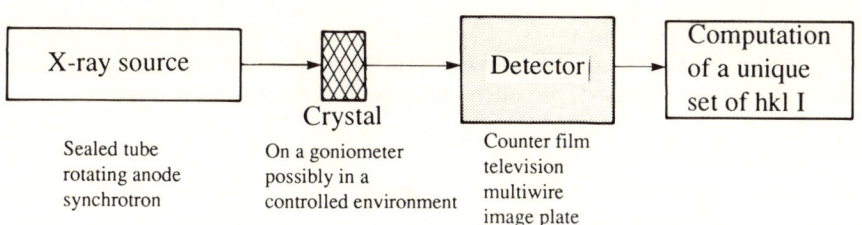

X-ray source		Crystal		Detector		Computation of a unique set of hkl I
Sealed tube rotating anode synchrotron		On a goniometer possibly in a controlled environment		Counter film television multiwire image plate		

Figure 2 The components of a data collection system.

3.2. DATA COLLECTION

Data collection is shown schematically in Fig. 2. For a small molecule, given good stable crystals, this is a fairly routine procedure, on a time-scale of hours to days. For proteins the situation is more complicated because of factors such as their generally weaker diffraction, temperature sensitivity, and susceptibility to X-ray damage, so that several crystals may be needed to collect a complete dataset. A very large volume of data must be stored and processed for the native protein and its derivatives. The time-scale here can be weeks or months; but recent developments at all stages of data collection are decreasing this rapidly. This section provides only a pointer to how developments are expanding the size and type of problems which can be tackled, with more intense X-ray sources, faster detectors and more computing power.

3.2.1. *The X-ray source*

This may produce either monochromatic or white radiation. Traditional methods use monochromatic radiation from a sealed tube or rotating anode source, and use movement of the crystal (and of the detector in a diffractometer) to obtain all the reflexions that are available. The new, more intense synchrotron sources[46] permit use of conventional diffractometers or oscillation cameras to collect data from smaller and much more weakly diffracting crystals,[47] and in a much shorter time. They also offer the ability to tune the radiation precisely to particular wavelengths so that anomalous diffraction can be measured near the absorption edges of heavy atoms. Several structures with intrinsic transition metal ions have been solved using multiple wavelength anomalous diffraction (MAD) methods.[48] Use of shorter wavelengths also reduces radiation damage to the crystal.

But synchrotron radiation also produces a good distribution of wavelengths in white radiation, which can be utilized in the *Laue method*. Here the sample and detector are stationary, and different wavelengths (say 0.2–2.5Å) satisfy the diffraction condition for different reflexions. Exposure times are very short (of the order of seconds or less) and only a few sample orientations are required to obtain most of the theoretically available data, particularly in high-symmetry space groups. The first electron density map of a protein crystal derived from Laue data was that of glycogen phosphorylase b,[49] and the short exposure times give the possibility of taking 'snapshots' of kinetic processes. Phosphorylase is catalytically active as a crystal (the whole crystal, not just the surface), and Laue photographs of various stages of the reaction have been interpreted.[50]

3.2.2. *The crystal*

This is either mounted on a fibre or in a capillary tube in contact with the mother liquor, and is then placed on an adjustable goniometer head which holds it in the X-ray beam. More specialized mountings may be used to control the temperature of the crystal by passing a stream of heated or cooled dry gas around it,[51] and even a pressure cell for proteins has been designed[52] although many protein crystals are extremely fragile. Flow cells have been used where the liquid surrounding the crystal is pumped from an external reservoir, so that it may be changed without disturbing the orientation of the crystal. The goniometer head is then fixed on the data collection equipment; either a camera or a diffractometer.

3.2.3. *The detectors*

These are either counters, such as a scintillation counter, on a diffractometer, which measures a single reflexion at a time; or film, such as that used on an oscillation camera, where many reflexions are recorded at one angular setting. Diffractometry is used for most small-molecule crystals, and for smaller proteins. The oscillation camera has been preferred, especially when used with a synchrotron source, for larger unit cell proteins, but new equipment offers more rapid and accurate data collection for these bigger problems. Electronic area detectors, such as the multiple wire proportional counter and the TV detector, collect many reflexions simultaneously, like the oscillation camera, but also allow on-line display of the diffraction pattern, and are more sensitive than film.[53] The imaging plate, used in place of film, has a larger dynamic range, reducing the necessity for scanning multiple film packs.

3.2.4. *Computers*

These are needed to control the equipment, to scan the films and to process the resultant raw data to give a unique set of *hklI* values. Data transmission and storage become more critical with rapid data collection methods and the large number of data required for big structures. New and more powerful computer hardware enables greater amounts of data to be processed; and developments in techniques, such as the Laue method, require new software to predict diffraction patterns, scale between films, apply geometric corrections, etc., so that this is a very active area of progress.

3.3. STRUCTURE SOLUTION

Structure solution is the process of obtaining a trial (approximate) structure which can then be refined. The main approaches are

- vector search methods (Patterson search or molecular replacement)
- direct methods

Early attempts to solve the phase problem were basically trial-and-error, combined with constraints from chemistry or symmetry; together with the limitations imposed by manual methods of calculation, this meant that only relatively small structures could be attempted.

The discovery of the Patterson function,[54] which contains peaks for each interatomic vector, with weights proportional to the product of the two atomic numbers, meant that if a structure contains a heavy atom then the associated peaks can be found, and for small molecules the whole structure can then usually be solved. Isomorphous replacement also uses a Patterson function to locate vectors between heavy atoms introduced by soaking into protein crystals. With a single isomorphous derivative (SIR) two possible values of the phases, enantiomorphic to the heavy atom vectors, are obtained; this is resolved by using also anomalous scattering (SIRAS), multiple derivatives (MIR), or both (MIRAS).

Molecular replacement methods[55] are used when a similar structure is known and can be employed as a search model, or when several copies of a molecule are present in the asymmetric unit and are related by non-crystallographic symmetry; they are generally applied to large structures and have had spectacular success in the field of virus crystallography. A recent possibility is that of solving crystal structures using as the search model a three-dimensional structure obtained from NMR interproton distance restraints, and this has been shown to work using crambin as a model system.[56]

Direct methods are those where assumptions about the electron density – that it will always be positive, and will have Gaussian peaks centred at the atomic positions – are incorporated in methods to derive the phases from the amplitudes alone. At first the method was restricted to centrosymmetric structures, but the extension to non-centrosymmetric structures has meant its wide application to small-molecule structures, many of which are routinely solved by automatic input into program packages. Knowledge of the geometry of a fragment of the structure or the location of a part of it can be incorporated into the direct methods algorithm. However, the presence of disorder, pseudosymmetry, or other usual features can still give rise to difficulties. A recent review of the history and future of direct methods was given by Woolfson.[57] Attempts to apply direct methods to macromolecules are less advanced, and cannot be done in quite the same way. However they are being developed for determination of heavy atom substructure, phase extension and refinement, and phasing of SIR and SAS data.[58]

A special technique should be mentioned for protein–substrate complexes, where the protein structure is already known; if the substrate can then be crystallized complexed to the protein, a difference map between the complex and the protein can show the substrate structure very clearly.

So, when is a structure solved? For a small molecule, most of the expected atoms are present, and in a chemically reasonable arrangement. It can then be refined (see below) and although sometimes there are alternative plausible solutions to be considered, this usually marks the end of structure solution. For a macromolecule, the heavy atoms' position or coordinates of the search molecule

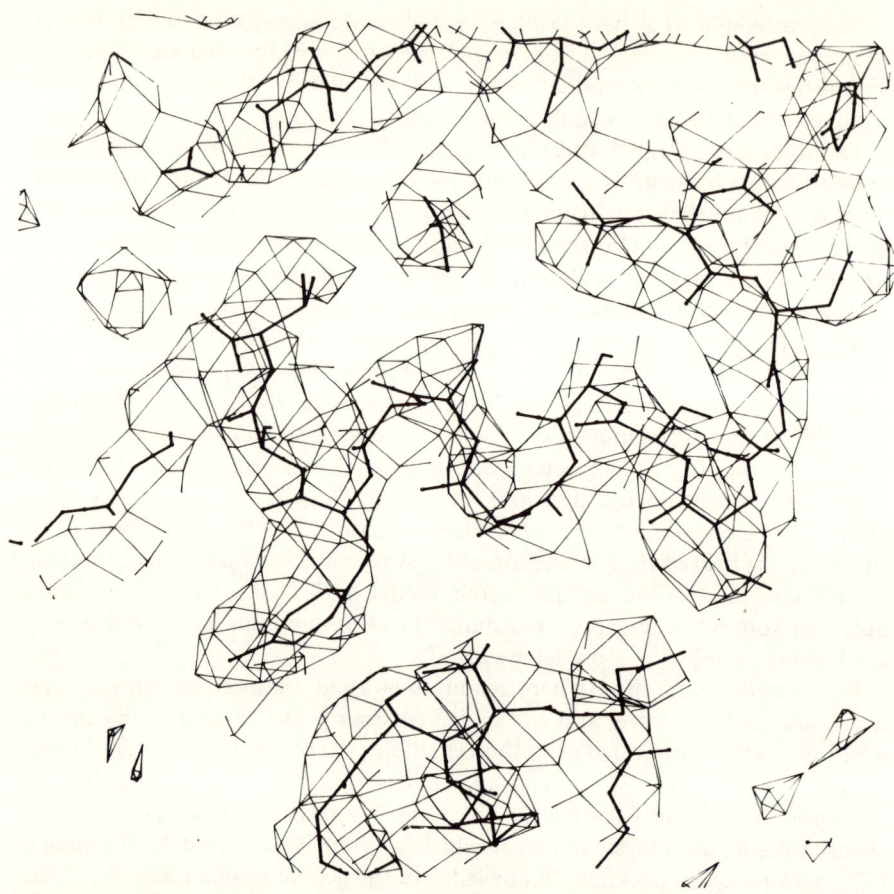

Figure 3 Monochrome plot from a colour graphics system of part of a typical region of electron density (light lines) at about 3.5 Å resolution with a polypeptide chain (heavy lines) being fitted into it. A segment of helix runs from right to left across the centre of the picture.

are used to give calculated phases for an electron density map, which will hopefully reveal the location of some or all of the macromolecule. A protein structure can only be considered solved if the map can be interpreted appropriately for the resolution of the data. At very low resolution this may simply be the boundary of the molecule, but at increasing resolution the fold, and then the direction of the main chain and location of side-groups, should be identifiable. The use of computer graphics for interpretation of electron density maps has developed enormously with programs such as FRODO.[59] A small portion of such a 'chicken-wire' map with a chain located in it is shown in Fig. 3. There are two facets to this; one is the ability to display the electron density and atoms simultaneously and to manipulate the atoms to best fit the density, and the other is in the automatic interpretation of the maps using pattern recognition and artificial intelligence methods.[60]

In a small proportion of problems, where the interpretation proves difficult, or impossible, then new derivatives may be necessary, or the molecular replacement solution may have to be reconsidered, possibly with a modified search model. Then large amounts of human and computer resources can be required and several years may elapse before the structure can be considered solved (if it is solved at all).

3.4 REFINEMENT AND ASSOCIATED TECHNIQUES

Refinement is the process of adjusting the initial atomic parameters so that the structure factors calculated from them (F_c) agree better with the observed structure factors (F_o).

A trial structure for a small molecule will have most atoms slightly displaced (say 0.3–0.5 Å), some may be completely misplaced (wrong chemical connectivity) or some may be missing. If most of the trial structure is approximately known, a difference map, with amplitudes ($F_o - F_c$) and phase angles from the trial structure will often show peaks at the positions of missing atoms and negative regions where an atom should not have been placed. Once most of the expected atoms have been located, a non-linear least-squares refinement is used to optimize the fit between F_o and F_c. The least-squares process is iterative, and has converged when the shifts in parameters are less than their standard deviations, and the R-factor is a minimum (for a well-behaved small molecule this might be of the order of 5% or less). Incorrectness in an initial model is revealed in least-squares in several ways. The R-factor may fail to decrease, atoms may move to chemically ridiculous positions, or the temperature factor of an atom may increase rapidly–the usual sign that it should not be where it has been placed. Disorder may be discovered if an atom or group of atoms fails to refine satisfactorily, and alternative positions may then be sought from a difference map. For least squares to work, the number of observations should exceed the number of parameters (for each atom x,y,z and one isotropic or if possible six anisotropic temperature factors), and in practice for small molecules is about 5–10 times greater. During least-squares refinement geometrical constraints can be introduced (e.g. standard planar phenyl rings) and this reduces the number of parameters required. So straightforward refinement of a small molecule might involve:

- an initial difference map to locate missing atoms
- several cycles of least squares
- a second-difference map to locate solvent, and possibly hydrogen atoms
- more least squares, possibly with geometric constraints and hydrogen atoms placed in geometrically expected positions, (e.g. on $-CH_2-$ groups)
- a final, hopefully featureless, difference map

Refinement programs are included in several software packages, and in uncomplicated cases are rapid and require little decision-making from the user.

In protein crystallography the aim is the same, but the initial model will generally not be so good as for a small molecule because of the lower resolution and the difficulties of map interpretation. Techniques for map improvement, such as symmetry averaging and density modification, are often effective, and can assist in extending the phases to higher resolution.

3.4.1. *Symmetry averaging*

This is useful if there are multiple copies of a molecule in the asymmetric unit (i.e. not related by crystallographic symmetry, but by an approximate symmetry axis).[61] The electron densities are superimposed and averaged, and then the averaged density transformed back to the locations of the individual subunits.

3.4.2. *Density modification*[62]

This works if molecules in the crystal are fairly well separated (usually over 50% solvent) so that an approximate molecular boundary can be determined. Electron density in the solvent region is then set to zero (or to an expected value lower than in the 'protein') the amplitudes and phases of the reflexions are calculated from this new map by Fourier inversion, and then a new electron density map is calculated with these new coefficients. This cycle is repeated until no further improvement is obtained. This technique has the advantage that no model is required; but the phases obtained may be combined with model phases if these are available.

The above techniques are combined with least-squares refinement in macro-molecular refinement. In the initial stages of refinement a fairly complete molecule, e.g. from a molecular replacement solution, may be refined as a rigid body, or each of several domains may be treated as a rigid body. At low-to-medium resolution the observation : parameter ratio is still low, and it would be wrong to refine individual atomic parameters. Each amino acid (or other well-defined group) may be constrained to a particular geometry and refined as a single unit. At rather higher resolution, distance restraints can be used to maintain the proper stereochemistry while individual atoms are free to move. Methods based on constraints, restraints, and combinations of the two have been compared by Sussman.[63]

The progress and correctness of the refinement is monitored, as in small molecules, by a falling R-factor; a correct protein structure may refine to $R < 20\%$. Calculation of difference density maps can reveal parts of the molecules missing from the initial model, and maps with coefficients $2F_o - F_c$ are used to fit or re-fit the model to electron density after each round of refinement. However, a structure may be too far displaced for least-squares refinement to improve it and there may be no clear indication from difference maps of how to improve it.

Several cycles of refinement and model-building into electron density on an interactive graphics display will normally be needed before a protein structure is complete, and the investment in computer time and human effort is considerable.

Least-squares refinement as thus described is capable of correcting only fairly small (< 1 Å) errors in atomic positions, and it is therefore easy for it to converge to a local minimum from which it can only be moved by manual rebuilding. A more efficient method of approaching this problem is to use molecular dynamics to explore the conformational space of the protein[64] while also carrying out crystallographic refinement.

3.4.3. Simulated annealing

This technique increases the temperature of the system to explore conformational space, then cools slowly, thus enabling higher energy barriers to be crossed than at room temperature. Programs implementing these methods have been developed and they can also incorporate, for example, restraints based on interproton distances from nuclear Overhauser effect (NOE) measurements (see Chapter 2) and the effects of crystallographic packing. This method can produce atomic movements in excess of 5 Å and is thus capable of escape from local energy minima where conventional least squares would be powerless.

3.5. INTERPRETATION OF RESULTS

The direct results of the crystal structure determination are cell dimensions, unit cell contents, structure factors, and a list of atomic coordinates and thermal parameters with their estimated standard deviations. However, listing of these quantities gives little immediate insight into molecular architecture, except that inspection of the atomic parameters will show atoms or groups where temperature factors are unusually high (because of thermal movement or disorder), or where a structure with disordered regions has alternative conformations included in the model. For most purposes derived quantities such as bond lengths and angles, and torsion angles are needed, together with pictures and models (Fig. 4).

Undoubtedly the most important development for proper interpretation of the results is the increased power and availability of interactive graphics hardware and software (e.g. FRODO,[59] HYDRA[65]) for display and manipulation of molecules. This enables us to display the atoms in the structure, and colour them according to parameters such as chemical type, B-value, electrostatic potential, accessibility, or hydrophobicity/philicity. Molecules can be superimposed, compared, and docked with one another, and the effects of mutations or torsion angle changes rapidly visualized.

4. Results

Examples of crystal structure results here have been chosen to illustrate what to expect from a small peptide and a protein, using enkephalin and insulin as

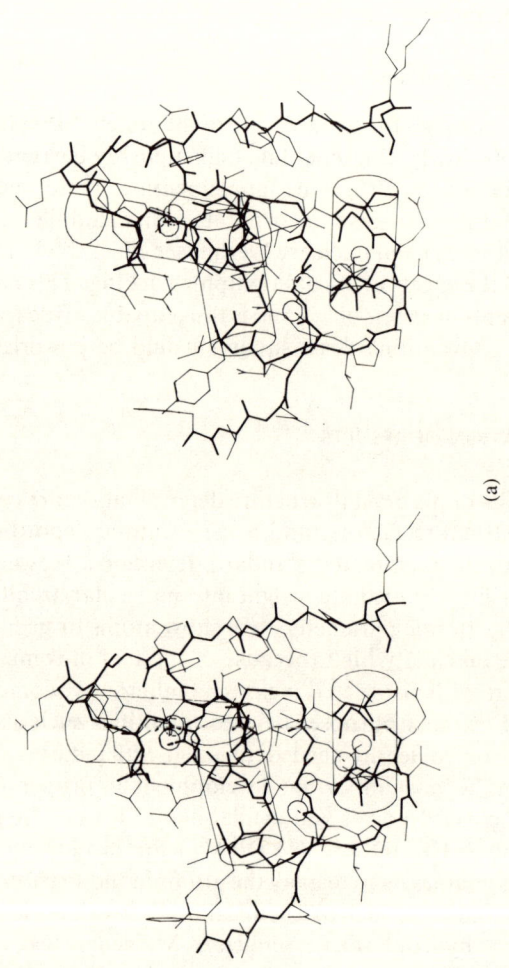

(a)

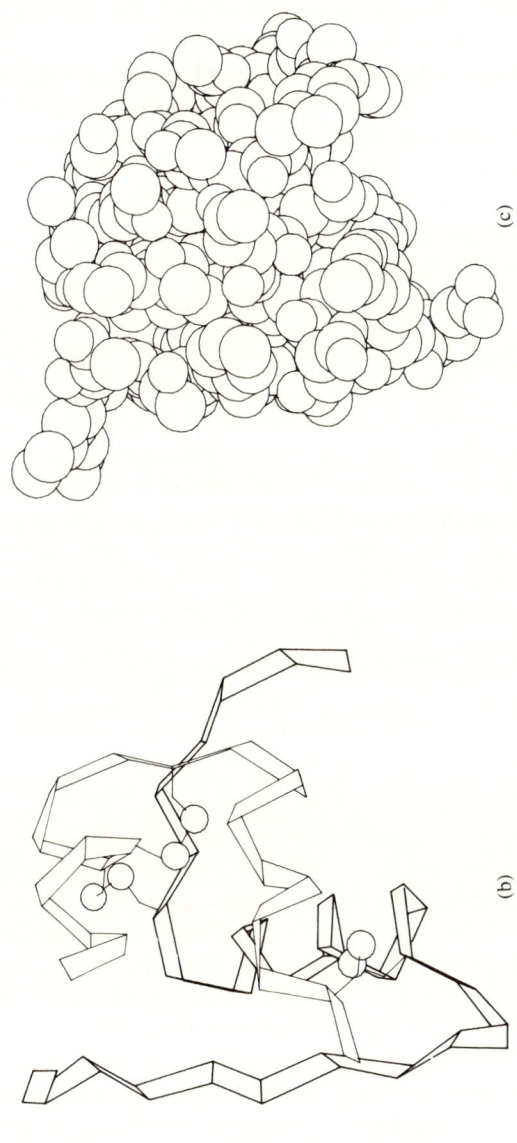

(b)

(c)

Figure 4 Different representations of an insulin molecule drawn using a program written by A. M. Lesk and K. D. Hardman (*Science* **216** (1982), 539–540): (a) stereo pair showing a stick model with the three helical regions outlined; (b) the main chains and disulphide bonds; (c) space-filling model.

examples. The details of crystallographic papers often seem inaccessible, and many chemists and modellers may prefer to be presented with a 'finished' three-dimensional picture of their molecule. But to use the data critically it is pertinent to be aware of other features, such as disorder, crystal packing, the existence of varied crystal forms, the quality of the data, and the stage which refinement has reached. For a small peptide the conformation(s) of the backbone and side-chains are usually well defined, but location of solvent of crystallization and the influence of crystal packing interactions are important in assessing the structure as a whole. For proteins, we are much more dependent on the resolution of the available data obtaining anything from a general chain-tracing to placing of all the protein atoms and at least some of the solvent. In many protein crystals the large solvent content (between 30 and 70%) means that crystal packing has much less influence on conformation than in small molecules, but there are specific contacts between protomers in an oligomeric protein.

4.1. ENKEPHALIN

The determination of the enkephalin structures illustrates the variability of small peptide structure and the difficulties that can arise even in solution of structures of rather small molecules. Leu- and Met-enkephalin are endogenous pentapeptides which act as potent analgesics binding to the opioid receptor(s). These small flexible peptides have little obvious similarity with the classic opiate morphine (1), but both contain the 'tyramine' moiety (dotted outline) – a phenolic ring separated by two carbon atoms from a nitrogen atom.

As might be expected, there is no single conformational minimum for the enkephalins, and models for the bound conformation based on various β-bend structures as well as an extended backbone have been proposed,[66] and different conformations are believed to bind to different subclasses of the receptor. Ten independent molecules of Leu-enkephalin in three crystal structures, and two independent molecules of Met-enkephalin in one structure, are now known (Table 1), and a survey of these shows that the two predominant backbone conformations are extended and a β-bend centred on Gly–Gly. Met-enkephalin exists only in an extended form, and there are no crystal structures with β-bends centred on Gly–Phe. Within each crystal structure the backbones are similar, i.e. there is no asymmetric unit containing both bent and extended conformers, but in each asymmetric unit there are differences in the side-chain orientations. The

Table 1 The different crystal forms of Leu- and Met-enkephalin

Solvent	Spacegroup/cell	Asymmetric unit	References
Leu-enkephalin (Tyr–Gly–Gly–Phe–Leu)			
Aqueous methanol	A2 $a = 31.937$, $b = 17.084$, $c = 24.861$ Å $\beta = 95.54°$ $Z = 16$	Four nearly identical conformers with a Gly–Gly β I′ bend; several water molecules	69, 70
N,N-Dimethyl formamide/ water	P2$_1$ $a = 18.720$, $b = 24.732$, $c = 20.311$ Å $\beta = 115.86°$ $Z = 8$	Four extended conformers forming antiparallel β-sheet; eight water, eight DMFA molecules plus some disordered solvent	68, 71
Ethanol	P2$_1$ $a = 11.549$, $b = 15.587$, $c = 16.673$ Å $\beta = 92.19°$ $Z = 4$	Dimers with the two extended molecules related by a pseudo twofold axis; one molecule of water per dimer	67
Met-enkephalin (Tyr–Gly–Gly–Phe–Met)			
Aqeous 1% pyridine and 0.05% acetic acid	P2$_1$ $a = 11.607$, $b = 17.987$, $c = 16.519$ A° $\beta = 91.24°$ $Z = 4$	Very similar to Leu-enkephalin from ethanol, but with 10.6 molecules per dimer	67

opiate receptor is also heterogeneous with at least three types being recognized; the enkephalins show mainly δ but also μ activity. Although these are 'only' pentapeptides, none of the structure determinations was automatic or straight-forward. The Leu-enkephalin from N,N-dimethylformamide (DMFA)/water in fact had an asymmetric unit of over 200 atoms (Fig. 5), and computerized direct-phasing multiple solution methods were unsuccessful. The structure was solved[72] by a combination of Patterson and direct methods, with side-chain and solvent atoms located from difference maps. Direct methods provided a small number of plausible solutions for the other form containing β-sheets,[69] but the chain termini could not be distinguished. Translational searches using a single planar pentapeptide also failed because a large number of approximate fits could be obtained. The structure was eventually solved with a search model containing a partial peptide backbone dimer, and refinement by tangent formula methods gave the remaining amino acid residues and tentative C_β positions. The initial structure determination of the β-bend Leu-enkephalin went relatively smoothly, by direct methods, but it was subsequently shown to have a unit cell four times larger, and refinement of this structure, in which there are small differences between four molecules related by pseudosymmetry, is still incomplete.

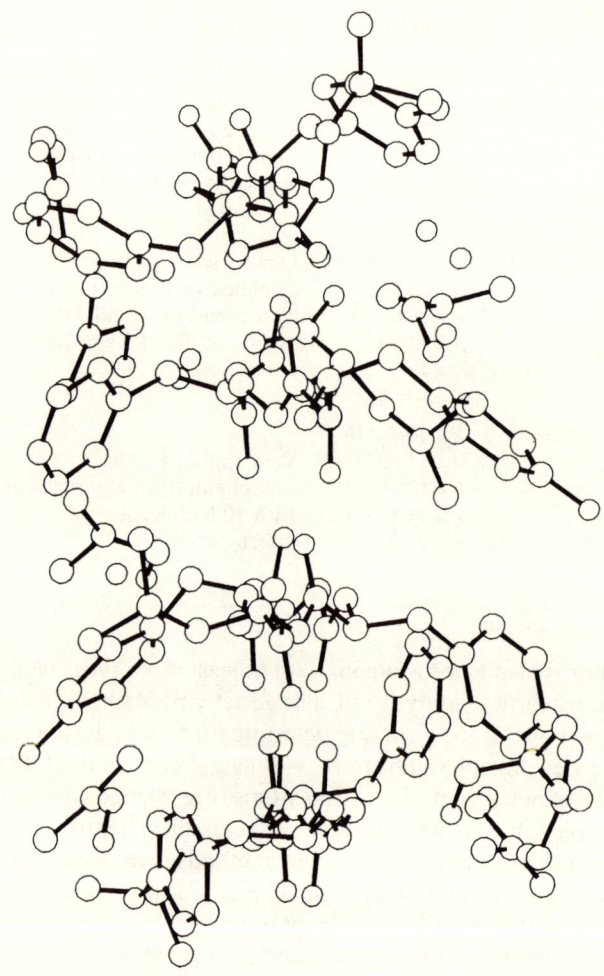

Figure 5 The asymmetric unit of Leu-enkephalin crystallized from DMFA/water.

4.2. INSULIN

Insulin is one of the proteins most extensively examined by X-ray crystallography. It is a relatively small protein, with A and B chains of 21 and 30 amino acids respectively, and three disulphide bonds linking A6–A11, A7–B7, and A20–B19. The chains are folded together to produce a compact globular core and surface regions involved in hydrophobic and hydrophilic interactions. Structures of insulins in various crystal forms and from various species have been determined, and some of them are listed in Table 2. The cell dimensions of bovine insulin were determined in the 1930s, and the structure of 2Zn porcine insulin

Table 2 Some of the published insulin crystal structures

Molecule	Crystal data	Assembly	Asymmetric unit	Reference
Pig insulin				
2 Zn	rhombohedral, R3 $a = 82.5$, $c = 34.0$ Å	hexamer	dimer	78
4 Zn	rhombohedral, R3 $a = 80.7$, $c = 37.6$ Å	hexamer	dimer	80
Zinc-free	cubic, $I2_13$ $a = b = c = 78.9$ Å	dimer	monomer	83
Human insulin				
2 Zn	rhombohedral, R3 $a = 82.85$, $c = 34.0$ Å	hexamer	dimer	79
4 Zn	rhombohedral, R3 $a = 80.953$, $c = 37.636$ Å	hexamer	dimer	81
Phenol	monoclinic, $P2_1$ $a = 61.36$, $b = 61.71$, $c = 47.95$ Å, $\beta = 110.8°$	hexamer	hexamer	82
Hagfish insulin				
	tetragonal, $P4_12_12$ $a = 38.4$, $c = 85.3$ Å	dimer	monomer	84
Des (B26-B30) pentapeptide insulin				
Beef	monoclinic, C2 $a = 52.74$, $b = 26.11$, $c = 51.64$ Å, $\beta = 93.41°$	monomer	2 monomers	85
Sheep	monoclinic, C2 $a = 53.90$, $b = 25.90$, $c = 25.60$ Å, $\beta = 93.80$	monomer	monomer	86
A1-B29 diaminosuberic acid cross-linked insulin				
Beef	rhombohedral, R3 $a = 83.2$, $c = 33.9$ Å	hexamer	dimer	87
Des (Phe-B1) insulin				
Beef	rhombohedral, R3 $a = 81.6$, $c = 34.0$ Å	hexamer	dimer	85

later determined to 2.8 Å[15,73] and 2.5 Å.[74] Phases were gradually extended to higher resolution,[75,76] and low-temperature (4°C) data collected.[77] Positions corresponding to 349 water sites, as well as the Zn and protein atoms, are included in the refined structure at 1.5 Å resolution.[78] The insulin monomer is shown in Fig. 4; in different crystal forms it associates to give dimers and hexamers. This parallels the aggregation with increasing concentration in solution; insulin is believed to exist as a monomer at physiological concentration in the blood, but to be stored as a hexamer. In the Zn-free form the monomers are related by an exact twofold axis to give dimers; the monomers in 2Zn, 4Zn, and phenol-insulin are related only by approximate twofold axes, and there are considerable differences in conformation between the protomers, particularly in the N-terminus of the B-chain. The transition between 2Zn and 4Zn insulin takes place both in solution[88] and reversibly in the crystal, and the transition is being explored using synchrotron radiation Laue diffraction data.[89]

Ever since the initial structure determination, the relationships between chemistry, structure, and function of insulin have been extensively investigated[90,91,92] and the mechanism of transmission of conformational change,[93] and identification of receptor-binding regions[94] are just two of many features for which a detailed knowledge of the structures is indispensable. The design of modified insulins with particular properties for clinical use is mentioned in the next section.

5. Discussion

There are many target systems for drug design, including viruses, compounds interacting with DNA, e.g. antibiotics[95] and anti-cancer drugs,[96] and proteins involved in hypertension[97,98] and anaemia.[99] A number of these systems have been elegantly reviewed by Hol.[100] A list of some recent structures which are relevant to drug design is given in Table 3.

The results of single crystal structure determinations are used in several ways, though these interact with each other, and with other computational and experimental methods. Rough categories might be as follows:

- determination/confirmation of chemical formula, conformation, connectivity and chirality, particularly of small molecules
- determination of the structural basis of mechanism and function, including interactions with substrates, receptors and co-factors; (a) modelling and design of peptides and other small-molecule substrates, and (b) modelling of proteins

Crystallography is widely used for small-molecule structures; but the problem of conformational variability of small peptides has already been stressed and illustrated by the conformations of enkephalin found in different crystal forms. An assessment of the usefulness of X-ray studies of conformation[101] suggests, inter alia, that conformers found in the crystal are likely to be within a few kcal mol^{-1} of a major (undisturbed) conformer, and if a similar conformation of a

Table 3 Some relevant macromolecular structures. The list is not exhaustive, but indicates the range of systems for which crystallographic determinations are available

Macromolecule	Nominal resolution	Reference
Genetically engineered porcine growth hormone	2.8 Å	133
Calmodulin	3.0 Å	134
Troponin C	2.8 Å	135
T-state haemoglobin	2.1 Å	136
Deoxyhaemoglobin	3.0 Å	137
Human c-H ras P^{21}	2.7 Å	138
Recombinant human renin	2.5 Å	139
Human serum albumin	6.0 Å	140
Interleukin 1-beta	3.0 Å	141
Rous sarcoma virus protease	3.0 Å	142
HIV-1 aspartyl protease	2.7 Å	143
Dihydrofolate reductase	1.7 Å	144
Carboxypeptidase A	1.54 Å	145
Phospholipase A2	2.5 Å	146
Mutant phospholipase A2	2.1 Å	147
Phospholipase C	1.5 Å	148
CheY	2.7 Å	149
Thymidylate synthetase	3 Å	150
Tumour necrosis factor	2.9 Å	151
Thermolysin-inhibitor complexes	1.6 Å	152
Porcine pancreatic elastase	1.65 Å	153
Beta-lactam inhibitor complex with elastase	1.84 Å	154
Beta-lactamase	2.5 Å	155
Colicin A	2.5 Å	156
Subtilisin Carlsberg–eglin c complex	1.2 Å	157
Foot and mouth disease virus	2.9 Å	158
Polio virus	2.9 Å	159
Mengo virus	3.0 Å	160
Human rhinovirus 14	3.0 Å	161
Antiviral agents complexed with human rhinovirus 14	3.0 Å	162
Influenza virus haemagglutinin	3.0 Å	163
Antigenic mutant of influenza virus haemagglutinin	3.0 Å	164
Influenza virus neuraminidase–antibody complex	3.0 Å	165
Photosynthetic reaction centre	3.0 Å	166
Yeast aspartyl tRNA	3.5 Å	167
cis-[PtCl$_2$(NH$_3$)$_2$] bound to Phe tRNA	6.0 Å	168
Phage 434 Cro–DNA complex	3.2–5.5 Å	169
Repressor–operator complex of bacteriophage 434	3.2–5.5 Å	170

molecule or fragment appears in several crystal structures then the average conformation is likely to be close to a global energy minimum. However, studies of isolated peptides give no direct evidence for receptor-bound or similarly interacting states. The conformations of cyclic peptides are considerably more restricted, as illustrated in a discussion of cyclic tetra- and pentapeptides,[102] and crystallography is even more valuable for determination of structures of natural metabolites such as (2) the cyclic peptide cyclosporin[103] (Fig. 6) and the related (3) non-peptide FK-506,[104] which also shows immunosuppressive activity. Conformational analysis of small molecules based on comparison of structures from the Cambridge Crystallographic Database can be very extensive, as in the work of Duax et al. on steroids.[105]

(2)

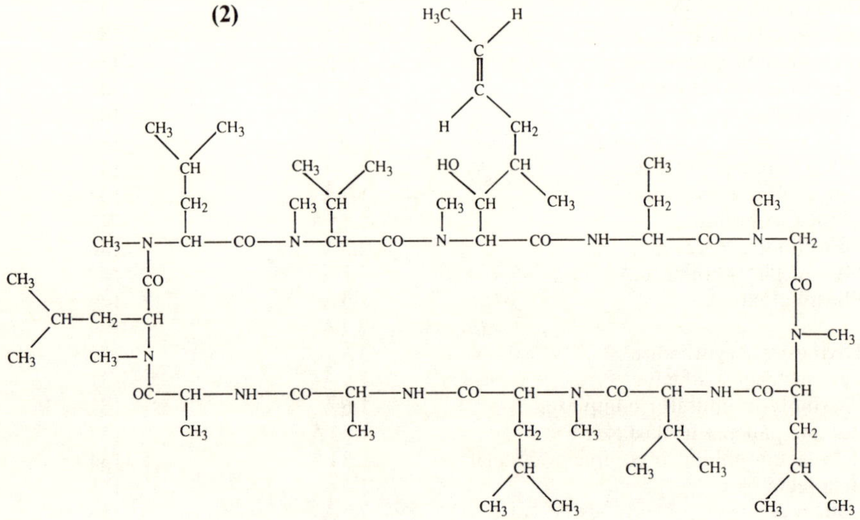

(3)

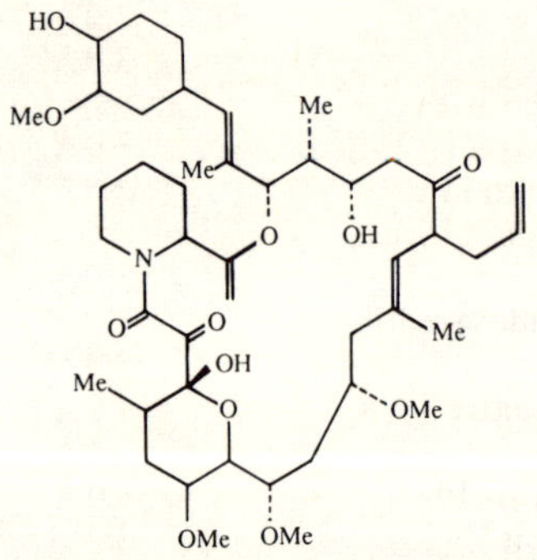

The future for crystallographic advances lies in two directions: more accurate determination of accurate electron density for small molecules; and solution of larger macromolecules, particularly as families, complexes, or other functionally significant assemblies. Protein crystallography has benefited more recently from advances in techniques of crystallization, data collection, and structure solution, and the results can be used to understand the functions of individual proteins, and families of proteins.

The importance of conformational changes has been repeatedly emphasized, e.g. in binding drugs to haemoglobin[106] and in interactions of immunoglobulins with antigens and effector molecules.[107] Such changes are thought to be crucial in signalling mechanisms, and as cell surface receptors are being expressed for a number of proteins,[108] the structures of these, and of the protein–receptor

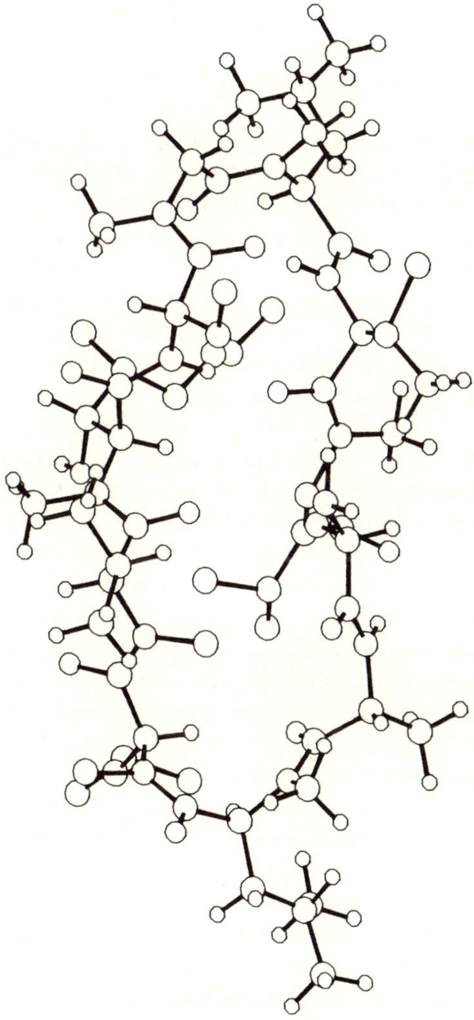

Figure 6 Structure of cyclosporin in the crystal.

complexes will need to be compared to investigate how extracellular ligand binding leads to intracellular response.

Pointers to the magnitude of structures which are becoming accessible are seen in the photosynthetic reaction centres[109] and human leucocyte antigen (HLA2); the structure of an extracellular fragment of HLA2[110,111] has shown the location of polymorphic residues, foreign antigen binding sites and the T-cell recognition region. Technical progress will allow solution of even more complicated assemblies, such as the ribosomal particles for which low-temperature single-crystal data at up to 4.5Å resolution have recently been collected.[112] A typical bacterial ribosome has a molecular weight of 1.45 or 2.3 million, and contains several different proteins and one or two RNA chains.

The use of groups of related structures is often necessary, as in that involving the Trp repressor, where the aporepressor,[113] repressor,[114] pseudorepressor,[115] and repressor/operator complex[116] have all been determined.

Structures of mutant and complexed forms of the same protein can be completed on a rapid time-scale, and such comparative data is very valuable. For example, the membrane glycoprotein of the influenza virus, haemagglutinin, is responsible for binding the virus to cell surface receptors during infection. Sialic acid from the receptor fills a conserved pocket surrounded by antibody binding sites (which are known to have mutated since 1968). Structures of the wild-type molecule and a receptor-binding mutant and their complexes with sialyllactose have been used to suggest an approach to antiviral drugs that block viral entry into the cell by designing an inhibitor targeted to the conserved residues.[117]

Good examples of the use of crystallographic data as a basis for designing modified proteins are the insulins of Brange et al.[118] and Markussen et al.[119-121] with improved therapeutic properties. In the first case the aim was to produce faster-acting insulin by reduction of self-association, which was suspected of limiting the rate of absorption. This required identification of residues involved in subunit interactions, particularly dimer formation, but also not too central to the putative receptor-binding region. Mutations were proposed to reduce dimer formation, e.g. by increasing the bulk to be accommodated at the dimer interface or by increasing charge repulsion. The results of these mutations were visualized by computer graphics, then the various mutants produced and their state of aggregation and biological activity measured. Insulins monomeric at pharmaceutical concentration were indeed obtained, and are rapidly absorbed following subcutaneous injection. In the second case the intention was to produce soluble, prolonged-acting insulins, by mutations which change the crystallization behaviour. Again, the three-dimensional structure of insulin was used to indicate useful residues for mutation, and molecular graphics employed in attempting to explain why addition of positive charge to the C-terminus of the B-chain promotes crystallizability.

Modelling of peptides and proteins based on crystallographic data is a rapidly expanding area which requires interaction with techniques of energy minimization and sequence alignment. A number of analogues of small peptides, such as enkephalin, oxytocin, somatostatin, and vasopressin are seen as targets, and a number of conformationally restricted or non-peptide analogues have reached the market.[122] Some principles for design of small peptides have been proposed by Kaiser;[123] novel small peptides can readily be constructed on an interactive graphics system, using programs such as QUANTA,[124] and energy-minimized and unknown non-peptides can be patched together from fragments in the Cambridge Structural Database to form initial models. Modelling of peptides into the binding sites in proteins has pointed towards more effective inhibitors of renin[125,126] and angiotensin converting enzyme.[98]

Modelling of proteins is necessary because there are far more sequences known (around 15 000) than there are three-dimensional structures, and the *ab initio* calculation of protein folding pathways is not yet feasible. The effects of even fairly site-specific mutations, insertions, and deletions are difficult to quantify, but rules for such changes are being developed.[127-129] If the tertiary structure of a homologous protein is known, it may be used as a basis for modelling other members of a family. If there is no crystal structure of a homologous protein available, modelling may still be possible if patterns of residues corresponding to known motifs in other proteins can be identified. It may then be possible to patch these together sufficiently well to give a model which can then be energy-minimized or used as a starting point for molecular dynamics simulation, in a similar way to that used for small molecules. A recent review of all these three aspects of modelling is given by Blundell *et al.*[130] The sort of tools required are a combined sequence structure database,[131] and alignment programs, the most sophisticated of which take account of topological equivalence, and possibly other amino acid properties as well as sequence similarity.[132]

In conclusion – there is a wide gulf between the design of an active molecule and its clinical use, because we cannot yet predict, in general, undesirable metabolic effects or delivery problems. But until protein folding pathways can be predicted with certainty from the sequence, and thus from the gene sequence, crystallography will still have a dominant role in drug design.

References

1. 'Structure of myoglobin' J. C. Kendrew, R. E. Dickerson, B. E. Strandberg, R. G. Hart, D. R. Davies, D. C. Phillips and V. C. Shore (1960) *Nature*, **185**, 422–427.
2. 'Structure of haemoglobin' M. F. Perutz, M. G. Rossmann, A. F. Cullis, H. Muirhead, G. Will and A. C. T. North (1960) *Nature*, **185**, 416–422.
3. 'Sequential resonance assignments as a basis for determination of spatial protein structures by high resolution proton nuclear magnetic resonance' K. Würtrich, G. Wider, G. Wagner and W. Braun (1982) *J. Mol. Biol.*, **155**, 311–319.
4. 'Three-dimensional nmr spectroscopy of a protein in solution' H. Oschkinat, C. Griesinger, P. J. Kraulis, O. W. Sørensen, R. R. Ernst, A. M. Gronenborn and G. M. Clore (1988) *Nature*, **332**, 374–376.
5. 'Macromolecular structure in solution' I. D. Kuntz (1987) *Protein Eng.*, **1**, 147–150.
6. 'Studies by ^{1}H nuclear magnetic resonance and distance geometry of the solution conformation of the alpha-amylose inhibitor tendamistat' A. D. Kline, W. Braun and K. Würtrich (1986) *J. Mol. Biol.*, **189**, 377–382.

7. 'Crystal structure determination, refinement and the molecular model of the alpha-amylase inhibitor Hoe-467A' J. Pflugrath, E. Wiegland, R. Huber and L. Vertesy (1986) *J. Mol. Biol.*, **189**, 383–386.
8. 'Comparison of the solution and X-ray structures of barley serine proteinase inhibitor 2' G. M. Clore, A. M. Gronenborn, M. N. G. James, M. Kjaer, C. A. McPhalen and F. M. Poulsen (1987) *Protein Eng.*, **1**, 313–318.
9. 'Determination of three-dimensional structures of proteins by simulated annealing with interproton distance restraints. Application to crambin, potato carboxypeptidase inhibitor and barley serine proteinase inhibitor 2' M. Nilges, A. M. Gronenborn, A. T. Brünger and G. M. Clore (1988) *Protein Eng.*, **2**, 27–38.
10. 'The crystal structure of cholesteryl iodide' C. H. Carlisle and D. Crowfoot (1945) *Proc. Roy. Soc.*, **A184**, 64–83.
11. 'X-ray crystallographic evidence on the structure of vitamin B_{12}' C. Brink, D. C. Hodgkin, J. Lindsey, J. Pickworth, J. H. Robertson and J. G. White (1954) *Nature*, **174**, 1169–1170.
12. 'The conformation of deamino-oxytocin: X-ray analysis of the 'dry' and 'wet' forms' J. Husain, T. L. Blundell, S. Cooper, J. E. Pitts, I. J. Tickle, S. P. Wood, V. J. Hruby, A. Buku, A. J. Fischman, H. R. Wyssbrod and Y. Mascarenhas (1990) *Phil. Trans. R. Soc. Lond.*, **B327**, 625–654.
13. 'X-ray analysis of glucagon and its relationship to receptor binding' K. Sasaki, S. Dockerill, D. A. Adamiak, I. J. Tickle and T. Blundell (1975) *Nature*, **257**, 751–757.
14. 'Anisotropic thermal motion and polypeptide secondary structure studied by X-ray analysis at 0.98 Å resolution' I. D. Glover, D. S. Moss, I. J. Tickle, J. E. Pitts, I. Haneef, S. P. Wood and T. L. Blundell (1985) *Adv. Biophys.*, **20**, 1–12.
15. 'Structure of rhombohedral 2 zinc insulin crystals' M. J. Adams, T. L. Blundell, E. J. Dodson, G. G. Dodson, M. Vijayan, E. N. Baker, D. C. Hodgkin, B. Rimmer and S. Sheat (1969) *Nature*, **224**, 491–495.
16. 'Common cold viruses' M. G. Rossmann, E. Arnold, J. P. Griffith, G. Karner, Ming Luo, T. J. Smith, G. Vriend, R. R. Rueckert, B. Sherry, M. A. McKinley, G. Diana and M. Otto (1987) *Trends Biochem. Sci.*, **12**, 313–318.
17. 'Characterization of a helical protein designed from first principles' L. Regan and W. DeGrado (1988) *Science*, **241**, 976–978.
18. 'Protein crystallography and its new revolution' G. G. Dodson (1986) *Trends Biochem. Sci.*, **11**, 309–310.
19. 'The anatomy and taxonomy of protein structure' J. Richardson (1981) *Adv. Protein Chem.*, **34**, 167–339.
20. 'The Cambridge Crystallographic Data Centre: computer-based search, retrieval, analysis and display of information' F. H. Allen, S. Bellard, M. D. Brice, B. A. Cartwright, A. Doubleday, H. Higgs, T. Hummelink, B. G. Hummelink-Peters, O. Kennard, W. D. S. Motherwell, J. R. Rodgers and D. G. Watson (1979) *Acta Cryst.*, **B35**, 2331–2339.
21. 'The protein data bank: a computer-based archival file for macromolecular structures' F. C. Bernstein, T. F. Koetzle, G. J. B. Williams, E. F. Myer, M. D. Brice, J. R. Rodgers, O. Kennard, T. Shimanouchi and M. Tasumi (1977) *J. Mol. Biol.*, **112**, 535–542.
22. 'Three dimensional structure of cholera toxin penetrating a lipid membrane' H. O. Ribi, D. S. Ludwig, K. L. Mercer, G. K. Schoolnik and R. D. Kornberg (1988) *Science*, **239**, 1272–1276.
23. 'Three dimensional structural analysis of tetanus toxin by electron crystallography' J. P. Robinson, M. F. Schmid, G. Morgan and W. Chiu (1988) *J. Mol. Biol.*, **200**, 367–375.
24. *Protein Crystallography* T. L. Blundell and L. N. Johnson (1976) Academic Press, New York.

25. *X-ray Structure Determination: A Practical Guide* G. H. Stout and L. H. Jensen (1968) Macmillan, London.
26. *Structure Determination by X-ray Crystallography* M. F. C. Ladd and R. A. Palmer (1977) Plenum Press, New York.
27. *X-ray Analysis and the Structure of Organic Molecules* J. D. Dunitz (1979), Cornell University Press, Ithaca.
28. *Crystal Structure Analysis*, 2nd edn J. P. Glusker and K. N. Trueblood (1985) Oxford University Press, New York.
29. *International Tables for X-ray Crystallography*, Vol. A Published for the International Union of Crystallography by D. Reidel, Dordrecht.
30. 'Structure and charge density of the 1 : 1 complex of thiourea with parabanic acid at 298°K' H-P. Weber and B. M. Craven (1987) *Acta Cryst.*, **B43**, 202–209.
31. 'Anomalous dispersion in phase determination for macromolecules' W. A. Hendrickson (1985). In: *Crystallographic Computing* **3** (G. M. Sheldrick, C. Kruger and R. Goddard, eds) Oxford University Press.
32. *Methods in Enzymology* Vols **114** and **115** (1985) Academic Press, New York.
33. *Preparation and Analysis of Protein Crystals* A. McPherson (1982) Wiley, New York.
34. 'Crystallization of proteins and nucleic acids: a survey of methods and importance of the purity of the macromolecules' R. Giege (1987). In: *Crystallography in Molecular Biology* (D. Moras, J. Drenth, B. Strandberg, D. Suck and K. Wilson, eds) Plenum Press, New York.
35. 'Cryocrystallography of biological macromolecules: a generally applicable method' H. Hope (1988) *Acta Cryst.*, **B44**, 22–26.
36. 'A biological macromolecule crystallization database: a basis for crystallization strategy' G. L. Gilliland (1988) *J. of Crystal Growth*, **90**, 51–59.
37. 'Automatic preparation of protein crystals using laboratory robotics and automated visual inspection' K. B. Ward, M. A. Perrozzo and W. M. Zuk (1988) *J. of Crystal Growth*, **90**, 325–339.
38. 'Experiments with automated protein crystallization' M. J. Cox and P. C. Weber (1987) *J. Appl. Cryst.*, **20**, 366–373.
39. 'Automated protein crystallization and a new crystal form of a subtilisin–eglin complex' H. A. Kelders, K. H. Kalk, P. Gros and W. G. H. Hol (1987) *Protein Eng.*, **1**, 301–303.
40. 'A new universally applicable procedure for wall-contact-free single crystal growth from a suspended droplet under conditions of microgravity' W. Littke (1988) *J. of Crystal Growth*, **90**, 344–348.
41. 'Macromolecular crystals' A McPherson (1989) *Scientific American*, **260**, 42–49.
42. D. Carvin, in preparation.
43. 'Multiple heavy-atom reagents for macromolecular X-ray structure determination: application to the nucleosome core particle' T. V. O'Halloran, S. J. Lippard, T. J. Richmond and A. Klug (1987) *J. Mol. Biol.*, **194**, 705–712.
44. 'Use of site-directed mutagenesis to obtain isomorphous heavy-atom derivatives for protein crystallography: cysteine-containing mutants of phage T4 lysozyme' S. Dao-Pin, T. Alber, J. A. Bell, L. H. Weaver and B. W. Matthews (1987) *Protein Eng.*, **1**, 115–123.
45. 'Crystallographic phases through genetic engineering: experiences with colicin A' A. D. Tucker, D. Baty, M. W. Parker, F. Pattus, C. Lazdunski and D. Tsernoglou (1989) *Protein Eng.*, **2**, 399–405.
46. 'New Opportunities in synchrotron X-ray crystallography' C. T. Prewitt, P. Coppens, J. C. Philips and L. W. Finger (1987) *Science*, **238**, 312–318.
47. 'Piperazine silicate (E419): the structure of a very small crystal determined with synchrotron radiation' S. J. Andrews, M. Z. Papiz, R. McMeeking, A. J. Blake, B. M. Lowe, K. R. Franklin, J. R. Helliwell and M. M. Harding (1988) *Acta Cryst.*, **B44**, 73–77.

48. 'Methods in MADness' K. Moffat (1988) *Nature*, **336**, 422–423.
49. 'Millisecond X-ray diffraction and the first electron density map from Laue photographs of a protein crystal' J. Hadju, P. A. Machin, J. W. Campbell, T. J. Greenhough, I. J. Clifton, S. Zurek, S. Gover, L. N. Johnson and M. Elder (1987) *Nature*, **329**, 178–181.
50. 'Catalysis in enzyme crystals' J. Hadju, K. R. Acharya, D. I. Stuart, D. Barford and L. N. Johnson (1988) *Trends Biochem. Sci.*, **13**, 104–109.
51. 'Crystal cooling for protein crystallography with synchrotron radiation' H. D. Bartunik and P. Schubert (1982) *J. Appl. Cryst.*, **15**, 227–231.
52. 'A fixture for X-ray crystallographic studies of biomolecules under high gas pressure' R. F. Tilton Jr. (1988) *J. Appl. Cryst.*, **21**, 4–9.
53. 'Data collection' J. R. Helliwell (1987). In: *Crystallography in Molecular Biology* (D. Moras, J. Drenth, B. Strandberg, D. Suck and K. Wilson, eds) Plenum Press, New York.
54. 'A Fourier series method for the determination of the component of interatomic distances in crystals' A. L. Patterson (1934) *Phys. Rev.*, **46**, 372–376.
55. 'Molecular replacement: the method and its problems' E. J. Dodson (1988). In: *Crystallographic Computing* **4** (N. W. Isaacs and M. R. Taylor, eds) Oxford University Press.
56. 'Solution of a protein crystal structure with a model obtained from nmr interproton distance restraints' A. T. Brünger, R. L. Campbell, G. M. Clore, A. M. Gronenborn, M. Karplus, G. A. Petsko and M. M. Teeter (1987) *Science*, **235**, 1049–1053.
57. 'Direct methods – from birth to maturity' M. M. Woolfson (1987) *Acta Cryst.*, **A43**, 593–612.
58. 'Direct methods applications to macromolecules' S. Fortier (1988). In: *Crystallographic Computing* **4** (N. W. Isaacs and M. R. Taylor, eds) Oxford University Press.
59. 'A graphics model building and refinement scheme for macromolecules' T. A. Jones (1978) *J. Appl. Cryst.*, **11**, 268–272.
60. 'Computer skeletonization and automatic electron density map analysis' J. Greer (1985). In: *Methods in Enzymology*, **115**, pp. 207–224. Academic Press, New York.
61. 'Methods and programs for direct-space exploitation of geometric redundancies' G. Bricogne (1976) *Acta Cryst.*, **A32**, 832–847.
62. 'Density modification methods' A. D. Podjarny (1987). In: *Crystallography in Molecular Biology* (D. Moras, J. Drenth, B. Strandberg, D. Suck and K. Wilson, eds) Plenum Press, New York.
63. 'Application of refinement constraints and restraints to protein and nucleic acids' J. L. Sussman (1984). In: *Methods and Applications in Crystallographic Computing* (S. R. Hall and T. Ashida, eds) Oxford University Press.
64. 'Crystallographic R factor refinement by molecular dynamics' A. T. Brünger, J. Kuriyan and M. Karplus (1987) *Science*, **235**, 458–460.
65. *Hydra* (A Graphics Software Package) R. E. Hubbard (1986) University of York.
66. 'Conformation–activity relationships of opiate analgesics' J. Martin and P. Andrews (1987) *J. Computer-Aided Molecular Design*, **1**, 53–72.
67. 'The crystal structures of [Met^5]enkephalin and a third form of [Leu^5]enkephalin: observations of a novel pleated beta sheet' J. F. Griffin, D. A. Langs, G. D. Smith, T. L. Blundell, I. J. Tickle and S. Bedarkar (1986) *Proc. Natl. Acad. Sci. USA*, **83**, 3272–3276.
68. '[Leu^5]enkephalin: four co-crystallizing conformers with extended backbones that form an antiparallel beta sheet' I. L. Karle, J. Karle, D. Mastropaolo, A. Camerman and N. Camerman (1983) *Acta Cryst.*, **B39**, 625–637.
69. 'Conformation of [Leu^5]enkephalin from X-ray diffraction: features important for recognition at opiate receptor' G. D. Smith and J. F. Griffin (1978) *Science*, **199**, 1214–1216.

70. 'Crystal structure of [Leu^5]enkephalin' T. L. Blundell, L. Hearn, I. J. Tickle, R. A. Palmer, B. A. Morgan, G. D. Smith and J. F. Griffin (1979) *Science*, **205**, 220.
71. 'Crystal structure of leucine enkephalin' A. Camerman, D. Mastropaolo, I. Karle, J. Karle and N. Camerman (1983) *Nature*, **306**, 447–450.
72. 'General and specific techniques for solving equal atom structures containing more than 200 atoms per asymmetric unit' I. L. Karle (1984). In: *Methods and Applications in Crystallographic Computing* (S. R. Hall and T. Ashida, eds) pp. 478–479. Oxford University Press.
73. 'Atomic positions in rhombohedral 2-zinc insulin crystals' T. L. Blundell, J. F. Cutfield, S. M. Cutfield, E. J. Dodson, G. G. Dodson, D. C. Hodgkin, D. A. Mercola and M. Vijayan (1971) *Nature*, **231**, 506–511.
74. 'Insulin's crystal structure at 2.5Å resolution' Peking Insulin Structure Research Group (1971) *Peking Rev.*, **40**, 11–16.
75. 'Studies on the insulin crystal structure: the molecules at 1.8 Å resolution' Peking Insulin Structure Research Group (1974) *Scientia Sinica*, **17**, 752–778.
76. 'The extension and refinement of the 1.9 Å spacing isomorphous phases to 1.5 Å spacing in 2Zn insulin by determinantal methods' C. de Rango, Y. Mauguen, G. Tsoucaris, E. J. Dodson, G. G. Dodson and G. Taylor (1985) *Acta Cryst.*, **A41**, 3–17.
77. 'Insulin structure in 1.2Å resolution: flexibility of local conformation and surrounding water molecules' N. Sakabe, K. Sakabe and K. Sasaki (1978). In: *Proinsulin, Insulin, C-peptide* (S. Baba, T. Kaneko and N. Yanaihara, eds) pp. 73–80. Elsevier North-Holland, Amsterdam.
78. 'The structure of 2Zn pig insulin crystals at 1.5Å resolution' E. N. Baker, T. L. Blundell, J. F. Cutfield, S. M. Cutfield, E. J. Dodson, G. G. Dodson, D. C. Hodgkin, R. E. Hubbard, N. W. Isaacs, C. D. Reynolds, K. Sakabe, N. Sakabe and M. Vijayan (1988) *Phil. Trans. Roy. Soc.*, **319**, 369–456.
79. 'The crystal structures of three non-pancreatic human insulins' S. A. Chawdhury, E. J. Dodson, G. G. Dodson, C. D. Reynolds, S. P. Tolley, T. L. Blundell, A. Cleasby, J. E. Pitts, I. J. Tickle and S. P. Wood (1983) *Diabetologia*, **25**, 460–464.
80. 'Structure of insulin in 4-zinc insulin' G. Bentley, E. Dodson, G. Dodson, D. Hodgkin and D. Mercola (1976) *Nature*, **261**, 166–168.
81. 'Structural stability in the 4-zinc human insulin hexamer' G. D. Smith, D. C. Swenson, E. J. Dodson, G. G. Dodson and C. D. Reynolds (1984) *Proc. Natl. Acad. Sci. USA*, **81**, 7093–7097.
82. 'Phenol stabilizes more helix in a new symmetrical zinc insulin hexamer' U. Derewenda, Z. Derewenda, E. J. Dodson, G. G. Dodson, C. D. Reynolds, G. D. Smith, C. Sparks and D. Swenson (1989) *Nature*, **338**, 594–596.
83. 'Zinc-free cubic pig insulin: crystallization and structure determination' E. J. Dodson, G. G. Dodson, A. Lewitova and M. Sabesan (1978) *J. Mol. Biol.*, **125**, 387–396.
84. 'Structure and biological activity of hagfish insulin' J. M. Cutfield, S. M. Cutfield, E. J. Dodson, G. G. Dodson, S. F. Emdin and C. D. Reynolds (1979) *J. Mol. Biol.*, **132**, 85–100.
85. 'The structure of des-Phe B1 bovine insulin' G. D. Smith, W. L. Duax, E. J. Dodson, G. G. Dodson, R. A. G. de Graaf and C. D. Reynolds (1982) *Acta Cryst.*, **B38**, 3028–3032.
86. 'Insulin's structure as a modified and monomeric molecule' R. C. Bi, Z. Dauter, E. Dodson, G. Dodson, F. Giordano and C. Reynolds (1984) *Biopolymers*, **23**, 391–395.
87. 'Crystal structure, aggregation and biological potency of beef insulin cross-linked at A1 and B29 by diaminosuberic acid' G. G. Dodson, S. Cutfield, E. Hoenjet, A. Wollmer and D. Brandenburg (1980). In: *Insulin: Chemistry, Structure and Function of Insulin and Related Hormones* (D. Brandenburg and A. Wollmer, eds) pp. 17–26. De Gruyter, Berlin.

88. 'Proton nmr studies of insulin. Reversible transformation of 2 zinc to 4 zinc insulin hexamer' V. Ramesh and J. H. Bradbury (1986) *Int. J. Peptide and Protein Res.*, **28**, 146–153.

89. 'Preliminary study of a phase transformation in insulin crystals using synchrotron radiation Laue diffraction' C. D. Reynolds, B. Stowell, K. K. Joshi, M. M. Harding, S. J. Maginn and G. G. Dodson (1988) *Acta Cryst.*, **B44**, 512–515.

90. 'Insulin: the structure in the crystal and its reflection in chemistry and biology' T. Blundell, G. Dodson, D. Hodgkin and D. Mercola (1972) *Adv. Protein Chem.*, **26**, 279–402.

91. 'Similarities and differences in the crystal structures of insulin' J. F. Cutfield, S. M. Cutfield, E. J. Dodson, G. G. Dodson, C. D. Reynolds and D. Vallely (1981). In: *Structural Studies on Molecules of Biological Interest* (G. Dodson, J. P. Glusker and D. Sayre, eds) pp. 527–546. Clarendon Press, Oxford.

92. 'Insulin's structural behaviour and its relation to activity' E. J. Dodson, G. G. Dodson, R. E. Hubbard and C. D. Reynolds (1983) *Biopolymers*, **22**, 281–291.

93. 'Transmission of conformational change in insulin' C. Chothia, A. M. Lesk, G. G. Dodson and D. C. Hodgkin (1983) *Nature*, **302**, 500–505.

94. 'Receptor-binding region of insulin' R. A. Pullen, D. G. Lindsay, S. P. Wood, I. J. Tickle, T. L. Blundell, A. Wollmer, G. Krail, D. Brandenburg, H. Zahn, J. Gliemann and S. Gammeltoft (1976) *Nature*, **259**, 369–373.

95. 'Non Watson–Crick GC and AT base pairs in a DNA–antibiotic complex' G. J. Quigley, G. Ughetto, G. A. van der Marel, H. van Boom, H. J. Wang and A. Rich (1986) *Science*, **232**, 1255–1258.

96. 'Binding of an antitumour drug to DNA: netropsin and C-G-C-G-A-A-T-T-BrC-G-C-G' M. L. Kopka, C. Yoon, D. Goodsell, P. Pjura and R. E. Dickerson (1985) *J. Mol. Biol.*, **183**, 553–563.

97. 'A rational approach to the design of antihypertensives: X-ray studies of complexes between aspartic proteinases and aminoalcohol renin inhibitors' J. Cooper, S. I. Foundling, T. L. Blundell, R. J. Arrowsmith, C. J. Harris and J. N. Champness (1988). In: *Topics in Medicinal Chemistry* (P. R. Leeming, ed.) pp. 308–313. Royal Society of Chemistry, London.

98. 'Design of potent competitive inhibitors of angiotensin-converting enzyme: carboxyalkanoyl and mercaptoalkanoyl aminoacids' D. W. Cushman, H-s. Cheung, E. F. Sabo and M. A. Ondetti (1977) *Biochemistry*, **16**, 5484–5491.

99. 'Haemoglobin as a receptor of drugs and peptides: X-ray studies of the stereochemistry of binding' M. F. Perutz, G. Fermi, D. J. Abraham, C. Poyart and E. Bursaux (1986) *J. Amer. Chem. Soc.*, **108**, 1064–1078.

100. 'Protein crystallography and computer graphics – towards rational drug design' W. G. J. Hol (1986) *Angewandte Chemie* (*Int. Ed.*) **25**, 767–852.

101. 'How useful are X-ray studies of conformation?' P. Murray-Rust (1982). In: *Molecular Structure and Biological Activity* (J. F. Griffin and W. L. Duax, eds) pp. 117–133. Elsevier Biomedical, New York.

102. 'Conformational flexibility and characteristics of cyclic tetra- and pentapeptides' I. L. Karle (1982). In: *Molecular Structure and Biological Activity* (J. F. Griffin and W. L. Duax, eds) pp. 215–227. Elsevier Biomedical, New York.

103. 'The conformation of cyclosporin A in the crystal and in solution' H-R. Loosli, H. Kessler, H. Oschkinat, H-P. Weber, T. J. Petcher and A. Widmer (1985) *Helv. Chim. Acta*, **68**, 682–686.

104. 'Structure of FK-506, a novel immunosuppressant isolated from streptomyces' H. Tanaka, A. Kuroda, H. Marusawa, H. Hatanaka, T. Kino, T. Goto, M. Hashimoto and T. Taga (1987) *J. Amer. Chem. Soc.*, **109**, 5031–5033.

105. 'Steroid conformation, receptor binding and hormone action' W. L. Duax, J. F. Griffin and D. C. Rohrer (1984). In: *X-ray Crystallography and Drug Action* (A. S. Horn and C. J. DeRanter, eds) pp. 405–426. Oxford University Press.

106. 'Physiological and X-ray studies of potential anti-sickling agents' D. J. Abraham, M. F. Perutz and S. E. V. Phillips (1983) *Proc. Natl. Acad. Sci. USA*, **80**, 324–328.
107. 'Elbow motions in the immunoglobulins involves a molecular ball- and socket joint' A. M. Lesk and C. Chothia (1988) *Nature*, **335**, 188–190.
108. 'Structure–function relationships of growth factors and their receptors' N. McDonald, J. Murray-Rust and T. Blundell (1989) *British Medical Bulletin*, **45**, 554–569.
109. 'Structure and function of bacterial photosynthetic reaction centres' G. Fehler, J. P. Allen, M. Y. Okamura and D. C. Rees (1989) *Nature*, **339**, 111–116.
110. 'Structure of the human class 1 histocompatibility antigen HLA A2' P. J. Bjorkman, M. A. Saper, B. Samraoui, W. S. Bennett, J. L. Strominger and D. C. Wiley (1987) *Nature*, **329**, 506–512.
111. 'The foreign antigen binding site and T cell recognition regions of class 1 histocompatibility antigens' P. J. Bjorkman, M. A. Saper, B. Samraoui, W. S. Bennett, J. L. Strominger and D. C. Wiley (1987) *Nature*, **329**, 512–518.
112. 'Cryocrystallography of ribosomal particles' H. Hope, F. Frolow, K. van Bohlen, I. Makowski, C. Kratky, Y. Halfon, H. Danz, P. Webster, K. S. Bartels, H. G. Wittmann and A. Yonath (1989) *Acta Cryst.*, **B45**, 190–199.
113. 'The crystal structure of trp aporepressor at 1.8 Å resolution shows how binding tryptophan enhances DNA affinity' R-G. Zhang, A. Joachimiak, C. L. Lawson, R. W. Schevitz, Z. Otowinowski and P. B. Sigler (1987) *Nature*, **327**, 591–597.
114. 'The three-dimensional structure of trp repressor' R. V. Schevitz, Z. Otwinowski, A. Joachimiak, C. L. Lawson and P. B. Sigler (1985) *Nature*, **317**, 782–786.
115. 'The structure of trp pseudorepressor at 1.65 Å shows why indole propionate acts as a trp "inducer"' C. L. Lawson and P. B. Sigler (1988) *Nature*, **333**, 869–871.
116. 'Crystal structure of trp repressor/operator complex at atomic resolution' Z. Otowinowski, R. W. Schevitz, R-G. Zhang, C. L. Lawson, A. Joachimiak, R. Q. Marmorstein, B. F. Luisi and P. B. Sigler (1988) *Nature*, **335**, 321–329.
117. 'Structure of the influenza virus haemagglutinin complexed with its receptor, sialic acid' W. Weiss, J. H. Brown, S. Cusack, J. C. Paulson, J. J. Skehel and D. C. Wiley (1988) *Nature*, **333**, 426–431.
118. 'Monomeric insulins obtained by protein engineering and their medical implications' J. Brange, U. Ribel, J. F. Hansen, G. Dodson, M. T. Hansen, S. Havelund, S. G. Melburg, F. Norris, K. Norris, L. Snel, A. R. Sørensen and H. O. Voight (1988) *Nature*, **333**, 679–682.
119. Soluble prolonged-acting insulin derivatives I' J. Markussen, P. Hougaard, U. Ribel, A. R. Sørensen and E. Sørensen (1987) *Protein Eng.*, **1**, 205–213.
120. 'Soluble prolonged-acting insulin derivatives II' J. Markussen, I. Diers, A. Engesgaard, M. T. Hansen, P. Hougaard, L. Langkjaer, K. Norris, U. Ribel, A. R. Sørensen E. Sørensen and H. O. Voigt. (1987) *Protein Eng.*, **1**, 215–223.
121. 'Soluble prolonged-acting insulin derivatives III' J. Markussen, I. Diers, P. Hougaard, L. Langkjaer, K. Norris, L. Snel, A. R. Sørensen and H. O. Voight (1988) *Protein Eng.*, **2**, 157–166.
122. 'Small peptides – new targets for drug research' A. S. Dutta (1989) *Chem. in Brit.*, (February), 159–162.
123. 'Design principles in the construction of biologically active peptides' E. T. Kaiser (1987) *Trends Biochem. Sci.*, **12**, 305–309.
124. *Quanta* (A graphics software package) Polygen Corporation.
125. 'High resolution X-ray analyses of renin inhibitor–aspartic proteinase complexes' S. J. Foundling, J. Cooper, F. E. Watson, A. Cleasby, L. H. Pearl, B. L. Sibanda, A. Hemmings, S. P. Wood, T. L. Blundell, M. J. Valler, C. G. Norey, J. Kay, J. Boger, B. M. Dunn, B. J. Leckie, D. M. Jones, B. Atrash, A. Hallett and M. Szelke (1987) *Nature*, **327**, 349–352.
126. 'Computer graphics modelling of human renin: specificity, catalytic activity and

intron–exon junctions' B. L. Sibanda, T. L. Blundell, P. M. Hobart, M. Fogliano, J. S. Bindra, B. W. Dominy and J. M. Chirgwin (1984) *FEBS Letts.*, **174**, 102–111.

127. 'Comparison of solvent inaccessible cores of homologous proteins: definitions useful for protein modelling' T. J. P. Hubbard and T. L. Blundell (1987) *Protein Eng.*, **1**, 159–171.

128. 'Structural and sequence patterns in the loops of beta-alpha-beta units' M. S. Edwards, M. J. E. Sternberg and J. M. Thornton (1987) *Protein Eng.*, **1**, 173–181.

129. 'Analysis of the relationship between sidechain conformation and secondary structure in globular proteins' M. J. McGregor, S. A. Islam and M. J. E. Sternberg (1987) *J. Mol. Biol.*, **198**, 295–310.

130. 'Three dimensional structural aspects of the design of new protein molecules' T. L. Blundell, D. Barlow, B. L. Sibanda, J. M. Thornton, W. Taylor, I. J. Tickle, M. J. E. Sternberg, J. E. Pitts, I. Haneef and A. M. Hemmings (1986) *Phil. Trans. Roy. Soc. London*, **317**A, 333–344.

131. 'A protein sequence/structure database' Protein engineering club database group (1988) *Nature*, **335**, 745–746.

132. 'A holistic approach to protein structure alignment' W. R. Taylor and C. A. Orengo (1989) *Protein Eng.*, **2**, 505–519.

133. 'Three dimensional structure of a genetically engineered variant of porcine growth hormone' S. Abel-Meguid, H-S. Shieh, W. W. Smith, H. E. Dayringer, B. N. Violand and L. A. Bentle (1987) *Proc. Nat. Acad. Sci. USA*, **84**, 6434–6437.

134. Three dimensional structure of calmodulin' Y. S. Babu, J. S. Sack, T. J. Greenhough, C. E. Bugg, A. R. Means and W. J. Cook (1985) *Nature*, **315**, 37–40.

135. 'Structure of the calcium regulatory muscle protein troponin C at 2.8 Å resolution' O. Herzberg and M. N. G. James (1985) *Nature*, **313**, 653–659.

136. 'Structure of the liganded T state of haemoglobin identifies the origin of cooperative oxygen binding' R. Liddington, Z. Derewenda, G. Dodson and D. Harris (1988) *Nature*, **331**, 725–728.

137. 'Refined crystal structures of deoxyhaemoglobin S. I Restrained least-squares refinement at 3.0Å resolution' E. A. Padlan and W. E. Love (1985) *J. Biol. Chem.*, 260, 8272–8279.

138. 'Three dimensional structure of an oncogenic protein: Catalytic domain of c-H rasP^{21}' A. de Vos, L. Tong, M. V. Milburn, P. M. Matias, J. Janoarik, S. Noguchi, S. Nishimura, K. Miura, E. Ohtsuka and S. H. Kim (1988) *Science*, **239**, 888–893.

139. 'Structure of recombinant human renin, a target for cardiovascular active drugs, at 2.5Å resolution' A. R. Sielecki, K. Hayakawa, M. Fujinaga, M. E. P. Murphy, M. Fraser, A. K. Muir, C. T. Carilli, J. A. Lewicki, J. D. Baxter and M. N. G. James (1989) *Science*, **243**, 1346–1351.

140. 'Three dimensional structure of human serum albumin' D. C. Carter, X-M. He, S. H. Munson, P. D. Twigg, K. M. Gernert, M. B. Broom and T. Y. Miller (1989) *Science*, **244**, 1195–1198.

141. 'Crystal structure of the cytokine interleukin 1-beta' J. P. Priestle, H-P. Schar and M. G. Grutter (1988) *EMBO Journal*, **7**, 339–343.

142. 'Crystal structure of a retroviral protease proves relationship to aspartic protease family' M. Miller, M. Jaskolski, J. K. M. Rao, J. Leis and A. Wlodawer (1989) *Nature*, **337**, 576–579.

143. 'X-ray analysis of HIV-1 proteinase at 2.7 Å resolution confirms structural homology among retroviral enzymes' R. Lapatto, T. Blundell, A. Hemmings, J. Overington, A. Wilderspin, S. Wood, J. R. Merson, P. J. Whittle, D. E. Danley, K. F. Geoghegan, S. J. Hawrylik, S. E. Lee, K. G. Scheld and P. M. Hobart (1989) *Nature*, **342**, 299–302.

144. 'Dihydrofolate reductase: the stereochemistry of inhibitor selectivity' D. A. Matthews, J. T. Bohn, J. M. Burridge, D. J. Filman, K. W. Volz and J. Kraut (1985) *J. Biol. Chem.*, **260**, 392–399.

145. 'Refined crystal structure of carboxypeptidase A at 1.54Å resolution' D. C. Rees, M. Lewis and W. N. Lipscomb (1983) *J. Mol. Biol.*, **168**, 163–179.

146. 'The refined crystal structure of dimeric phospholipase A2 at 2.5Å. Access to a shielded catalytic center' S. Brunie, J. Bohn, D. Gewirth and P. B. Sigler (1985) *J. Biol. Chem.*, **260**, 9742–9749.

147. 'Enhanced activity and altered specificity of phospholipase A2 by deletion of a surface loop' O. P. Kuipers, M. M. G. M. Thunnissen, P. de Geus, B. W. Dijkstra, J. Drenth, H. M. Verheij and G. H. de Haas (1989) *Science*, **244**, 82–85.

148. 'High resolution (1.5 Å) crystal structure of phospholipase C from *Bacillus cereus*' E. Hough, L. K. Hansen, B. Birknes, K. Jynge, S. Hansen, A. Hordvik, C. Little, E. Dodson and Z. Derewenda (1989) *Nature*, **338**, 357–360.

149. 'Three-dimensional structure of CheY, the response regulator in bacterial chemotaxis' A. M. Stock, J. M. Mottonen, J. B. Stock and C. E. Schutt (1989) *Nature*, **337**, 745–749.

150. 'Atomic structure of thymidylate synthetase: target for rational drug design' L. W. Hardy, J. S. Finer-Moore, W. R. Montfort, M. O. Jones, D. V. Santi and R. M. Stroud (1987) *Science*, **235**, 448–455.

151. 'Structure of tumour necrosis factor' E. Y. Jones, D. I. Stuart and N. P. C. Walker (1989) *Nature*, **338**, 225–228.

152. 'Structures of two thermolysin-inhibitor complexes that differ by a single hydrogen bond' D. E. Tronrud, H. M. Holden and B. W. Matthews (1987) *Science*, **235**, 571–574.

153. 'Structure of native porcine pancreatic elastase at 1.65Å resolution' E. Meyer, G. Cole, R. Radhakrishnan and O. Epp (1988) *Acta Cryst.*, **B44**, 26–38.

154. 'Crystallographic study of a beta-lactam inhibitor complex with elastase at 1.84Å resolution' M. A. Navia, J. P. Springer, T-Y. Lin, H. R. Williams, R. A. Firestone, J. M. Pisano, J. B. Doherty, P. E. Finke and K. Hoogsteen (1987) *Nature*, **327**, 79–82.

155. 'Tertiary structural similarity between a class A beta lactamase and a penicillin-sensitive D-alanyl carboxypeptidase-transpeptidase' B. Samraoui, B. J. Sutton, R. J. Todd, P. J. Artymuik, S. G. Waley and D. C. Phillips (1986) *Nature*, **320**, 378–380.

156. 'Structure of the membrane-pore forming fragment of colicin A' M. W. Parker, F. Pattus, A. D. Tucker and D. Tsernoglou (1989) *Nature*, **337**, 93–96.

157. 'Refined 1.2 Å crystal structure of the complex formed between subtilisin Carlsberg and the inhibitor eglin c. Molecular structure of eglin and its detailed interaction with subtilisin' W. Bode, E. Papamokos, D. Musil, U. Seemueller and H. Fritz (1986) *EMBO Journal*, **5**, 813–818.

158. 'The three-dimensional structure of foot and mouth disease virus at 2.9 Å resolution' R. Acharya, E. Fry, D. Stuart, G. Fox, D. Rowlands and F. Brown (1989) *Nature*, **337**, 709–716.

159. 'Three-dimensional structure of polio virus at 2.9 Å resolution' J. M. Hogle, M. Chow and D. J. Filman (1985) *Science*, **229**, 1358–1365.

160. 'The atomic structure of mengo virus at 3.0 Å resolution' M. Luo, G. Vriend, G. Kamer, I. Minor, E. Arnold, M. G. Rossmann, U. Boege, D. G. Scraba, G. M. Duke and A. C. Palmemberg (1987) *Science*, **235**, 182–191.

161. 'The structure determination of a common cold virus, human rhinovirus 14' E. Arnold, G. Vriend, M. Luo, J. P. Griffith, G. Kamer, J. W. Erickson, J. E. Johnson and M. G. Rossmann (1987) *Acta Cryst.*, **A43**, 346–361.

162. 'Structural analysis of a series of antiviral agents complexed with human rhinovirus 14' J. Badger, I. Minor, M. J. Kremer, M. A. Oliveira, T. J. Smith, J. P. Griffith, D. M. A. Guerin, S. Krishnaswamy, M. Luo, M. G. Rossmann, M. A. McKinley, G. D. Diana, F. J. Dutko, M. Fancher, R. L. Rueckert and B. A. Heinz (1988) *Proc. Nat. Acad. Sci. USA*, **85**, 3304–3308.

163. 'Structure of the haemagglutinin membrane glycoprotein of influenza virus at 3 Å

resolution' I. A. Wilson, J. J. Skehel and D. C. Wiley (1981) *Nature*, **289**, 366–373.

164. 'Three dimensional structure of an antigenic mutant of the influenza virus haemagglutinin' M. Knossow, R. S. Daniels, A. R. Douglas, J. J. Skehel and D. C. Wiley (1984) *Nature*, **311**, 678–680.

165. 'Three dimensional structure of a complex of antibody with influenza virus neuraminidase' P. M. Colman, W. G. Laver, J. N. Varghese, A. T. Baker, P. A. Tulloch, G. M. Air and R. G. Webster (1987) *Nature*, **326**, 358–363.

166. 'Structure of the protein subunits in the photosynthetic reaction center of *Rhodopseudomonas viridis* at 3 Å resolution' J. Deisenhofer, O. Epp, K. Miki, R. Huber and H. Michel (1985) *Nature*, **318**, 618–624.

167. 'Three dimensional structure of Yeast tRNA Asp. 1 Structure determination' M. B. Comarmond, R. Giege, J. C. Thierry and D. Moras (1986) *Acta Cryst.*, **B42**, 272–280.

168. 'Binding of the antitumour drug *cis*-[PtCl$_2$(NH$_3$)$_2$] to crystalline tRNA Phe at 6.0 Å resolution' J. C. Dewan (1984) *J. Amer. Chem. Soc.*, **106**, 7239–7244.

169. 'Structure of a phage 434 Cro/DNA complex' C. Wolberger, Y. Dong, M. Ptashne and S. C. Harrison (1988) *Nature*, **335**, 789–795.

170. 'Structure of the repressor–operator complex of bacteriophage 434' J. E. Anderson, M. Ptashne and S. C. Harrison (1987) *Nature*, **326**, 846–852.

—— *Chapter 4* ————————————————————

Theoretical approaches to peptide drug design

D. J. Ward, A. M. Brass, J. Li, E. Platt, Y. Chen and B. Robson

I. Introduction

There is currently considerable interest in the design of new drugs based on endogenous peptide hormones, neuropeptides, and proteins.[1-5] The production of highly specific drugs, which by implication would be expected to be non-toxic, would be substantially aided by knowledge of the major solution conformers and of the active form of the *native* peptides. We are interested in exploring protocols for computer modelling of peptide hormone conformation and dynamics for application in *rational* drug design.

Application of these protocols requires first, some confidence that low-energy conformers have been identified, and for this we compare predicted structures with those from experiment; second, that we have techniques for distinguishing the active or bound form of the peptide, from a range of stable and metastable conformers, when there is no detailed knowledge of the receptor binding site, and this chapter describes several ways in which this can be attempted; third, some knowledge of the biological effects of the endogenous peptide, together with data for analogues, especially those for which small structural changes cause large differences in biological effects. Accumulation of such data permits structure–activity relationships (SAR) to be proposed, which is a first step in suggesting activities for, as yet, unsynthesized molecules; this is *rational* drug design.

The principal aim of this chapter is to identify methods and protocols for modelling the conformational behaviour of peptides, and consideration is given to thyrotropin releasing hormone (TRH), the enkephalins, and oxytocin, to show how predicted conformational preferences can be related to biological actions and consequently be of value in the rational design of peptide-based drugs. For *de novo* studies, there is an enormous number of possible conformers for even relatively small peptides (oxytocin has nine residues, enkephalin has five), and this initially demands simplified interatomic potentials with regard to solvent and other environmental effects.

In view of the importance placed in the previous chapters on molecular dynamics (MD) in the refinement of conformers from experimental conformational analysis, as well as its general importance for molecular modelling, the technique will be described in some detail. To obtain some insight into the simulation processes we are also interested in the new theories on *chaos*, as a means of investigating and describing dynamical systems. Descriptions for dynamical trajectories in phase space and of chaos theory will be outlined, and it is shown that molecular dynamics exhibits deterministic chaos.

I.I. HISTORICAL ASPECTS

Although it is only recently that efforts have been made to design novel proteins, the design of novel drugs has proceeded for more than a century. A revolution in drug design came from the studies of Hansch and co-workers,[6] in quantitative structure–activity relationship (QSAR) studies,[7] who modelled drug partitioning by chemical modification of polarity. The drug should be non-polar enough to cross membranes, reach, and interact with cells, while being polar enough to avoid non-specific interactions. In this respect, the receptor becomes a simple phase, amorphous and unstructured.

Relating chemical modification to biological action leads to SAR,[8] which are often qualitative descriptions of the effect of changing chemical constituents, rather than being a *quantitative* assignment (as in QSAR) and providing an additive score to the change. Only recently has QSAR analysis for peptides been reported.[9] Computer-aided design of *peptide-based* drugs utilizes techniques of SAR, QSAR, generalized computer-aided drug design, methods developed for secondary[10] and tertiary structure prediction of proteins,[11-15] information from databases of sequence and conformational information,[16-18] and molecular graphics, as well as developments in computer architecture.[19,20]

Methods of energy calculation have immediate application in peptide conformational analysis for several related reasons. Most experimental data point towards peptides as having flexible, or random coil, structure, so secondary structure predictions on peptides are rarely used. Since there is no well-defined secondary structure there is not the problem in modelling the relative orientation of β-sheets/α-helices. Therefore, for the ensemble of interconverting conformers, the relative energy of any *allowed* peptide conformer has application.

Non-peptide molecules are mainly used in drug design,[21] and have formed the basis for most SAR/QSAR analysis because they can be relatively very rigid. They often have similar *molecular weight* to amino acids or very small peptides, and are therefore one and two orders of magnitude smaller than most peptides and proteins, respectively. For these different sizes, different approaches are taken, and in the search for predictive techniques for protein secondary and tertiary structure, the analysis is of *sequence–structure relationships* (SSR). For peptides, with large numbers of possible conformers, we seek *conformation–activity relationships* (CAR).[5]

One way in which QSAR techniques can be applied as an aid in peptide drug design is investigated in Section 4, and involves the orientation of the *pharmacophore* (the essential functional groups). This arrangement is likely to be different for binding and antagonist action and for agonist action, and differing again for different biological actions.[22] Taking the view that the *active* conformation of the drug and receptor are mutually induced, then increasing the potency of the drug depends on enhancing the *energetic* stability of the *active* form over other conformations. Similarly, design of an antagonist concentrates on stabilizing *bound* conformations, which are compatible with *inactive* conformations of the receptors.[3,5] The dimensionality of description for the pharmacophore orientation can be reduced, permitting comparison across a range of analogues.[5] Separate considerations for this method include enhancing the strength of interaction between the ligand and the receptor, as well as ways of improving a drug once a chemical modification has locked it in its active conformation. Detailed knowledge of the pharmacophore, as well as overlay of predicted conformers across a series of analogues, can allow such effects, as well as steric considerations, to be included.

Different techniques and approaches are usually used in the conformational analysis of proteins, reflecting the differences in physical properties. For example, secondary structure prediction depends on having well-defined preferred conformations, which proteins do but peptides do not. Nonetheless, local information, for example secondary structure, provides useful starting points from which to start energy calculations for peptides.[5] In the simulation of interaction between a peptide hormone and a receptor, or for either of them in isolation, a key consideration is the *size* of the 'system' and the level of detail for the study. In general, the larger the system, the lower the level of detail and therefore presumably accuracy.

Interaction of a peptide with a cell surface receptor will require a different protein to propagate the response, and it is this protein which responds to and therefore determines the active conformation of the receptor. Unfortunately, if the protein with which a receptor acts defines the active form of the receptor, so may any subsequent interaction in the chain of events from the initial ligand–receptor interaction. The system which must be simulated is therefore potentially enormous, perhaps constituting the whole biological cell. A system of solvent, ligand–receptor complex, associated protein or proteins, and possibly also membrane and counter-ions, remains beyond current simulation capabilities. It is not so much a question of memory, but the time requirement for simulation which is roughly proportional to the square of the number of atoms present. At the other extreme, the simplest system to study is the isolated peptide, *in vacuo*, for which active and bound forms must be deduced.[3,5]

2. Theory and methods

2.1. ENERGY MINIMIZATION

In an energy minimization simulation, the energy is only allowed to decrease and therefore the molecule can never escape from a minimum in which it is trapped. Using techniques such as the SIMPLEX method,[15,23] originally designed for fitting rough, discontinuous data, coarse steps are taken and energy barriers can be crossed. In an N-dimensional problem (with N variables), $N+1$ conformations are generated, and the one of highest energy is moved downhill in energy by attempting to reflect its position through the centroid of the positions of the remaining points. Repeated application of this operation, as well as extending the reflected point and contracting the pattern of points, are consistent with the original method. The GLOBEX procedure[15] uses the SIMPLEX method itself to predict the positions of *new minima* from those of *old minima* on the energy surface. The old minima could equally be experimental structures, which can be used to impose a net trend.

The potential energy of a peptide can be minimized as a function of conformation. In *in vacuo* calculations, if we are using *rigid* geometry and every atom is being considered (*all-atom* mode)[24] then, excluding a term for hydrogen bonds, the intramolecular energy E_{TOT} is given by equation (1), where A and B are Van der Waals terms and C is the electrostatic interaction:[15]

$$E_{TOT} = \sum \frac{A_i A_j}{r^9} + \frac{B_i B_j}{r^6} - \frac{C_i C_j}{r} \tag{1}$$

This is the *non-bonded* contribution in flexible geometry calculations (see Section 2.2.1), and the r^9 term can be replaced by r^{12}. In *united-atom* mode,[15] again with rigid geometry, the potential energy of a molecule with n rotatable bonds ($E_n(s)$) can be defined in equation (2)[15]:

$$E_n(s) = \sum_{i=1}^{i=j} \sum_{j=2}^{j=n} e_{ij}(s) + \sum_{k=1}^{k=m} I(\alpha_k) \tag{2}$$

where

$$e_{ij}(s) = A_{ij}s^9 - B_{ij}s^6 + C_{ij}s$$

and $I(\alpha_k)$ is the internal rotational potential for bond α_k (for more details see reference 15). The use of s rather than r for the distance between pairs of atoms is novel.[15] s is defined according to equation (3):

$$p = (x_i - x_j)^2 + (y_i - y_j)^2 + (z_i - z_j)^2 = r_{ij}^2$$
$$q = 1 + 0.25p$$
$$s = 2q/(q^2 + p) \tag{3}$$

With the rigid geometry model, planar *trans* peptide bonds are employed, and rotation is allowed around all single bonds. The difference between all-atom and united-atom is that the all-atom parameters include all aliphatic hydrogens, each with partial charge $+0.11$. The carbon has a complementary negative charge, with the value depending on whether one, two, or three hydrogens are attached. In the rigid geometry studies here, all bond lengths and valence angles of the same type are assigned the same value throughout the molecule, e.g. all carbonyl $C=O$ bond lengths are $1.24\,\text{Å}$.

The interaction potentials used in the *united-atom* simulations have been calibrated for the hydrophobic effect[15] and a dielectric constant of 3.5 is included to reduce charge–charge interactions, as would be the case in the presence of water.

2.2. MOLECULAR DYNAMICS

A single protein molecule may have many thousands of degrees of freedom coupled by complicated anharmonic potentials. At the same time, peptide and protein molecules lack any high degree of symmetry, so techniques used in statistical physics to describe simple crystalline solids such as lattice dynamics are inappropriate. It is only since the advent of modern high-speed digital computers that it has been possible to model macromolecules at the atomic level and one of the most successful computer techniques used is the molecular dynamics (MD) algorithm. Given an analytic form for the various interatomic distances, bond stretch, valence angle, and torsional potentials and the positions of all the atoms in the molecule (the conformation), Newton's equations of motion can be solved numerically for all the particles in the simulation.

2.2.1. The molecular dynamics algorithm

The first MD simulations were carried out by Alder and Wainwright[25] who simulated a system of elastic hard spheres, whilst the first MD simulations of a system of particles interacting via a realistic potential were carried out by Rahman[26] on liquid argon.

Consider a system described by the Hamiltonian:

$$H = K + \Phi \tag{4}$$

where K represents the kinetic energy and Φ the potential energy, of the system. For a peptide or protein, Φ typically has the form[13]

$$\Phi = \tfrac{1}{2}\Sigma k_b(b-b_0)^2 + \tfrac{1}{2}\Sigma k_\theta(\theta-\theta_0)^2 + \tfrac{1}{2}\Sigma k_\varphi[1+\cos{(n\varphi-\delta)}]$$
$$+ \Sigma\ (A/r^{12} - B/r^6 + q_1 q_2/Dr) \tag{5}$$

The first term in equation (5) models the contribution to the total energy from bond stretch vibrations. Every covalent bond in the molecule is treated as a Hooke's law spring with a characteristic length (b_0) and spring constant (k_b). This approximation will be valid at biological temperatures where bond stretch motions are small. The second term models the fluctuations from valence angles, and the third term models the energy required for twisting atoms along the axis of the covalent bond between them. The first three terms model the covalent bonding energy of the peptide or protein. The final term models the non-bonded interactions (Van der Waals and electrostatic interaction) between atoms in the molecule separated by a distance r. The bonding terms are naturally short-ranged, as are the Van der Waals terms (potentials going as $1/r^{12}$ and $1/r^6$ are well modelled by short-range interactions). The electrostatic term is more complex to handle, because it is so long-ranged (decaying only as $1/r$). The electrostatic terms means that peptide and protein simulations naturally fall into the class of N-body problems (a given charged atom or group will interact with every other charge in the simulation). The number of computer operations required to model such systems goes as N^2 (where N is the number of charged species in the simulation), rather than as N (as would be the case for systems with short-ranged interactions). This problem is partially overcome by artificially cutting off the electrostatic term at some finite (but large) radius by multiplying it by a smoothing function.[27]

Extra terms can be added to the potential energy to incorporate distance constraints obtained from NMR spectra or to close cyclic peptides:

$$\Phi_{dc} = 1/2 \, k_{dc}(r_{ij} - r_{ij}^0)^2$$

where r_{ij}^0 is the desired separation between atoms i and j, r_{ij} is the actual separation between atoms i and j, and k_{dc} is the magnitude of the forcing potential.[28]

Given the total potential energy of the system, the force on particle i is given by

$$\mathbf{F}_i(\{\mathbf{r}\}) = -\sum \partial \Phi(\mathbf{r})/\partial \mathbf{r}_j \tag{6}$$

where the notation $\mathbf{F}_i(\{\mathbf{r}\})$ is used to emphasize the fact that the force on particle i can depend on the position of all other particles in the system. Using equation (6) we obtain the forces on all the atoms in the simulation. The particle trajectories at some later time can be obtained by numerically integrating Newton's equation of motion:

$$(d^2\mathbf{r}_i/dt^2) = (1/m_i)\mathbf{F}_i(\{\mathbf{r}\})$$

The numerical integration of the equations of motion will give us the total configuration of the simulation at successive timesteps from which time-dependent properties of the simulation can be obtained explicitly, whilst equilibrium properties can be calculated by averaging over many configurations.

2.2.2. Integrating the equations of motion

Many algorithms are available for numerically integrating Newton's equations of motion, although typically they are linear multistep equations relating function values at discrete time levels. Mathematically, an updating step can be described as a mapping $T(\delta)$ acting on points in the phase space $(\mathbf{r}, \mathbf{v})$:

$$T(\delta):(\mathbf{r},\mathbf{v}) \rightarrow (\mathbf{r}',\mathbf{v}')$$

where $(\mathbf{r},\mathbf{v})$ is the position in phase space of the system at time t and $(\mathbf{r}',\mathbf{v}')$ is the position in phase space of the system at time $t+\delta t$. It is convenient to split $T(\delta)$ into two separate mappings:

$$T_r(\delta):(\mathbf{r},\mathbf{v})\rightarrow(\mathbf{r}',\mathbf{v})$$
$$T_v(\delta):(\mathbf{r}',\mathbf{v})\rightarrow(\mathbf{r}',\mathbf{v}')$$

Consider, for example, the well-known leapfrog integration algorithm.[29] This algorithm consists of three stages:

Step 1. Calculate all the accelerations, $\mathbf{a}_i(t)$, on the atoms at time t.
Step 2. Calculate the velocity at time $t+\delta t/2$ using:
$$\mathbf{v}_i(t+\delta t/2)=\mathbf{v}_i(t-\delta t/2)+\mathbf{a}_i(t)\delta t$$

Step 3. Calculate the positions at time $t+\delta t$ using:

$$\mathbf{r}_i(t+\delta t)=\mathbf{r}_i(t)+\mathbf{v}_i(t+\delta t/2)\delta t$$

The path through phase space of the atoms can be calculated by repeating Steps 1–3 many times. The mapping $T_r(\delta)$ is represented by Step 2, and the mapping $T_v(\delta)$ by Step 3. Writing the MD updating step in terms of mappings, it can be seen that the MD algorithm belongs to the class of problems in which an iterative non-linear mapping is applied to points in some space (the mapping is non-linear because $\mathbf{a}_i$ is a non-linear function of the positions). We will show later that MD simulations exhibit many of the properties typical of such systems.

The above algorithms can be shown to be accurate to order δt^2, i.e. the size of the errors in the simulation should be proportional to δt^2. It is possible to construct algorithms which use a higher-order difference equation and are theoretically more accurate in δt, but they are inherently unstable. Problems arise when the number of solutions to the difference equation exceeds the number of characteristic solutions of the differential equation being solved.[29] In this case the extra 'parasitic' solutions to the difference equation swamp the true solution to the differential equation for all but the smallest values of δt. This effect is well demonstrated by the results quoted by Beeman.[30]

2.2.3. *Initial conditions*

The potential energy surface of a peptide or protein can be very complex, with many deep metastable minima. If the simulation is started in one of these, the time needed to leave and move to the area of the global minimum could be many orders of magnitude larger than it is possible to simulate on the computer. In practice, this means that it is difficult to ensure that the simulation is properly sampling phase space, or to ensure that the results obtained over a few tens or hundreds of picoseconds of an MD simulation, from some initial conformation, bear any resemblance to those that would be over a run of several seconds (were such simulations possible to run). Indeed, as the iso-energy surface is potentially disconnected, it is perfectly possible to start in a region of phase space for which there is no connected iso-energy trajectory to the global minimum. In such cases the system is non-ergodic and the simulated molecule would never reach the area of the global minimum.

In fact, there is no *a priori* reason to believe that a global minimum in the potential energy surface will exist for a protein, which can be thought of as a many-body system with competing interactions. The ground state of such systems can be a 'glassy' phase[31] in which there are a large number of different minima all of equivalent energy and every attempt to minimize the structure will find a different final conformation. An example of such a system would be a protein in denaturing conditions.[32]

In order to minimize the time taken for the system to equilibrate, it is important that the initial conformation should be as close as possible to the global minimum. Ideally a simulation should be started with the atoms of the peptide or protein in the positions predicted from an X-ray structure, if available.[33] If an X-ray structure does not exist, other possible starting conformations include positions obtained from a molecular mechanics simulation or the results of a secondary structure prediction.[34] For small peptides good results have recently been obtained in which atom–atom distances from NMR experiments are used to constrain the MD simulation.[35]

The velocities of the atoms in the molecule are chosen from a Gaussian distribution with a width appropriate to the required temperature of simulation. If necessary, a constant velocity should be added to every atom in the system to cancel out any net linear momentum.

Typically an MD simulation will include several hundred to several thousand degrees of freedom; anything significantly larger than this is prohibitively expensive in computer time. Modelling small isolated systems, such as a peptide or protein, in vacuum is now routine, but in order to simulate bulk systems, such as solvent, we apply some form of *boundary* conditions. One of the commonest choices is to use periodic boundary conditions (equivalent to running the simulation on a *torus*, see Section 3). Under these conditions, particles which leave one side of the periodic cell re-enter through the opposite face, thereby minimizing edge effects, though not reducing finite-size effects. The simulation cell should be chosen to be large enough that the individual atoms/molecules do not interact significantly with their periodic images.

2.2.4. *MD ensembles*

As the simulation is modelling the Hamiltonian described by equation (4), it should share the same conserved quantities, i.e. total energy should be constant and the linear and rotational momenta should be constant (there will be no rotational momentum for systems with periodic boundary conditions). These conserved quantities can be used to check that the simulation is working properly; at every timestep the total energy and momenta can be calculated. Although these quantities will fluctuate from timestep to timestep, their averages should be constant. In practice they are surprisingly constant, the drift in energy per timestep typically being less than the round-off error of the computer. The algorithm described above simulates a system in the micro-canonical ensemble, i.e. a system with a constant number of particles, a constant energy, and a constant volume.

Sometimes it is desirable to perform simulations at constant pressure or temperature (or both). This can be achieved by adapting the Hamiltonian of the system (equation 4) so that the quantity which is required to be constant is a conserved quantity of the system. For example, Nosé and Klein[36] describe a Hamiltonian in which temperature was a conserved quantity. Parinello and Rahman[37] describe a Hamiltonian in which the unit cell vectors are dynamical variables of the Hamiltonian and which conserves pressure and stress (this algorithm is particularly useful in modelling structural phase transitions in crystals, as it allows the unit cell to change shape). A similar technique is described by Anderson[38] in which pressure is conserved by isotropically changing the volume of the unit cell.

2.2.5. *Equilibration*

As mentioned above, it is difficult to ensure that a peptide or protein simulation is in the global minimum. However, it *is* possible to ensure that the molecule reaches thermal equilibrium, even if the 'equilibrium' state is merely a long-lived metastable minimum.

When the simulation is started it will be far away from thermal equilibrium, in particular the balance between the potential energy and kinetic energy will not be the same as in a well-equilibrated system. For a simulation to be run as described above, the final temperature of the simulation would be unpredictable – it would be the equilibrium temperature for a system with the same total energy as the initial system. In order for the simulation to equilibrate at some predetermined temperature, energy must be added and subtracted from the simulation. This is normally done by rescaling the velocities such that at every timestep the average energy per degree of freedom is $(1/2)kT_0$, where k is Boltzmann's constant and T_0 is the required temperature. It should be noted that any simulation in which the temperature is being scaled is no longer in the microcanonical ensemble and it will not share the same conservation laws as the original Hamiltonian.

In order for the system to equilibrate in the microcanonical ensemble the temperature rescaling must be switched off, and this is usually done when the amount of energy being added or removed from the simulation becomes small (i.e. when the average potential energy of the system becomes constant). The system must then be run further to equilibrate in the microcanonical ensemble. It is difficult to know precisely when a simulation has reached equilibrium, but one of the most stringent tests is to monitor the fluctuations in the potential energy and temperature. When the size of these fluctuations has reached a constant value (equivalent to the system having a well-defined specific heat – see below) the system is said to be equilibrated. The problem is knowing whether the system has equilibrated in the global minimum of the potential surface, and one way to test is to heat it up and then slowly cool it down to T_0. If it repeatedly returns to the same minimum, then that is some indication that it corresponds to the global minimum.

2.2.6. Measurement of equilibrium properties

The typical quantities determined in an MD simulation are average potential energy, specific heat, various spatial and dynamical distribution functions (e.g. radial distribution functions, X-ray or neutron scattering functions, pressure, average temperature, etc.).

In statistical physics the average value of a quantity, Q, in the microcanonical ensemble (partition function Ω) is defined as its weighted average over all possible conformations, where the weight $\exp[-\beta H]$ gives the probability of finding an ensemble of energy H in phase space. If Z is the canonical partition function

$$Z = \int d\mathbf{r}\, d\mathbf{v}\, \exp[-H/kT]$$

where k is Boltzmann's constant, then Ω is defined via

$$Z = \int dE\, \Omega(E) \exp[-E/kT] \tag{7}$$

and the configurational average of Q, $\langle Q \rangle$, is defined by

$$\langle Q \rangle = \Omega^{-1} \int d\mathbf{r}\, d\mathbf{v}\, Q\delta(E-H)$$

To measure thermodynamic quantities from the simulation we therefore need to be able to calculate the integral defined by equation (7). We could naively calculate the configurational average by running many independent simulations from different points in phase space, but this method would be very expensive in computer resources. The assumption made in MD simulations is that configurational averages can be replaced by time averages, i.e. that over a long MD run the simulation will sample many points in phase space and that the average of Q measured over the MD run will approximate $\langle Q \rangle$. It is possible to show that this approximation will be valid providing that the system is ergodic and that data is collected for at least twice the period of the slowest oscillation in the system.[39] In cases where the ergodicity requirement is not met (for example, systems with disconnected iso-energy surfaces) then it is important to run several independent simulations starting in different regions of phase space in order to sample all of phase space properly.

The temperature of an MD simulation is defined via the total kinetic energy (K). Assuming that the equipartition theorem holds then the temperature of a system of N particles with N_{df} degrees of freedom is defined by[40]

$$T = 2\langle K \rangle / N_{df} k$$

By examining the fluctuations in the kinetic energy it is possible to calculate the specific heat at constant volume:

$$\langle K^2 \rangle - \langle K \rangle^2 = (3/2)Nk^2T^2(1-(3/2)k/C_v)$$

The pressure in the system is calculated via the virial:

$$p = \rho k \langle T \rangle - (1/3V)\langle \sum \mathbf{F}_{ij}.\mathbf{r}_{ij} \rangle$$

where V is the volume of the system.

X-ray and neutron scattering functions can be calculated from the density fluctuation correlation function:[41]

$$F(\mathbf{q},t) = \langle \rho(\mathbf{q},t)\rho(\mathbf{q},0)^* \rangle$$

where $\mathbf{q}$ is a reciprocal vector. $\rho(\mathbf{q},t)$ is the Fourier transform of the particle density:

$$\rho(\mathbf{q},t) = (1/\sqrt{N})\sum \exp[i\mathbf{q}.\mathbf{r}_i(t)]$$

where $\mathbf{r}_i(t)$ is the position of particle i at time t.

2.2.7. *Accuracy of the MD algorithm*

Because of the finite nature of computer arithmetic, any MD simulation will suffer from the effects of round-off. Consider a simulation of N identical particles of mass m. Adapting a technique from Hockney and Eastwood[29] it is possible to investigate how these errors will propagate under the mapping $T(\delta)$. Define $\mathbf{x}^n$ to be a $3N$ vector describing the positions of the particles after n applications of $T(\delta)$. Define $\mathbf{X}^n$ to be the positions of the particles after n iterations if the computer arithmetic was done at infinite precision. Assume that $T(\delta)$ is exact except at the second timestep when a small rounding error, ε^2, is introduced:

$$\varepsilon^2 = \mathbf{x}^2 - \mathbf{X}^2$$

and that from this step onwards all arithmetic is done at infinite precision. What we would like to know is whether the error introduced at Step 2 increases or decreases as the simulation precedes. Applying $T(\delta)$ on $\mathbf{X}^2$ and Taylor expanding we get the error propagation equation:

$$\varepsilon^{n+1} - 2\,\varepsilon^n + \varepsilon^{n-1} = -\mathbf{A}^2(\delta t)^2\,\varepsilon^n \tag{8}$$

where

$$A^2 = (1/m)|\partial \mathbf{F}/\partial x|_{\max}$$

Solving equation (8), it can be shown that the error introduced at the second timestep will decay if:

$$\delta t < \delta t_c = 2/A \tag{9}$$

If the results quoted above can be extended to a real MD simulation (where round-off errors will be introduced at every timestep), then for values of the timestep less than some critical value, the effect of round-off should not be cumulative – i.e. the conserved quantities of the Hamiltonian should be constant. For values of the timestep larger than some critical value the simulation should become unstable – i.e. conserved quantities of the Hamiltonian will no longer be

Table 1 The values of the drift in total energy per timestep as a function of timestep for a run of 7000 steps of a system of 4096 particles interacting via a Lennard–Jones potential in two dimensions. The simulation was run with $\rho^* = 0.92$ and $t^* = 0.43$. The simulations were run using single-precision arithmetic.

Timestep (δt^*)	Drift in the total energy per timestep
0.006	1.5×10^{-10}
0.012	1.1×10^{-9}
0.024	4.9×10^{-10}
0.030	1.9×10^{-8}
0.036	2.2×10^{-7}
0.042	3.2×10^{-6}

constant. In order to test this we ran a series of simulations on two-dimensional Lennard–Jonesium in which we measured the drift in the total energy per timestep as a function of the size of δt (Table 1). From this graph it can be seen that the drift per timestep stays approximately constant for a range of timestep values from $\delta t = 0.006$ to $\delta t = 0.024$. However, for values of the timestep greater than this the drift per timestep grows rapidly with timestep until the simulation becomes completely unstable. It can be shown that the value of δt at which the simulation becomes unstable agrees surprisingly well with the value predicted from equation (9).

2.2.8. MD and chaos

One signal of a chaotic system is sensitivity to initial conditions. To investigate the stability of MD trajectories to perturbation we calculated the velocity cross-correlation function for two simulations of two-dimensional Lennard–Jonesium, both starting at exactly the same point in phase space, but for which we used two different timesteps, both much less than the critical timestep (δt_c) we calculated in the previous section. The velocity cross-correlation function is defined by:

$$C(t) = \sum \mathbf{v}_i^1(t) \cdot \mathbf{v}_i^2(t) / (|\mathbf{v}_i^1(t)||\mathbf{v}_i^2(t)|)$$

where $\mathbf{v}_i^1(t)$ is the velocity of particle i in simulation 1 at time t.

Figure 1 shows the velocity cross correlation function for simulations of two-dimensional Lennard–Jonesium at $\rho^* = 0.92$ and $T^* = 0.42$. From this graph it appears that the two systems remain well correlated up until approximately 3 ps, at which point the correlation function decays rapidly to zero. This behaviour appears strange for a system obeying Newton's equation of motion. We have already shown that rounding errors decay in the simulation and that energy is conserved, yet two simulations starting at the same point in phase space but using a different timestep rapidly diverge from each other in phase space. This divergence is a sign of an 'instability' implicit in any MD simulation. To demonstrate this instability, consider oxygen at normal temperature and

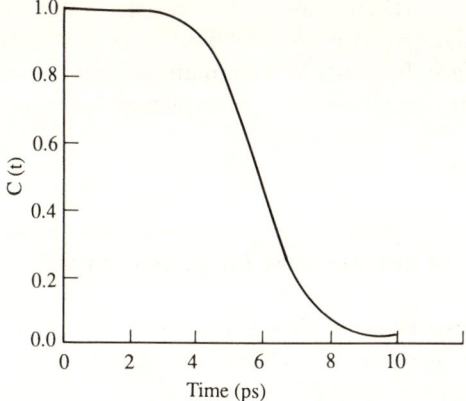

Figure 1 The velocity cross-correlation function, $C(t)$, plotted against time for a simulation of 4096 particles via a Lennard–Jones potential in two dimensions. The simulation was run with $\rho^* = 0.92$ and $T^* = 0.43$.

pressure (NTP). Neglecting the gravitational interaction of an electron at the edge of the known universe introduces an error in the angle of motion of any oxygen atom that has undergone only 56 collisions of one radian.[42] Calculating the velocity cross-correlation function for a system of interacting hard spheres we obtain the graph shown in Fig. 2 – the form of the graph is similar to that obtained from two-dimensional Lennard–Jonesium. If we repeat the Lennard–Jonesium simulations using double precision, then the time taken for $C(t)$ to drop to zero doubles. Powles[43] has shown that the divergence of MD trajectories grows exponentially with round-off error.

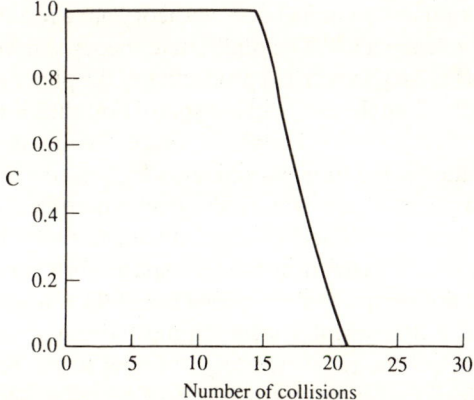

Figure 2 The calculated value of the velocity cross-correlation function, C, plotted against the number of collisions for a system with $\rho^* = 0.92$ and $T^* = 0.43$.

Round-off error in MD simulations has the effect of introducing exponential divergence in simulation trajectories in phase space, but does not move them off the iso-energy surface. In short, MD simulations provide a classic example of a system with a chaotic Hamiltonian. The implications of this are discussed in the next section.

3. Development of descriptions for phase space

This section will describe some basic geometric structures and characteristics of dynamical systems. A dynamical system[44] is defined by

$$\dot{\mathbf{x}} = \mathbf{F}(\mathbf{x},t) \tag{10}$$

where $\mathbf{x} = (x_1, x_2, \ldots, x_n)$ describes the state of the system, and $\dot{\mathbf{x}}$ is the derivative with respect to time t. The solution of (10) at time t identifies a point in an abstract n-dimensional space, which is called the phase space of the system. The evolution of the system for particular time series traces out a trajectory (or orbit) in phase space. The set of all possible trajectories is called *phase flow*, and for lower-dimensional systems, the possible trajectories can be plotted as a diagram known as a *phase portrait* of the phase flow.

It is known that dynamical system theory is the theoretical basis for molecular dynamics simulation, and many issues studied by dynamical system theory have direct significance for molecular dynamics simulation. Consider for example, the problem of structural stability, which is: if a dynamical system X has a known phase structure P, and is then perturbed to a slightly different system X' (this usually takes the form of changing coefficients in the differential equations), then is the new phase structure P' close to P in some topological sense? It is obvious that this problem has direct importance in practice, since the qualitative information obtained for P is applied not only to X, but to some nearby system X'. Studies of this kind in dynamical system theory have resulted in the so-called *qualitative theory of dynamics*.[45] The qualitative theory started with the work of Poincaré, and is based on geometrical properties of the phase portrait: the family of trajectories, which fill up the entire phase space. For questions such as stability, it is necessary to study the entire phase space, including the behaviour of solutions for all values of the time parameter, which means that it is essential to consider the entire phase space as a single geometric object.

It is known that for a mechanical system with angular variables or constraints, the phase space can be a general non-linear space; for example, a generalized cylinder. Thus, the notion of a differentiable manifold was used by Poincaré, as the phase space for a dynamical system. In mechanical systems, this manifold always has a special geometric structure pertaining to the occurrence of phase variables in canonically conjugate pairs. We now describe a few common regular geometrical structures encountered in dynamical system study.

3.1. GEOMETRICAL STRUCTURES OF PHASE SPACE

3.1.1. *Circle*

Consider an ordinary differentiable equation

$$\dot{x} = c, \quad x \; \varepsilon \; (0, 2\pi)$$

where $c > 0$ and is a constant. The phase space of this equation is the interval $(0, 2\pi)$, with 0 and 2π identified. Thus, the phase space has the structure of a circle of length 2π which is usually denoted as S^1 (the superscript refers to the dimension of the phase space).

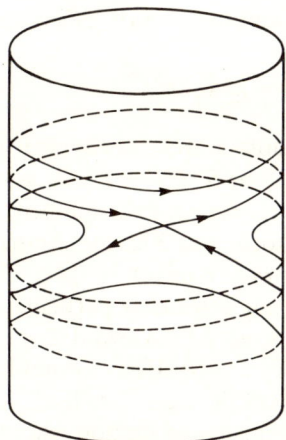

Figure 3   Schematic diagram of the trajectories of a moving pendulum, whose phase space can be represented by a cylinder.

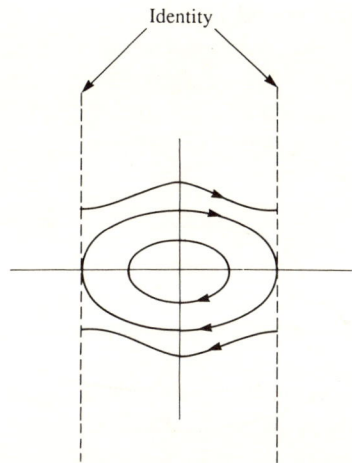

Figure 4   The two-dimensional representation of the trajectories described in Fig. 3.

3.1.2. Cylinder

The dynamical motion of a free undamped pendulum can be described by the following differential equations:

$$\dot{x} = y$$
$$\dot{y} = -\sin(x)$$

where y can take on any value in $\mathbb{R}$, but since the motion is rotational, the position x is periodic with period 2π. Hence, the phase space for this system is the cylinder, which is denoted by $\mathbb{R}^1 \times S^1$ (see the diagrams in Figs 3 and 4).

3.1.3. Torus

Consider the following system of ordinary differential equations:

$$\dot{x} = C_1$$
$$\dot{y} = C_2 \qquad C_1, C_2 \; \varepsilon \; (0, 2\pi) \tag{11}$$

where C_1 and C_2 are constants. x and y are angular variables, which defines the phase space of the system to be $S^1 \times S^1 = T^2$, i.e. a two-dimensional torus. For this particular system, if C_1/C_2 is a rational number, the trajectories of (11) spiral around the surface of the torus (see Fig. 5); alternatively, they densely fill the surface if C_1/C_2 is an irrational number.

The concept of motion on a torus is particularly useful in that it can be generalized to systems with more than two degrees of freedom. The n-torus is usually written as $T^n = S^1 \times \ldots \times S^1$. Each constant motion reduces the dimensionality of the phase space of the trajectory by 1, such that for an n-degree-of-freedom system with n constants of motion, the dynamical evolution in $2n$-dimensional phase space is reduced to that of an n-dimensional surface or manifold on which n angular variables run. The topological properties of the surface are those of an n-torus, i.e. n phase variables are orthogonal to one another and periodic with period 2π.

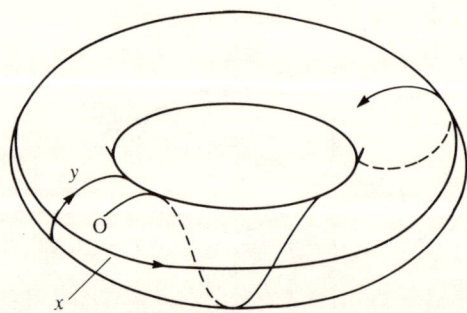

Figure 5 Pictorial representation of dynamical motion confined on a two-dimensionaltorus.

3.1.4. *Sphere*

The *n*-sphere of radius R is denoted by S^n and defined as

$$S^n = \{x \; \varepsilon \; \mathbb{R}^{n+1} \quad |x| = \mathbb{R}\} \tag{12}$$

One example is the dynamical motion of a rigid body,[46] whose equations of motion are given by:

$$\dot{x}_1 = \frac{I_2 - I_3}{I_2 \, I_3} \cdot x_2 \, x_3$$

$$\dot{x}_2 = \frac{I_3 - I_1}{I_3 \, I_1} \cdot x_1 \, x_3 \qquad (x_1, x_2, x_3) \; \varepsilon \; \mathbb{R}^1 . \mathbb{R}^2 . \mathbb{R}^3$$

$$\dot{x}_3 = \frac{I_1 - I_2}{I_1 \, I_2} \cdot x_1 \, x_2$$

where $I_1 \geqslant I_2 \geqslant I_3$ are moments of inertia about the principal body fixed axes and $x_i = I_i \, \omega_i$, $i = 1,2,3$, where ω_i is the angular velocity about the *i*th principal axis. Equation (12) has two equations of motion:

$$H = \frac{1}{2} \sum_{i=1}^{3} \frac{x_i^2}{I_i}$$

$$L^2 = \sum_{i=1}^{3} x_i^2$$

Thus, the trajectories of (12) are given by the intersection of the ellipsoids, $H = $ constant; with the sphere, $L^2 = $ constant.

3.2. ASPECTS OF CHAOS

Research in dynamical systems over the past two decades has shown that nonlinear dynamical systems can behave in a random fashion. Even more surprising is the fact that systems with only two or three degrees of freedom exhibit such unpredictable behaviour.[47,48] Such random phenomena in dynamical systems have generally been termed *chaos*. Chaotic behaviours in dynamical systems are not pathological or exceptional. In fact, most Hamiltonian systems exhibit chaotic behaviour.[49] Therefore, it is important to be aware of issues in chaos theory when using molecular dynamics to simulate dynamical behaviour of molecular systems.

There is no generally accepted definition of chaos. However, it is usually considered to be bounded steady-state behaviour that is not an equilibrium point, not periodic, and not quasi-periodic. Some examples are given below.

3.2.1. *Discrete system*

The logistic iterative equation

$$x_{n+1} = 4ax_n \, (1 - x_n)$$

where $0 < a < 1$ and $0 < x < 1$, has been used to study population dynamics for many years. The iterative mapping has provided the most famous example of a simple dynamical system exhibiting very complex behaviour.[50] For small a, one finds that all iterates converge onto a single fixed point. This behaviour persists until a passes 0.75. The simple fixed point, for large a, bifurcates into a pair of fixed points. As a is increased further, the two fixed points break into four fixed points, which subsequently break into eight fixed points, and so on. The a values at which the bifurcations occur (a_1, a_2, ...) become ever closer, converging geometrically to a critical value a_∞ (about 0.892). At this point, the orbit becomes aperiodic. Beyond a_∞, both chaotic orbits and odd-period limit cycles appear. At $a = 1$, the motion is formally ergodic on the unit interval (0,1). This process shows the so-called period-doubling route to chaos.

One remarkable feature of this mapping is the existence of universality.[51] It is found that

$$\delta = \lim_{n \to \infty} \frac{a_{n+1} - a_n}{a_{n+2} - a_{n+1}} = 4.6692016\ldots$$

exist for a class of mapping, and it has also been observed in a variety of experiments. This is where the excitement about order within chaos arises.

3.2.2. Continuous system

A classic example for chaotic behaviour of continuous system is Duffing's equation:[48]

$$\dot{x} = y$$
$$\dot{y} = -x^3 - ay + b \cos(t)$$

This system has been used to illustrate how a non-autonomous system dynamically behaves as system parameters change. Within certain regions of parameter space of a and b, stable periodic motions prevail. For small values of a, as b is increased, a periodic solution bifurcates into a pair of symmetric solutions, which then undergo a period-doubling cascade into a chaotic solution. A further increase in b gives a return to a symmetric pair of period solutions that undergo period doubling into chaos, which then leads into a new periodic solution.

3.3. CRITERIA FOR THE ONSET OF CHAOS

3.3.1. Lyapunov exponents[52,53]

From the examples presented above, it is clear that an important characteristic of chaotic motion is the great sensitivity of the motion to small changes in initial conditions. Closely neighbouring trajectories are found to diverge exponentially, whereas regular trajectories are found to separate only linearly in time. One of the methods to quantify the rate of divergence is to compute Lyapunov exponents, which measure the mean rate of exponential separation of neighbouring trajectories. In fact, Lyapunov exponents are an extremely useful way of characterizing dynamical systems. They provide a computable, quantitative measure of the degree of stochasticity for a trajectory. The following are some brief descriptions of Lyapunov exponents.

Consider an autonomous system,

$$\dot{x}_i = F_i(x_1, x_2, \ldots, x_n) \quad i = 1, \ldots, n \tag{13}$$

In order to examine the stability of the system, the system is linearized about any reference $(\bar{x} = \bar{x}_1, \bar{x}_2, \ldots, \bar{x}_n)$ to yield the tangent map:

$$\frac{d\delta x_i}{dt} = \sum_{j=1}^{n} \delta x_j \left(\frac{\delta F_i}{\delta x_j}\right)_{x = x(t)} \tag{14}$$

The norm

$$d(t) = \sqrt{\sum_{i=1}^{n} \delta x_i^2(t)} \tag{15}$$

provides a measure of the divergence of two neighbouring trajectories, that is, the reference trajectory $\bar{x}$ and its neighbour with initial condition $\bar{x}(0) + \delta x(0)$. The mean rate of exponential divergence is defined as

$$\sigma = \lim_{\substack{t \to \infty \\ d(0) \to 0}} \left(\frac{1}{t}\right) \ln \left(\frac{d(t)}{d(0)}\right) \tag{16}$$

where

$$d(0) = \sqrt{\sum_{i=1}^{n} \delta x_i^2(0)} \tag{17}$$

It can be shown that there exists a set of n such quantities, $\sigma_i, i = 1, \ldots, n$. These σ_i are called the *Lyapunov characteristic exponents* and they can be ordered by size, that is $\sigma_1 \geqslant \sigma_2 \geqslant \ldots \geqslant \sigma_n$.

In order to understand these ideas in more detail, it is useful to discuss the Lyapunov exponents of mappings. The simplest case is a one-dimensional map of the form

$$x_{n+1} = f(x_n)$$

where $f(x)$ is some non-linear function of x, e.g. $f(x) = 4ax(1-x)$, as in the logistic map we discussed before. The tangent map for a one-dimensional map is

$$x_{n+1} = f'(x_n) \, \delta x_n = \prod_{i=0}^{n} f'(x_i) \, \delta x_o$$

where $f'(x_i)$ is the derivative of $f(x)$ evaluated at each point x_i along the given trajectory. The associated Lyapunov exponent is deduced by using equation (15):

$$\delta = \lim_{n \to \infty} \frac{1}{n} \ln \left[\prod_{i=1}^{n} f'(x_i) \, \delta x_o\right] = \lim_{n \to \infty} \frac{1}{n} \sum_{i=0}^{n} \ln \left|f'(x_i)\right|$$

The exponent σ is independent of the initial point x_0.

In the case of multidimensional mappings,

$$\mathbf{x}_{n+1} = \mathbf{F}(\mathbf{x}_n)$$

where $\mathbf{x}$ and $\mathbf{F}$ are n-dimensional vectors, there will be a set of n characteristic exponents corresponding to the n eigenvalues of the associated tangent map. Introducing the eigenvalues $\lambda_i(n), i = 1, \ldots n$, of the matrix

$$(\mathbf{TM})_n = (\mathbf{M}\ (\mathbf{x}_n)\ \mathbf{M}\ (\mathbf{x}_{n+1}) \ldots \mathbf{M}\ (\mathbf{x}_1))^{1/n}$$

where $\mathbf{M}\ (\mathbf{x}_i)$ is the linearization of $\mathbf{F}$ at $\mathbf{x}_i$, the exponents are defined as

$$\sigma_i = \lim_{n \to \infty} \ln \left| \lambda_i\ (n) \right|, \ i = 1, \ldots, n$$

For area-preserving maps and Hamiltonian flows, the sum of the exponents must be zero in order to ensure that phase volume is preserved.

For the case of flows governed by (13), the tangent map is

$$\frac{d\delta \mathbf{Z}}{dt} = \mathbf{M}\delta \mathbf{Z}$$

where $\delta \mathbf{Z} = (\delta x_i, \ldots, \delta x_n)$ and $\mathbf{M}$ is the linearized matrix with elements

$$\mathbf{M}_{ij} = \left(\frac{\delta \mathbf{F}_i}{\delta x_i} \right)_{\mathbf{X} = \mathbf{X}(t)}$$

There will exist a set of basis vectors $e_i\ (i = 1, \ldots, n)$ such that

$$\delta \mathbf{Z} = \sum_{i=1}^{n} a_i e_i$$

The stretching (or contracting) rates in each of the direction e_i provides the set of exponents $\sigma_i\ (i = 1, \ldots, n)$, which can be ordered as in (17). As time evolves, a small volume element will be stretched most in the direction e_i with the largest exponent. Thus, in practice, (16) will yield this exponent (σ_i). For the Hamiltonian system with n degrees of freedom, the vector $\delta \mathbf{Z}$ becomes $2n$-dimensional (i.e. $\delta \mathbf{Z} = (\delta q_1, \ldots, \delta q_n, \delta p_1, \ldots, \delta p_n)$, and there will be $2n$ exponents. However, now there is a special symmetry between σ_i, namely

$$\sigma_i = -\sigma_{2n-i+1}$$

Thus any stretching in one direction is cancelled by contraction in another, thus ensuring Liouville's theorem. If the exponents are calculated on a given energy shell, the space is $2n - 1$ dimensional. Thus it follows from (14) that two (or more) of σ_i must be zero.

The actual computation of exponents for n-dimensional flows is non-trivial. Consider, for example, working with definition (16). If the norm $d(t)$ increases exponentially, there will be the risk of computer overflow. To avoid this problem, some special techniques are required. One starts with the initial norm $d(0)$ normalized to unity and computes the divergence over some interval τ, which is then renormalized back to a norm of unity. In this way, one computes a sequence of norm

$$d_i = \| \, \delta \mathbf{x}^{\,(i+1)}(\tau) \, \|$$

where $\|\cdot\|$ denotes the Euclidean norm, and

$$\delta x_{(0)}^{(i)} = \frac{\delta x_{(0)}^{(i=1)}}{d_i}$$

which is computed from (14), with its initial values along the reference trajectory $\bar{\mathbf{x}}$ from $\bar{\mathbf{x}}(i\tau)$ to $\bar{\mathbf{x}}((i+1)\tau)$, and thus exponents are defined as

$$\sigma_n = \frac{1}{n\tau} \sum_{i=1}^{n} \ln d_i$$

Furthermore, if τ is not too large, it can be shown that the limit $n \to \infty$ exists and is independent of τ. Indeed one can show that

$$\lim_{n \to \infty} \sigma_n = \sigma_1$$

where σ_1 is the largest of the set of exponents. The computation of the complete spectrum of Lyapunov exponents $\sigma_1, \ldots \sigma_n$ requires more sophisticated techniques.

3.3.2. Power spectra[54]

The calculation of Lyapunov characteristic exponents has provided a quantitative method of identifying dynamical behaviours of trajectories. Although theoretically it is a clearly defined method, the practical calculation of Lyapunov exponents for actual systems requires very sophisticated numerical techniques. Power spectra characterization uses a spectral analysis method to differentiate trajectories. The power spectra, $I(\omega)$, of trajectory $x(t)$ is related to the Fourier transform of autocorrelation function $C(t)$:

$$I(\omega) = \frac{1}{2\pi} \int_{-\infty}^{+\infty} C(t)\, e^{-i\omega t}\, dt$$

where

$$C(t) = <x(0)\, x(t)>$$

i.e. $C(t)$ is an average of $x(t)$ over an appropriate ensemble which is determined by the problem.

This power spectrum $I(\omega)$ is related to a certain time average of t:

$$I(\omega) = \frac{1}{2\pi} \lim_{T \to \infty} \frac{1}{2T} < \left| \int_{0}^{2T} x(t)\, e^{-i\omega t}\, dt \right|^2 >$$

By calculating $I(\omega)$ this way, it is possible to observe the differences between regular and irregular motions of dynamical systems. For regular motion, the spectrum is a clear curve.

For irregular motion, it is still possible to compute $I(\omega)$, using a simple trajectory. One finds that the spectrum of an irregular trajectory is much more complicated than for a regular one. Typically, one sees some dominant peaks surrounded by a lot of 'grass'. On the basis of numerical evidence alone, it is not clear whether this grassy portion of the spectrum is truly continuous for the irregular trajectories of generic Hamiltonian systems. Nonetheless, the difference in spectrum between regular and irregular motion is usually striking and again provides a valuable characterization of dynamical systems. Indeed, there are a number of important rigorous results available which tell us that a system will only have a discrete spectrum if it is 'ergodic', whereas to have a continuous spectrum it must be 'mixing'.

So far, the discussions of chaos have been of a local nature, in that they have concentrated only on the chaotic behaviour of individual trajectories as well as the means of identifying them and quantifying their behaviour. Useful as these techniques are, it would clearly be valuable if there were also methods for estimating when – as a function of energy or some non-linear coupling parameter – the bulk of the trajectories become chaotic. This is what is meant by the term *widespread chaos*. This is very important in even the most practical sense. For example, in theories of unimolecular decomposition, it can indicate the validity of statistical, rather than dynamical, theories. There are two well-known methods for this purpose.[55] The first is the method of overlapping resonances, which is capable of giving crude analytical estimates of the onset of widespread chaos. The other is Green's method, which is able to predict when individual tori will break down.

3.3.3. *Method of overlapping resonances*

In the study of dynamical systems it is often observed that a sharp transition occurs for increasing perturbation strength between regions of the phase space in which stochastic motion is closely bounded by Kolmogrov–Arnold–Moser (KAM) surfaces and regions for which the stochastic motion is interconnected over large portions of the space. In the former region, the excursion in the chaotic region is limited to the order of separatrix width. In the latter region, the motion may chaotically wander over the entire phase space. Clearly, a quantitative determination of the parameters for which this transition takes place yields important information concerning the system's behaviour.

Overlap resonance analysis is the earliest procedure to determine the transition to global or widespread chaos. In its simplest form, it postulates that the last KAM surface between two lowest-order resonances is destroyed when the sum of the half-widths of the two separatrices formed by the resonances, but calculated independently of one another, just equals the distances between the resonances. This criterion has an intuitive appeal, since it is known that regions near separatrix are stochastic. Rigorously, however, the overlap criterion is neither necessary or sufficient. One can imagine the last KAM surface breaking up well before isolated regions overlap, due to the interaction of slowly varying terms outside the separatrices. In fact, many results have indicated that the overlap criterion is too severe a condition for chaos. Detailed discussion of overlapping resonances analysis is outside the scope of this section, and interested readers are referred elsewhere.[56]

3.3.4. Green's method [57]

Green's method is an important method for predicting the onset of chaotic motion based on the stability properties of closed orbits. It is based on the hypothesis that the dissolution of an invariant curve (torus) can be associated with the sudden change from stability to instability of nearby closed orbits. To see this more precisely, imagine a weakly perturbed integrable system. According to the KAM theorem, those invariant curves with 'sufficiently' irrational winding number are preserved. The neighbouring rational curves break up in the manner described by the Poincaré–Birkhoff theorem, that is, into equal number of elliptic (stable) and hyperbolic (unstable) fixed points. Green's method is based on the observation that when the perturbation is made sufficiently strong (or the energy high enough), the set of stable fixed points also becomes unstable. The contention is that this then signals the dissolution of an invariant curve 'close' to that set of fixed points. The closeness of a closed orbit to a given invariant curve is estimated by expressing that curve's winding number in the form of a continued fraction, i.e.

$$a = a_0 + \cfrac{1}{a_1 + \cfrac{1}{a_2 + \cfrac{1}{a_3 + \ldots}}}$$

where $a_0, a_1, \ldots$ are positive integers. Thus the successive truncations of this representation of an irrational winding number yield the winding numbers of the closed orbits that became even 'closer' to that chosen invariant curve. By following the stability properties of these sequences of closed orbits, as they 'close in' on an invariant curve, it is possible to predict the breakup of that curve.

The two essential ingredients of Green's method are (a) finding the closed orbits and (b) determining their stability characteristics. Green's method represents what is probably the correct approach to predicting the onset of chaotic motion, i.e. concentrating on the destruction of individual tori. In this way, one avoids the problems associated with predicting the onset of chaos, such as defining the measure of 'widespread'. Furthermore, by allowing the breakup of a carefully chosen torus or invariant curve, it is possible to obtain a great deal of information about the transition to 'widespread' instability.

4. Development and testing of protocols for peptide modelling

Information transfer in peptide hormone–receptor systems involves several distinct steps (Fig. 6), which include recognition (binding), transduction

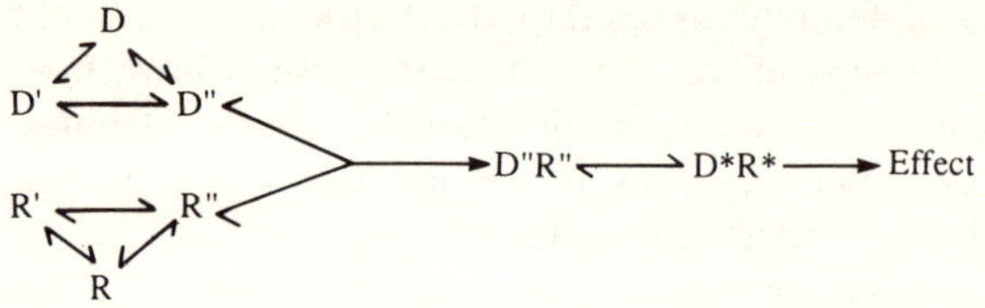

Figure 6 Schematic diagram for a model of drug (D)–receptor (R) interaction (D″R″) leading, via mutual conformational induction, to D*R* and subsequently to biological effect.

(biological activation), and reversal (displacement). Peptide hormones often also act as neuropeptides, and if this variety of biological actions involves different conformations, then a number of the multitude of the possible conformations of peptides are of biological relevance.[58]

For neuropeptides and peptide hormones, activity may occur in a non-aqueous environment, and a relatively high energy conformer may be stabilized by receptor interaction. To meet both eventualities, as well as compensate for the likely stabilizing effect of water, a conformer energy range of about 15 kcal/mol from the proposed global minimum has been selected for consideration in most of the cases described below.

4.1. THYROTROPIN RELEASING HORMONE

4.1.1. *Introduction*

The hypothalamic tripeptide, thyrotropin releasing hormone,[59] (TRH, Glp–His–ProNH$_2$) shown in Fig. 7, was originally identified as stimulating anterior pituitary secretion of thyrotropin (thyroid stimulating hormone, TSH), but has since been found to co-exist with neurotransmitters in nerves in the central nervous system (CNS).[60] The finding that the TRH analogue (3MeHis)–TRH (see Fig. 7), has about eight times the activity of TRH at the pituitary, whilst being of similar activity in the brain, supports the view that structural changes can lead to enhanced activity at selected sites. Attempts have also been made to synthesize TRH analogues selective to CNS activity.[61]

The variable activity of (3MeHis)–TRH could possibly be due to one of several reasons; firstly, that central and peripheral TRH receptors are different;[60] secondly, that the receptor environment is different in each area, and this can more easily induce particular conformers; and thirdly, that the structural change gives selective resistance to enzymes in different areas. Low-energy conformers of TRH and (3MeHis)–TRH are shown in Fig. 8.

4.1.2. *TRH analogues*

Other analogues which have been synthesized include RX77368, RX74355, CG3703, CF3509, and MK771, which have been developed as potential drugs.[59]

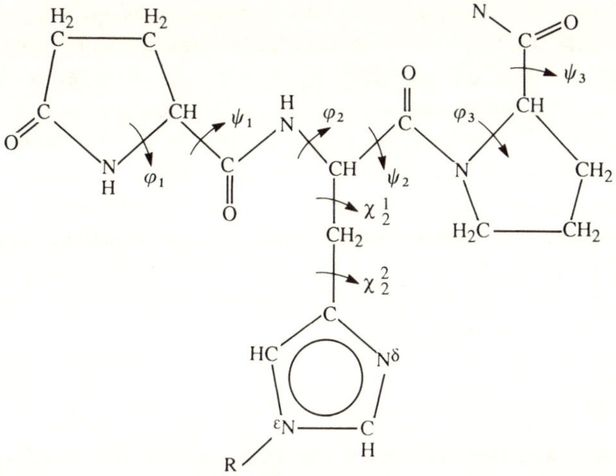

Figure 7 Chemical structure of thyrotropin-releasing hormone (TRH), when R is hydrogen. The structure is of (3MeHis)–TRH when R is a methyl group.

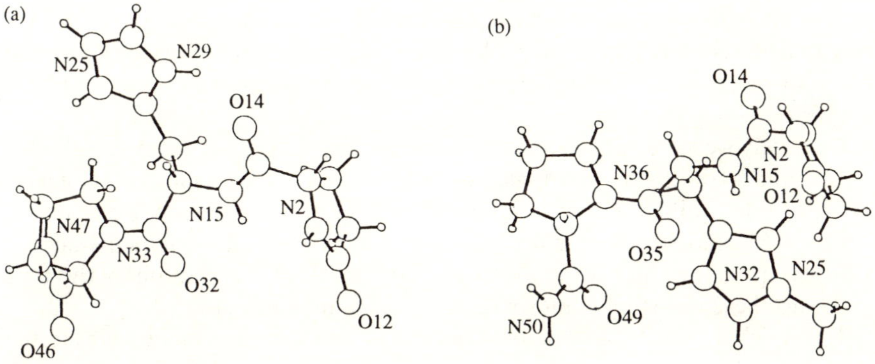

Figure 8 Predicted low-energy conformers of (a) TRH and (b) (3MeHis)–TRH.

Some have very enhanced stability to enzymes over TRH, for example in one study, 60% of RX77368 was found to be intact when excreted.[62] There is a variety of assays for TRH action including: antinociception, reversal of ethanol hypothermia, reversal of barbituarate sleeping time, and wet dog shakes. The potency of each analogue is quite variable across the tests, though only in the degree of activity. Amongst applications of the analogues, RX77368 has been used in treating traumatic cervical spinal injury and amyotropic lateral sclerosis,[63] and CG3703 has been used in trauma.[64]

The structure–activity data obtained can be useful in suggesting which functional groups are required for particular biological effect,[65] and to complement this we have studied *conformational preferences* for a series of analogues, whose biological activities range from high to low activity, and also analogues which are inactive.[66] The groups chosen for the TRH pharmacophore are the *lactam* group of pyroglutamate, the *imidazole* ring of histidine, and the C-terminal *amide* group of proline–amide.[65] As each of these is intimately associated or directly attached to a ring system, the centroid of each ring has been taken to represent the functional group. This does not take into account that different orientations will give the same centroid, though for a tripeptide it is a reasonable approximation.

4.1.3. *Starting conformers and solvent modelling*

For drug design purposes, we can use any methods to help us discover stable and metastable conformers of each peptide, so the starting conformations can include angles from experimental and theoretical published structures.[65,67] A systematic search, by definition, would not require consideration of starting conformations, but systematic search in small steps for even a small peptide is computationally prohibitive.

Although several other theoretical studies on TRH consider it to have only six rotatable bonds (ψ_1, φ_2, χ_1, χ_2, ψ_2, and ψ_3) we allow both the pyroglutamate and proline rings to vary. Although crystallographic data shows the pyroglutamate ring to be essentially planar, in proline the C_γ atom is either above or below the plane formed by the other four atoms, in configurations termed *exo* ($\varphi = -67.5°$) and *endo* ($\psi = -75°$).

Generally we assign all residues the same starting values, for example, all 180°, all 60°, and so on, so that the angles are independent of the amino acid sequence. We then select 'allowed' and 'disallowed' regions of the φ/ψ map, for example, $-65°$, $-55°$ and 90°, 90° for the backbone angles, with all side-chain angles of 180°, except χ_1 for which we choose $-60°$, 60°, or 180°. Combinations of this type make up the 25 starting conformations of each analogue studied.[66]

To model the effects of the solvent, we added the term in equation (18) to the intramolecular energy of each conformer. This term is modelled on Onsager's reaction field:[65,68]

$$RF = -14.4 \frac{\mu^2}{R^3} \frac{e-1}{2e-1} \quad (\text{kcal mol}^{-1}) \tag{18}$$

4.1.4. *Drug–receptor interaction*

The description of the pharmacophore *pattern* is in terms of the relative orientation of the essential functional groups, which can be the distance between them. The pharmacophore is deduced from SAR data,[65,66] and its preferred pattern for a series of analogues can be calculated by energy minimization and/or molecular dynamics. In general, this type of analysis will involve more than two

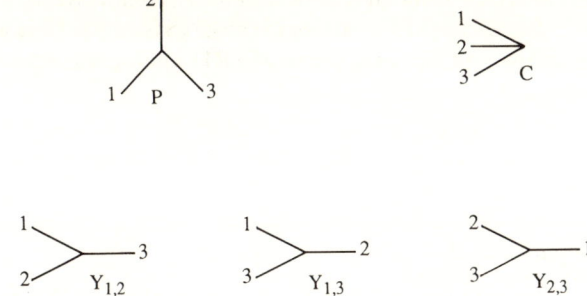

Figure 9 Generalized description of the conformer class classification used for TRH and its chemical analogues.[66]

distances, so the description can be as complex as interpreting a three-dimensional model. The method of principal component analysis has been applied to conformers of TRH and its analogues[66] in the process of investigating its general applicability, to reduce the dimensionality of description, whilst highlighting the degree of similarity between conformers.[66] Similar techniques have been used by others,[69,70] though based on systematic search rather than using energy-minimized conformers.

Using the nomenclature from Fig. 6, and working with statistical weights w for the conformers of each analogue a and the native form n, we can derive[66] equation (19) to relate the relative experimental potency of the analogue and native peptide to their relative energetic ease of attaining the active form D^*:

$$\log \left(\frac{Kp^a}{Kp^n}\right) = \log \left(\frac{w^a (D^* R^*) . z^n (D)}{w^n (D^* R^*) . z^a (D)}\right) \tag{19}$$

4.1.5. *Potency prediction*

Analysis of the groupings of conformers of TRH and its analogues developed into a generalized classification of conformers as $Y_{1,2}$, $Y_{1,3}$, $Y_{2,3}$ C, and P, across the analogue range, and are shown in Fig. 9. From the energy-minimization data, we have an energy for each conformer of each analogue. We can therefore calculate the relative energy of each conformer class for each analogue, and this is shown

Table 2 Conformational class preferences for TRH and (3MeHis)-TRH. * indicates that no conformer in this class was found within 12 kcal/mol of the lowest-energy conformer for that particular peptide. 0 refers to the lowest energy conformers for each molecule, and other values in each case are relative energies.

	P	Y_{12}	Y_{13}	Y_{23}	C
TRH	0	*	$+1$	$+3$	$+3$
(3MeHis)TRH	0	$+5$	*	$+4$	*

for TRH and (3MeHis)–TRH in Table 2. The P conformer is active at the pituitary, and the $Y_{2,3}$ conformer as active in the CNS is compatible with the data in Table 2 and on the analysis of a series of TRH analogues, across a range of biological activities.[66]

4.2. ENKEPHALINS

4.2.1. Introduction

The enkephalins, dynorphins, and β-endorphin share the same N-terminus, Tyr–Gly–Gly–Phe, and for the enkephalins this is followed by Met or Leu.[71] The enkephalins are therefore pentapeptides and they bind preferentially to δ receptors, whilst morphine is almost entirely μ-selective. β-endorphin also binds to μ receptors, whereas the dynorphins are κ-specific.[72] The actions of opioids include analgesia, respiratory depression, tolerance and physical dependence, euphoria, hypothermia, effects on gut motility, and cardiovascular effects.[72]

4.2.2. Comparison with experimental information

A number of X-ray crystallography studies to characterize the conformational preferences of Met- and Leu-enkephalin have been reported,[73-75] all have enkephalin in the extended conformation. Theoretical simulation of the crystal conformers shows them to be higher energy than other predicted conformers.[76] Spectroscopic data on the enkephalins includes average intramolecular (Tyr–Trp, inter-ring) distances in the Trp^4 analogue and comparison with the Trp^4 analogue of dynorphin 1–13. Results[77] suggested that in dynorphin the N-terminal tetrapeptide is almost completely extended – a helical configuration is favoured by other experimental[78] and theoretical[79] studies – whereas in enkephalin the same sequence is folded in most cases. Coupling constants for enkephalins have also been obtained in water[80] and are listed in Table 3, together with values based on the dihedral angles of stable (see Fig. 10) and metastable conformers predicted by molecular mechanics;[81] the aromatic side-chains are in close proximity, in agreement with similar studies on a conformationally-constrained specific δ-ligand.[82]

Table 3 A comparison between experimentally determined coupling constants for Leu- and Met-enkephalin,[80] and values averaged from predicted stable and metastable conformers[81] modelled with the inclusion of the reaction field component.[65]

		Present study	NMR in water[80]
Met-enkephalin	Met	7.1	7.6
	Phe	7.3	7.4
Leu-enkephalin	Leu	7.8	7.8
	Phe	7.5	8.0

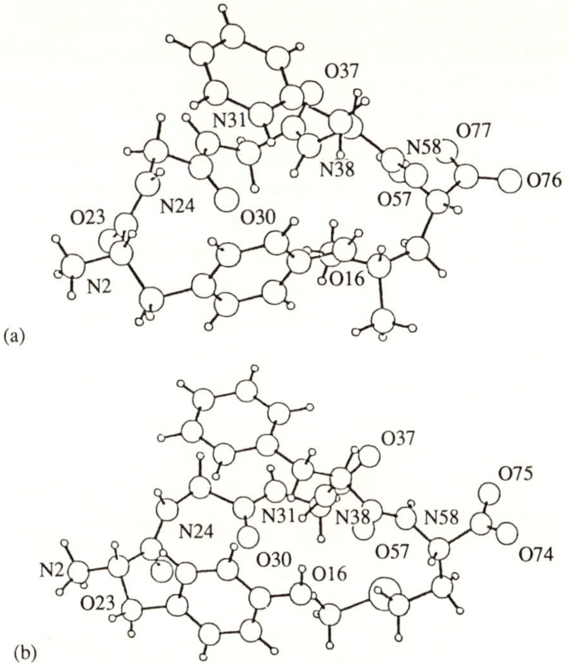

(a)

(b)

Figure 10 Lowest-energy conformers of (a) Leu-enkephalin and (b) Met-enkephalin calculated by molecular mechanics, incorporating the reaction field model for solvent effects.

4.2.3. *Agonist residue substitutions*

The number of possible enkephalin conformations is far too large to investigate, theoretically, the basis for SAR, so it is on predicted stable and metastable conformers that experimental data has been incorporated, even in the absence of knowledge of the receptor site. The reasoning is as follows: a structural change which increases potency (increased rate of effect, not just increased stability to enzyme degradation), should be consistent with incorporation into the predicted active conformation. The opposite should be true when the structural change abolishes activity. Antagonism presents a different case: substitutions might be accepted which lead to increased rigidity as suggested for [Pen^1]–oxytocin,[93] so that binding is not followed by mutual conformational change (for ligand and receptor) which would lead to biological effect. Alternatively, antagonists might act through different mechanisms to agonists,[58] for example via an allosteric-type mechanism as suggested for the peptide diazepam-binding inhibitor (DBI) and its fragments.[83]

The energies before and after structural changes are not comparable because the molecules are different, therefore a range of conformers is necessary. This procedure counters the possibility that a structural change might appear to be easily accommodated, but is so only at a large energy cost. Attempts at correlating experimental activities and conformation encompass *conformation–activity relationships*.

4.2.4. *Comparison with analogues and other opioid peptides*

From Fig. 11 it can be seen that D-Ala^2 substitution into the lowest energy Leu-enkephalin conformer (Fig. 10), and subsequent minimization, retains the main features especially the proximity of the aromatic rings. The same procedure for Aib^2 substitution (Fig. 12) yields similar results, which are compatible with these substitutions providing agonist analogues. However, D-Ala^3 substitution and minimization (Fig. 13) causes the breakup of the structure, consistent with [D-Ala^3]enkephalin being inactive. By way of comparison, Fig. 14 shows a predicted favoured conformer of dynorphin 1–17, which has a more extended arrangement at the N-terminus.[79]

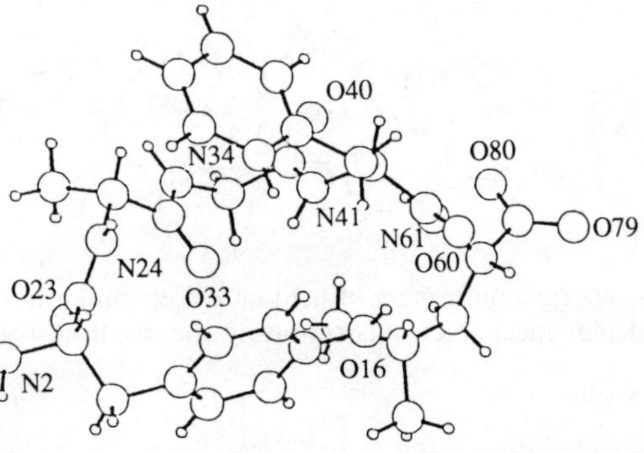

Figure 11 Conformation of [D-Ala^2]enkephalin, the result of substituting D-Ala for Gly^2 in the Leu-enkephalin conformer in Fig. 10 (a), and then energy-minimizing. The major features of the conformation are unchanged.

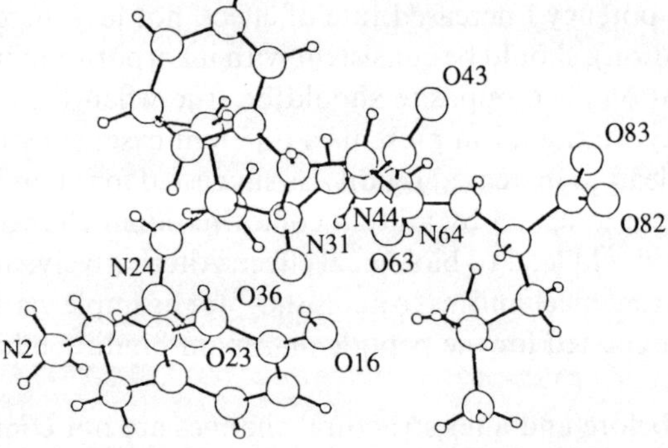

Figure 12 Conformation of [Aib^2]enkephalin, the result of substituting Aib for Gly^2 in the Leu-enkephalin conformer in Fig. 10 (a), and then energy-minimizing. The major features of the conformation are again unchanged.

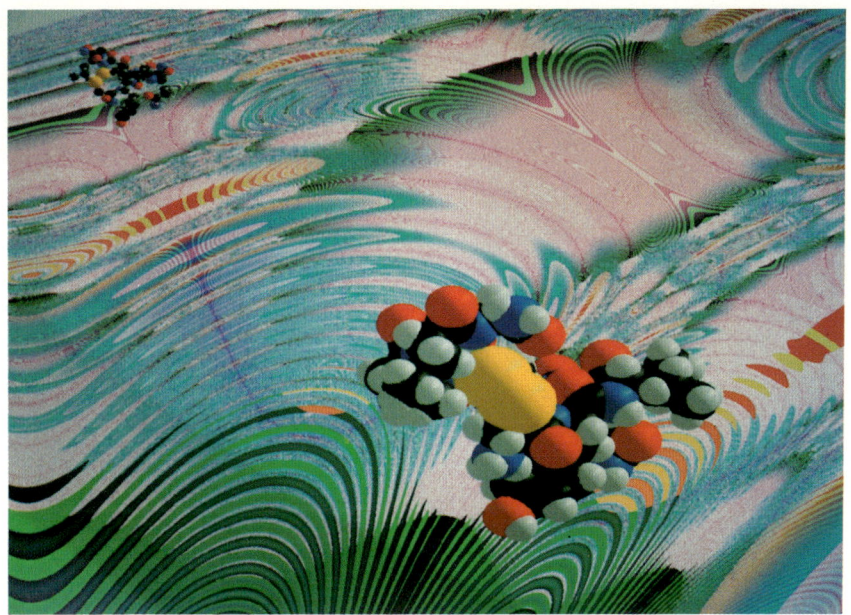

Plate 1
Space-filling and ball-and-stick representations of oxytocin conformations, overlying a graphics image of a dynamical system. Although the combination of images is symbolic, it emphasises our interest in relating aspects of chaos theory to the modelling of molecular behaviour.
(Software: Jin Li)

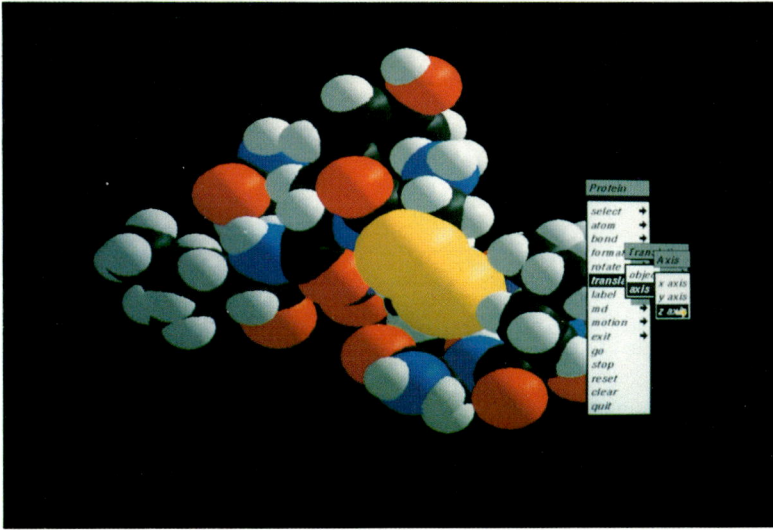

Plate 2
Space-filling representation of an oxytocin conformation, with a menu illustrating some of the program features available. The example selected is a translation along the z-axis. Colour code: oxygen – red; hydrogen – white; nitrogen – blue; carbon – black; sulphur – yellow.
(Software: Jin Li and Andrew Brass)

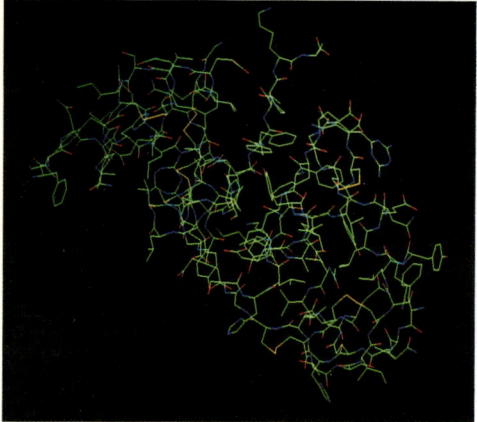

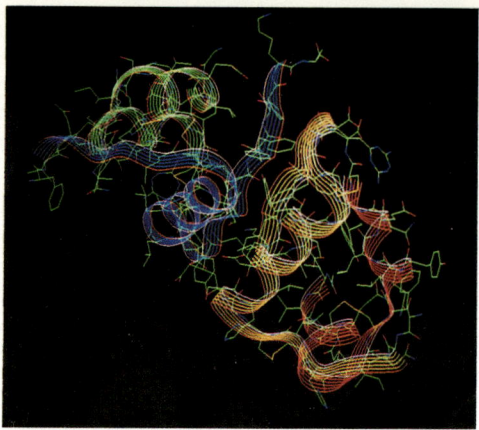

Plate 3
Stick representation of the
X-ray structure of an
insulin dimer from the
Brookhaven database.
Colour code: oxygen – red;
nitrogen – blue; carbon –
green; sulphur – yellow; no
hydrogens shown.
(Software: Polygen Inc.)

Plate 4
As Plate 3, but with a
'ribbon' overlay showing
the secondary structure (i.e.
backbone conformation).
(Software: Polygen Inc.)

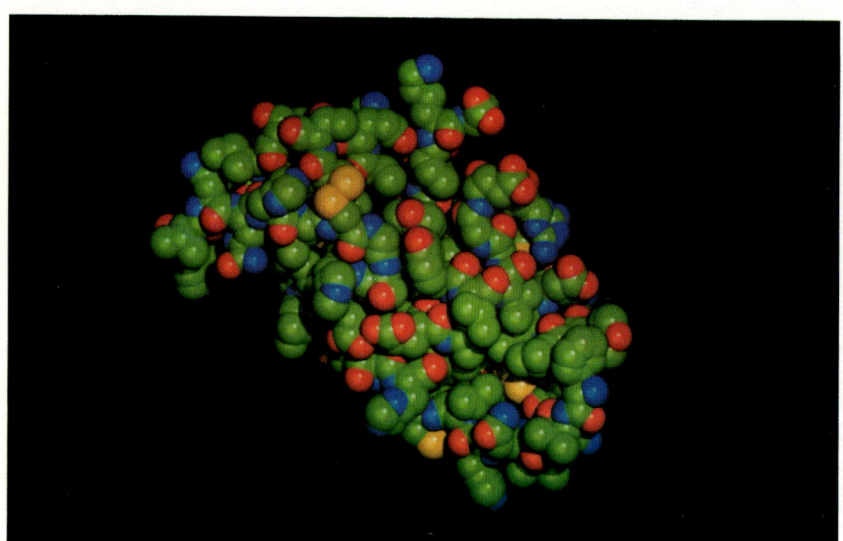

Plate 5
High-definition ray-tracing image of the insulin dimer conformation shown in
Plates 3 and 4. Colour codes as before. The image takes a few seconds to
generate and, in contrast to the representation shown in Plate 2, does not allow
interactive rotation.
(Software: Polygen Inc.)

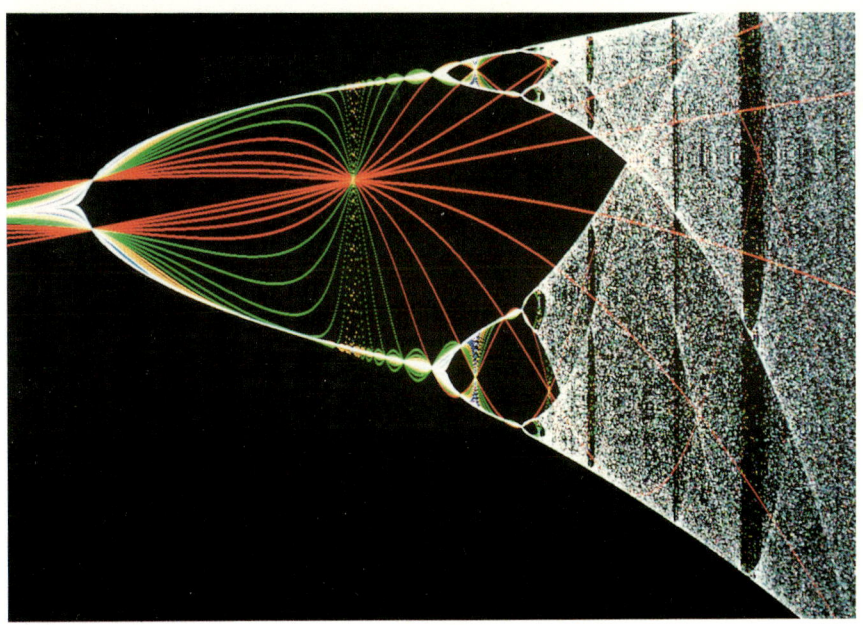

Plate 6
Feigenbaum diagram of the logistic map $f(x) = \lambda x(1-x)$. This non-linear mapping was originally introduced to model the population changes of certain species, and has been used to provide insight into the behaviour and complexity of dynamical systems. The diagram shown was produced by plotting the iterated valves of $f(x)$ against λ. As λ increases, the behaviour of $f(x)$ becomes more complicated and eventually steps into a chaotic region. This kind of diagram is usually used to illustrate the route to chaos. (Software: Jin Li)

Plate 7
Phase portrait of Duffing's equation, which was formulated to model the motions in a mechanical oscillator. There are periodic and chaotic solutions to Duffing's equation depending on the parameters used. One of the most important characteristics of chaotic dynamics is the sensitive dependence on initial conditions. This is illustrated in the photograph by plotting the trajectories of Duffing's equation for 5 very close starting positions (5 different colours). Because of exponential divergence in chaotic regions, these trajectories can be distinguished.
(Software: Jin Li)

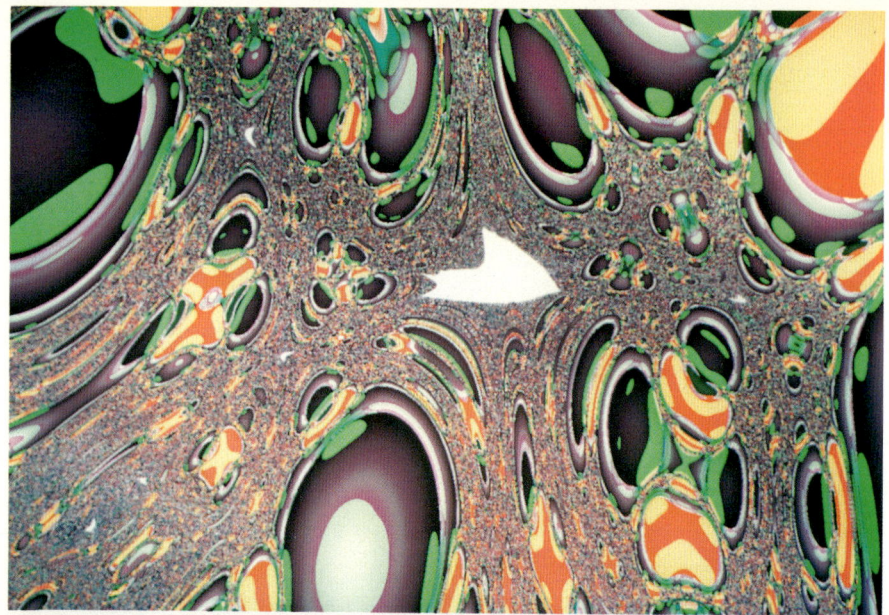

Plates 8 and 9
One of the most puzzling features in chaotic dynamical systems is that very often
mathematically-simple systems can exhibit extremely complicated behaviour. One well
known example is the Mandelbrot set. Shown in these colour plates are the images of
some non-linear maps which are similar to the map producing the Mandelbrot set.
Although the physical significance of these images is not known, they and the other
colour plates illustrate the importance of graphics in displaying the results of millions
of calculations.
(Software Jin Li)

All calculations performed and displayed on the Silicon Graphics 4D/240 GTX
workstation. All photography by Jin Li.

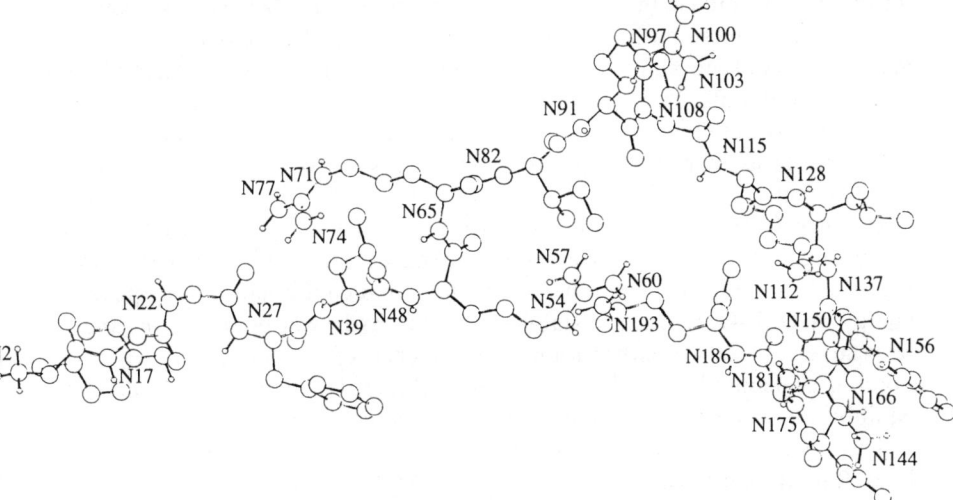

Figure 13 Conformation of [D-Ala^3]enkephalin, the result of substituting D-Ala for Gly^3 in the Leu-enkephalin conformer in Fig. 10 (a), and then energy-minimizing. The conformation is radically different to Figs 10–12, consistent with the D-Ala^3 substitution abolishing activity.

Figure 14 Plot of a predicted preferred conformation of dynorphin 1–17 obtained by molecular mechanics simulation.[79] Attention is drawn to the N-terminal four residues which are in a more extended conformation than predicted for enkephalin (see Fig. 10).

4.2.5. *Summary*

In contrast to the extended state of enkephalin in the crystal, most other data suggests folded conformers are preferred. Predicted stable and metastable conformers of Leu- and Met-enkephalin achieved in rigid geometry molecular mechanics simulation and with the inclusion of the reaction field, have folded conformations. Predicted coupling constants for this range of conformers is in agreement with experiment and selected residue substitutions are not 'accepted' in the *conformational model* if they abolish activity.

The proximity of the aromatic ring of Tyr^2 and Phe^4 are in agreement with similar calculations[82] on the conformationally-restricted and δ-selective DPDPE, and with average inter-ring distances from spectroscopic measurement.[77] As part of the investigation into the conformational basis for opioid receptor selectivity, we have also studied dynorphin 1–8 and 1–17 conformational behaviour[79] and find stable forms with the N-terminal 1–4 residues in a helical arrangement, as suggested by experiment.[78]

4.3. OXYTOCIN

4.3.1. Introduction

The neurohypophyseal hormone oxytocin (Cys^1–Tyr^2–Ile^3–Gln^4–Asn^5–Cys^6–Pro^7–Leu^8–Gly^9–NH_2), see Fig. 15 for the structural formula, is synthesized in the brain, and stored in the posterior pituitary. Interaction between oxytocin and receptors on the smooth muscle cell surfaces of uterus, blood vessels, and mammary gland can lead to uterine contraction during labour, increase in blood pressure, and milk ejection respectively. Oxytocin is a precursor for other behaviourally-active neuropeptides,[84] and affects various brain functions.[85,86]

On the basis of proton nuclear magnetic resonance (NMR) spectroscopy measurement in dimethylsulphoxide (DMSO), a preferred conformation of oxytocin was proposed having two β-turns, with hydrogen bonds fn and qe stabilizing the first and sw and ou stabilizing the second[87] (see Fig. 15). It was also suggested that oxytocin could have three favoured conformations of equivalent energy in DMSO, which are stabilized by the hydrogen bonds jn, fn, and dn respectively.[88] Others either found no evidence of intramolecular hydrogen bonds in aqueous solution,[89] or proposed that there is a 'highly mobile' dynamic equilibrium between folded and highly solvated conformers.[90]

The torsion angles for the disulphide are mostly within $30° \pm 90°$ according to laser Raman spectroscopy in water and DMSO,[91] and the cysteinyl–prolyl peptide bond is *trans* according to carbon-13 NMR studies.[92] NMR studies suggest that the antagonist analogue [Pen^1]–oxytocin has a more rigid conformation than oxytocin,[93] favouring a right-handed disulphide torsion angle greater than $110°$. The two conformations defined in the crystal of deamino-oxytocin,[94] have disulphide torsion angles of $+76°$ and $-101°$, respectively, and are similar to one proposed by NMR in DMSO,[87] with each having a type II turn in the ring, stabilized by hydrogen bonds fn and qe, and a type III turn in the tail, stabilized by hydrogen bond sw (see Fig. 15).

4.3.2. Search procedure and structural comparison

In terms of structural comparisons, an assumption is made that the crystal structure of oxytocin corresponds to that of deamino-oxytocin *plus* an NH_3^+ group. This hypothesis draws partly on the fact that deamino-oxytocin has similar biological activity to the native hormone.[22]

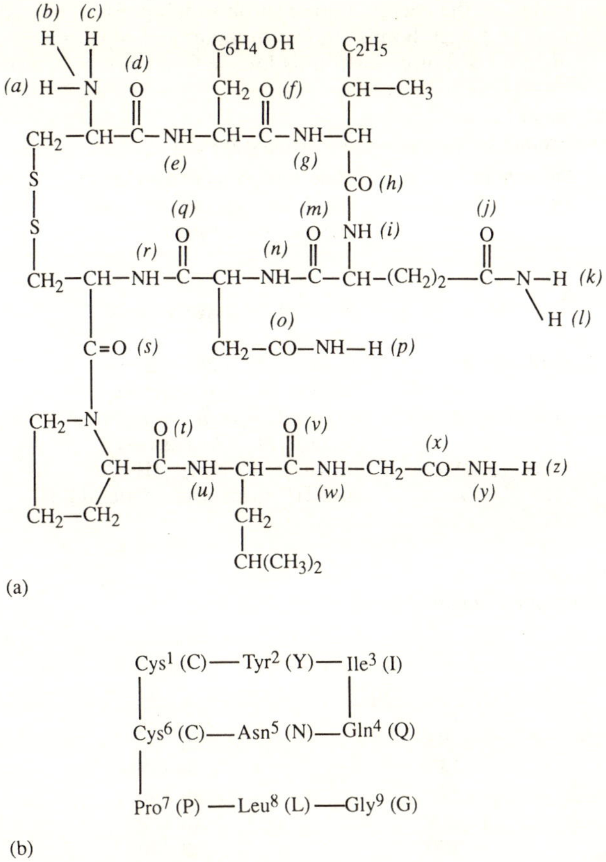

(a)

(b)

Figure 15 Structural formula for oxytocin; (a) italicized letters refer to groups/atoms involved in putative hydrogen bonds (see text), and (b) three-letter and one-letter codes for the residues in oxytocin.

For *de novo* simulations on oxytocin, the starting angles used are listed in Table 4. The backbone and side-chain angles given apply to all residues except proline, which was started in every case with $\varphi = -65°$, $\chi^1 = -20°$, $\chi^2 = 20°$, $\chi^3 = -20°$. The proline ψ value was as for the other residues. Backbone angles were chosen in a variety of locations on the φ/ψ map, and with side-chain angles set to either 180° or, on occasion, with χ^1 set to $-60°$ (see Table 4). As turns are often suggested as favoured conformations for peptides, a series of runs was carried out with backbone angles of $-135°,140°$ and separately types I, I′, II, II′, III, and III′ turn around Ile^3–Gln^4. The justification is these residues are in the middle of the disulphide-bonded ring. With linear peptides of up to about 15 residues, the central residues of the whole sequence could be chosen, or those suggested from

Table 4 Starting angles for the energy minimization simulations on oxytocin (i) with distance constraints for hydrogen bonds (shown in italics and separated by semicolon) and (ii) without distance constraints for the formation of hydrogen bonds.

Conformer starting angles
(using distance constraints for formation of hydrogen bonds)

Backbone angles	*Side-chain angles*	*Hydrogen bonds (see Fig. 1) which were attempted using distance constraints*
φ ψ	χ^1 $\chi^2,\chi^3 \ldots$	
Extended conformer 1	$\ll$all $180°\gg$	*fn,sw;fn;fn,qe,sw;qe,sw;hj*
Extended conformer 2	$\ll$all $180°\gg$	*fn,sw;nd;fn;fn,qe,sw;qe,sw; hj;fn,fr;fn,di;fn,qe,sw,ou*
$-90°, -90°$	$\ll$all $-90°\gg$	*fn,sw;nd;fn;fn,qe,sw;qe,sw; hj;fn,fr;fn,di;fn,qe,sw,ou*
$-135°, 140°$	$-60°$ $180°$	*fn* with I, I′, II, II′, III and III′ turns around Ile^3–Gln^4

Conformer starting angles
(no constraints except for disulphide bond)

Backbone angles		*Side-chain angles*	
φ ψ	χ^1	$\chi^2,\chi^3,\ldots$	
$90°, 90°$	$90°$	all $90°$	
$180°, 180°$	$180°$	all $180°$	
$-180°, -180°$	$-180°$	all $180°$	
$-90°, -90°$	$-60°$	all $180°$	
$-65°, -55°$	$180°$	all $180°$	
$-90°, 90°$	$-60°$	all $180°$	
$180°, -180°$	$180°$	all $180°$	
$60°, 60°$	$60°$	all $60°$	
$-60°, -60°$	$180°$	all $180°$	
$-120°, 120°$	$180°$	all $180°$	
$120°, 60°$	$180°$	all $180°$	
$-135°, 140°$ (except Type I, I′, II, II′, III, III′, VI and VI′ turns around Ile^3–Gln^4)	$-60°$	all $180°$	
$-135°, 140°$	$-60°$	all $180°$	

analysis of protein structures,[95] or peptide structures,[96,97] to be sites for turns. In the attempts to form hydrogen bonds, suggested from published studies, all angles were generated at random, on two separate occasions, in the range $180° \pm 10°$, and used as the starting points for simplex minimization. Although hydrogen bonds are usually defined as CO–HN interaction between 1.9 Å and 2.2 Å, interactions of this type are included for predicted oxytocin conformers up to 2.375 Å. This is because the third hydrogen bond in the crystal structure[94] (N→O distance) is 3.37 Å.

In order to simulate the standard disulphide bond geometry for peptides, in which the torsion angle of C_β^1–S–S–C_β^6 is around $\pm 90°$ and the valence angles for C_β^1–S–S and C_β^6–S–S are equal to $104°$, three distance constraints have been defined. The C_β–C_β distance is set at $3.893\,\text{Å}$ with a force constant of $1.08\,\text{kcal/mol Å}^{-2}$, which was chosen to allow the disulphide torsion angle to vary within $30°$ of $\pm 90°$; and the two S–C_β distances were constrained at $3.075\,\text{Å}$ each with force constants of $67\,\text{kcal/mol Å}^{-2}$, chosen to keep the valence angles fixed at $104°$. To simulate hydrogen bonds, which are suggested from studies on oxytocin by NMR spectroscopy, X-ray crystallography, and energy minimization, additional distance constraints (force constant of $13.0\,\text{kcal/mol Å}^{-2}$) have been included (see Table 4 and Fig. 15). The energy contribution due to the distance constraints is equal to the force constant multiplied by the absolute value of $((D_{ij}{}^{\text{expt}})^2 - (D_{ij}{}^{\text{model}})^2)$, where D_{ij} is the distance for the constrained atoms in the experimental (expt) structure and the modelled (model) structure.

A mapping procedure has been developed for energy-minimized conformers, which involves systematic rotation of two variables in $20°$ steps, and then systematic rotation around the lowest energy point in $10°$ steps. There is some similarity between this method and the *build-up* procedure of Scheraga and co-workers[98] in that φ/ψ maps are used to optimize conformations; however the current method is used for 'energy-minimized' conformers rather than *de novo* ones. The φ/ψ torsion angles for each residue have been varied together, and then χ_1/χ_2, χ_2/χ_3, and so on along the side-chain, except proline, in which just φ and ψ are varied. Energy minimization is then applied until convergence, here defined as an energy difference between the start and the end of a run of less than $0.05\,\text{kcal/mol}$. The mapping procedure is then iteratively repeated for conformers within about $15\,\text{kcal/mol}$ of the most stable conformer.

In the structural comparison between predicted conformers and the crystal, a root mean square (RMS) deviation value of 0 is unlikely, because all bond lengths and valence angles of the same type have the same value in the rigid geometry studies. Because of the equation used in the 'fitting' procedure,[15] and because the classical methods of assessing RMS is dependent on the degree of overlay between structures, whereas modelled structures may deviate strongly, the following equation has been used to calculate RMS deviations:

$$\text{RMS} = \sqrt{\sum_{ij} \frac{(d_{ij}{}^{\text{expt}} - d_{ij}{}^{\text{model}})^2}{n}}$$

where the square of the difference in distance (d) between two atoms (i and j) in two different structures (experimental and modelled), divided by the number of interactions (n) in the molecule, is summed over all pairs of atoms. Note that the RMS calculations use the crystal coordinates in the database, so the NH_3 group in the modelled conformers is excluded.

In the energy minimization stage, on average, there was an energy drop of the order of several hundred kcal/mol, following formation of the disulphide bond. Of the 51 starting conformations, about half the energy-minimized conformers had standard disulphide geometry and only these have been subjected to the mapping method. After mapping in $20°$ intervals, the energy of the conformation was reduced by, on average, 20 kcal/mol. After mapping in $10°$ intervals, the energy was reduced by up to a further 5 kcal/mol. After further energy minimization, there were eight conformations within 16 kcal/mol of the predicted global minimum. The energy reductions during mapping resulted mainly from side chain changes in the ring.

Fitting oxytocin, in the rigid geometry representation, to the crystal structure of deamino-oxytocin, without regard for the energy, tells us the best fit possible, and this was 0.28 Å RMS. With the assumption that the crystal structure is likely to be stable or metastable,[94] a fit with energy and structural deviation minimized, provides a 'ball-park' energy against which to assess the folded state of *de novo* simulations,[97] this gave a 0.61 Å RMS deviation, a relative energy of $+22.76$ kcal/mol and caused the disulphide to change from left-handed to right-handed helicity.

Table 5 shows the final energies, RMS values and hydrogen bonds for each of the eight conformations and the 'refined (minimized RMS and energy) "X-ray"' structure. All the conformers have standard disulphide geometry, as defined above. Conformers 1 and 3 were started from published angles of a theoretical study,[99] but have diverged substantially from the starting angles. Conformer 3 is very similar to Conformer 2, which was started from φ/ψ angles of $-90°,90°$, with $\chi^1 = -60°$ and the rest at $180°$. This, together with the similarities between Conformers 2 and 3, implies a powerful search of the potential surface. The

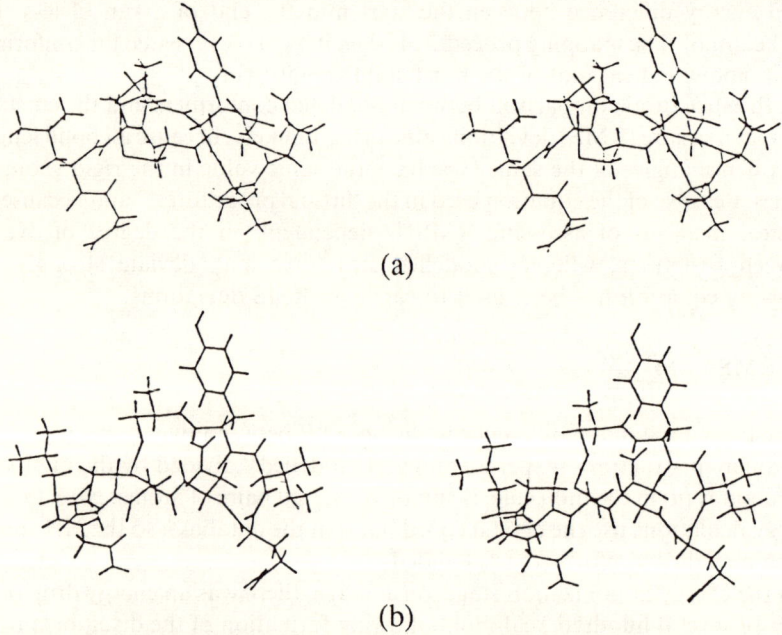

(a)

(b)

Figure 16 Stereo views of (a) the lowest-energy conformer of oxytocin predicted by molecular mechanics calculation, and (b) the X-ray structure of *deamino*-oxytocin, plus NH_3. The X-ray structure is taken from the Cambridge Structural Database.[107]

torsion angles of the eight low energy conformers, together with those of the X-ray structure and the 'refined "X-ray"' structure, are listed in Table 6. Stereo plots of the X-ray structure of deamino-oxytocin (plus NH_3) and the predicted lowest energy conformation of oxytocin, are shown in Fig. 16.

Conformer 1 (Tables 5 and 6) has the lowest energy and the best fit free of constraints (1.34 Å RMS) to the crystal structure of deamino-oxytocin.[94] It has a type III turn in the tocin ring around Ile^3 and Gln^4, whereas in the crystal there is a type II turn at the same point. Examination of the two structures (Fig. 16) shows a major difference in the orientation of the Gln^4 side-chain. The angles χ^2 and χ^3 were changed to those in the crystal and the conformer minimized, only very minor changes were found and the resulting conformer was about 5 kcal/mol higher than Conformer 1. Both the crystal structure and Conformer 1 have a type III turn in the tail, around Pro^7–Leu^8.

All the conformers have a β-turn or γ-turn, with hydrogen bonds which are listed in Table 5. Conformer 1 has three hydrogen bonds; *fn*, *qe*, and, *sw*, which stabilize the two β-turns. Hydrogen bond *fi* which makes a γ-turn around residue Ile^3, appears in Conformers 4 and 5 and was identified from NMR studies on oxytocin.[58] All hydrogen bonds (listed in Table 5) in all conformers except Conformer 7, have been formed during energy minimization without any precondition for their formation. Conformer 7 had distance constraints for two hydrogen bonds (*qe* and *sw*), proposed from a theoretical study,[100] included throughout energy minimization. A further hydrogen bond (*qa*) was formed without precondition (Table 5).

As shown in Table 6, the proline ring is *endo* (φ about $-75°$) in the crystal, but *exo* (φ less than $-70°$) in all predicted conformers except Conformer 5. All conformers were started in the *exo* conformation. The predicted low-energy conformers (Table 3) show a preference of 180° for the Cys^1 χ^1 value, which is the Cys^6 value in the most stable structure, and the crystal structure favours 180° in both cases. For the crystal, Tyr^2 χ^1 is $-60°$, and this is found in the predictions, though only for a relatively high energy conformer. More favoured is $\pm 20°$ and 180° (Table 6). For Ile^3, the crystal has χ^1 of 180°, whilst the predictions all favour either $+$ or $-60°$, in line with data from polypeptides and proteins.[16–18] For Gln^4, χ^1 values of $\pm 60°$ were predicted, and $-60°$ is favoured for the low-energy conformers in line with the crystal data for deamino-oxytocin and proteins.[16] By contrast, Asn χ^1 is generally $-60°$ in proteins[16] whilst the predicted favoured value of 180° in oxytocin agrees with that of deamino-oxytocin. Unlike the value of 180° for Ile^3 χ^1, the crystal value for Leu^8 is $-60°$, as in the predictions.

4.3.3. *Antagonist residue substitutions and the active form*

Data from NMR and structure–activity studies[58,101] support the idea that the active form of oxytocin has the aromatic ring of tyrosine positioned over the ring, with possible interaction with the side-chain of asparagine. The suggestion that this is prevented by the Pen^1 substitution[93] is investigated by energy calculations. Penicillamine^1 [Pen^1] substitutions have been imposed in the eight stable and

Table 5 Relative energy, disulphide bond helicity, hydrogen bond-type interactions up to a distance of 2.375 Å, and RMS deviations from the crystal structure of deamino-oxytocin, for stable and metastable conformers of oxytocin predicted by molecular mechanics simulation. Also shown are relative energies after substitution with penicillamine[1] and subsequent energy minimization. # refers to the C-terminal amide group and * refers to a side-chain amide group.

Conformer	Relative energy in kcal/mol (helicity of disulphide)	Hydrogen bonds with residue number superscripted. Distance between the groups is in Å. The italicized letters are shown in Fig. 1		— RMS Å — whole	ring	tail	Relative energy of Pen[1] conformers (kcal/mol)
1	0 (L)	CO^2-NH^5	(fn) 2.000	1.34	1.56	0.66	0
		CO^5-NH^2	(qe) 2.095				
		CO^6-NH^9	(sw) 2.335				
2	3.66 (R)	CO^1-NH^4	(di) 1.959	2.12	1.94	0.91	1.31
		CO^1-NH^5	(dn) 2.001				
		CO^6-NH^6	(sr) 2.360				
3	5.27 (R)	CO^1-NH^4	(di) 1.960	2.10	1.79	0.92	2.68
		CO^1-NH^5	(dn) 1.984				
		CO^6-NH^6	(sr) 2.360				
4	8.59 (L)	CO^2-NH^4	(fi) 1.947	2.38	2.59	0.50	17.74
		CO^2-#NH^9	(fy) 2.254				
		CO^4-NH^6	(mr) 2.289				
		CO^4-NH^4	(mi) 2.301				
		CO^1-NH^1	(dc) 2.363				
5	12.37 (R)	CO^2-NH^4	(fi) 2.199	2.67	2.58	0.94	9.72
		CO^2-#NH^9	(fy) 2.307				
6	12.91 (R)	CO^2-NH^5	(fn) 1.956	1.97	1.63	0.76	19.35
		CO^6-NH^6	(sr) 2.251				
		CO^8-#NH^9	(vz) 2.281				
		CO^9-NH^6	(xr) 2.291				
7	15.31 (L)	CO^5-NH^2	(qe) 1.979	2.95	1.55	0.65	26.81
		CO^6-NH^9	(sw) 2.165				
		CO^5NH^1	(qa) 2.216				
8	16.11 (L)	*CO^4-NH^5	(jn) 2.133	2.22	2.49	0.63	12.24
		CO^4-NH^5	(jp) 2.141				
		CO^4-NH^6	(mr) 2.155				
		CO^4-NH^4	(mi) 2.197				
		CO^8-#NH^9	(vz) 2.342				
		CO^9-NH^6	(xr) 2.359				
Reference 'X-ray' structure	22.76	CO^2-NH^5	(fn) 2.171	0.61	0.42	0.79	
		CO^5-NH^2	(qe) 2.329				

metastable conformers identified by molecular mechanics in the present study. The penicillamine is equivalent to cysteine plus two methyl groups on its C_β atom. The methyl groups were first rotated to identify the most stable starting point and then the conformers have been energy-minimized using molecular mechanics. Subsequent energy minimization caused some rearrangement of conformers in terms of their relative energies (Table 4) and some structural changes.

Conformer 6 of oxytocin (Fig. 17) has conformational features consistent with the proposed active form, it has the Tyr^2 aromatic group over the tocin ring, a turn at Ile^3–Gln^4 and Pro^7–Leu^8, and the Asn^5 side-chain in a position where it could interact with the Tyr^2 side-chain. The disulphide torsion angle has right-handed helicity. All this is consistent with a model for the biologically active conformation of oxytocin at uterine receptors (Table 2 in reference 58). Although with the [Pen^1] substitution these features are maintained, the relative energy rises appreciably and is probably outside what could be induced by the receptor. This supports the idea that the Pen^1 substitution prevents the active conformation from being attained,[93] and it is therefore suggested that Conformer 6 is the bioactive form.

4.3.4. *Molecular dynamics simulations*

A total of 28 oxytocin conformers have been identified at 4 K. Conformer relative energies, hydrogen bonds, RMS values relative to the crystal structure of the whole molecule, and for the ring and the tail separately, are listed in Table 7. For simulation purposes, an NH_3^+ group has been added at the N-terminus of the deamino-oxytocin crystal structure with the disulphide torsion of $-101°$, and the resulting structure used as an initial conformation for refinement. After a 13 ps

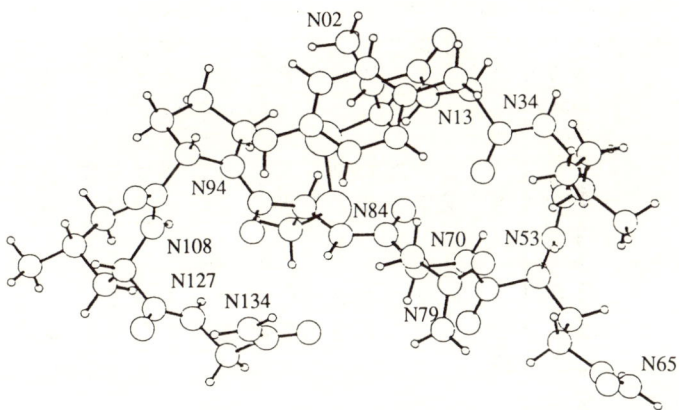

Figure 17 Ball-and-stick plot of the predicted bioactive conformation of oxytocin from molecular mechanics simulation. The conformer has the Tyr^2 side-chain over the tocin ring, a turn at Ile^3–Gln^4 and around Pro^7–Leu^8, and a right-handed disulphide helicity.

Table 6 Dihedral angles (in degrees) for the X-ray structure of deamino-oxytocin, for a refined (minimization of RMS and energy, in torsion angle space) 'X-ray' structure, and the conformers shown in Table 5, predicted by energy minimization.

	X-ray	'X-ray'	1	2	3	4	5	6	7	8
Φ_1		-60.33	159.80	170.74	179.78	-26.22	-179.83	-144.06	-75.49	-72.88
Ψ_1	101.54	95.16	99.47	-49.66	-49.49	157.13	116.50	137.66	-56.58	139.73
χ^1	177.89	-121.79	163.71	178.34	178.47	-119.38	-175.34	-69.74	-41.24	-171.51
Φ_2	-125.64	-115.72	-126.99	-54.91	-55.06	-117.91	-70.10	-115.99	-107.63	-70.32
Ψ_2	164.99	172.52	170.04	-34.93	-35.08	-177.95	-174.63	126.80	-106.92	134.04
χ^1	-55.05	-50.30	-19.83	60.02	20.01	-160.05	-147.72	20.09	-79.84	179.85
χ^2	98.47	100.06	96.01	-90.02	-99.92	69.93	62.60	-80.14	99.93	79.89
χ^3	180.00	171.06	167.90	179.95	179.16	179.78	179.49	163.99	179.50	179.92
Φ_3	-67.14	-62.46	-58.95	-44.82	-45.21	-71.74	-75.46	64.33	-76.27	-39.05
Ψ_3	125.12	132.95	-35.89	-35.08	-34.92	74.33	81.97	-114.65	-63.21	110.04
χ^1	-171.10	-170.66	-68.97	50.02	40.02	-50.20	-62.09	-49.61	-60.16	-50.17
χ^2	162.10	160.78	146.12	170.03	160.06	159.95	129.93	150.23	139.83	99.62
χ^3	-179.94	172.60	-176.00	-40.16	179.98	-178.62	157.93	179.91	-88.05	-135.21
χ^4	-179.57	-160.03	170.96	-59.90	179.25	-178.72	-172.28	-60.12	-84.60	-51.63
Φ_4	56.33	53.65	-58.67	-105.06	-104.99	-156.29	-131.70	-112.70	-105.71	163.08
Ψ_4	28.88	26.04	-47.01	-68.55	-68.65	-168.00	-131.35	5.54	-84.70	-176.74
χ^1	-63.63	-67.86	-59.12	-99.84	-99.97	60.07	56.21	70.37	-59.80	-110.16
χ^2	-51.14	-57.80	101.45	79.86	79.99	-179.93	-154.96	-159.79	-159.43	-99.96
χ^3	-47.04	-44.28	-119.25	91.85	91.88	179.24	48.48	159.44	-106.50	70.02

Table 6 (cont'd)

	X-ray	'X-ray'	1	2	3	4	5	6	7	8
Φ_5	−158.27	−160.89	−165.59	−168.21	−168.27	−67.47	−61.16	−95.67	−104.21	−63.35
Ψ_5	66.86	36.93	121.96	131.66	132.02	88.85	−32.02	−175.72	−140.42	85.57
χ^1	−179.61	177.20	178.00	−160.04	−179.79	−60.12	62.89	−49.52	−40.57	−79.44
χ^2	42.80	37.96	−94.23	−89.96	−99.83	89.91	−109.86	110.57	101.60	−80.88
Φ_6	−129.25	−147.33	−138.09	−165.67	−165.96	−119.80	−52.39	−161.32	−97.01	−94.85
Ψ_6	97.59	97.67	120.08	141.06	140.12	150.36	143.96	158.51	−57.10	151.89
χ^1	−179.64	167.63	174.73	−123.89	−123.66	−77.81	−77.87	−86.03	−97.92	−64.89
Φ_7	−72.84	−80.63	−58.18	−42.76	−41.70	−49.61	−70.88	−40.52	−40.81	−44.72
Ψ_7	−11.94	−35.47	−37.94	−41.95	−43.09	−38.21	−55.97	−41.31	−30.91	−38.15
χ^1	28.34	20.84	−19.30	−21.26	−20.98	−20.20	21.43	−21.51	−20.73	−21.55
χ^2	−40.13	−17.60	19.58	16.62	17.29	18.59	−16.10	15.73	7.50	15.65
χ^3		8.56	−12.96	−5.18	−7.13	−9.66	3.77	−4.10	14.27	−2.90
Φ_8	−76.70	−58.76	−79.26	−120.97	−121.25	−110.07	−124.02	−89.88	−50.98	−119.79
Ψ_8	−33.41	−25.83	−25.89	127.34	127.90	−54.51	137.05	−58.02	−55.12	63.05
χ^1	−65.36	−67.63	−57.95	−68.87	−69.49	−67.83	−68.07	−59.48	−92.87	−64.51
χ^2	176.29	120.65	104.94	155.36	156.87	153.99	154.36	102.94	−63.73	156.42
χ^3	178.58	58.47	177.35	−62.97	177.90	177.72	57.54	−178.38	−33.29	178.27
χ^4	71.86	−175.48	61.27	−177.91	−175.86	−177.41	−61.80	−178.74	47.85	−176.83
Φ_9	−176.44	−142.03	87.62	89.92	88.17	175.46	90.96	−85.82	88.88	82.0
Ψ_9	−22.13	73.05	−88.29	−29.47	−21.39	147.90	−30.84	71.04	166.68	−80.17

cooling period, from the crystal coordinates, using molecular dynamics simulation, the RMS value relative to the native structure is 1.37 Å RMS and the relative molecular dynamics energy is $+7.7$ kcal/mol, giving the 'relaxed "X-ray"' structure. The 1.37 Å RMS deviation is slightly larger deviation than was obtained by 'annealing' from the predicted lowest-energy conformer from molecular mechanics, which had a relative molecular dynamics energy of $+3.4$ kcal/mol and an RMS deviation of 1.36 Å (Conformer A in Table 5). However, the RMS difference between Conformer A and the 'relaxed "X-ray"' structure is 1.83 Å, so they are quite different conformations. This, together with the data from molecular mechanics, supports the idea that the X-ray structure may not be the lowest-energy conformation for a small peptide,[78] but that it could be a metastable conformer.[94]

Conformer A (in Table 7) is the result of cooling to 4 K the coordinates of the lowest-energy conformer from energy minimization. Conformers S, T, U, V are the result of the same procedure for Conformers 2, 4, 5, and 6, respectively. Conformers B–F were located by the regular cooling to 4 K during a 50 ps simulation at 310 K from Conformer A, whereas conformers G–K were similarly obtained over a 92 ps simulation at 3000 K. It is only after simulating at 3000 K, then cooling, that the 'global minimum' is obtained, suggesting that Conformer A was in a very deep minimum. The lowest-energy conformer identified has proline in the *endo* configuration, the same as the X-ray structure of deaminooxytocin. Conformers L–R were obtained by repeated heating up (to 310 K and 3000 K) and cooling down (to 4 K), as a continuous process, starting from Conformer K.

Conformers W, X, and Y result from cooling to 4 K over 13 ps, starting with backbone angles of $(180°,180°)$, $(-135°,140°)$ and $(-90°,90°)$ respectively and all other angles 180°. Conformers Z and Z1 were identified at the mid-point and end of a 350 ps simulation at 3000 K, for a conformer started from all angles 180°, and Z2 after a further 150 ps at 310 K. After this exhaustive search, no lower-energy conformer was found than 'W', which was located after 13 ps at 4 K, from the same starting conformation. The total simulated time in this study was greater than 1600 ps.

In the present study, therefore, we have used molecular dynamics in two ways: firstly as a type of flexible geometry gradient minimizer, when cooling to 4 K; secondly, as a probe of phase space, raising the temperature so that energy barriers may be crossed. Running at temperatures much higher than would be tolerated physiologically serves to 'age' the simulation, with the molecule 'moving' much further than it would in the same time at 310 K (37°C).

4.3.5 Summary

In the rigid geometry studies, all hydrogen bonds except one, in all the preferred conformers from energy minimization, formed without any precondition for their formation. It may be that additional constraints (beyond those for disulphide formation) cause the molecule to be 'pulling' in too many directions simulta-

Table 7 Relative energy, disulphide bond helicity, hydrogen bond-type interaction distances and the groups involved (residue number superscripted) and RMS deviations from the X-ray structure of deamino-oxytocin, for 28 conformers predicted by molecular dynamics simulation (cooled to 4 K). # refers to the C-terminal amide group, and * refers to a side-chain amide group.

Conformer	Relative energy in kcal/mol	Hydrogen bonds with residue number superscripted. Distance between the groups is in Å. The italicized letters are shown in Fig. 1.			$---RMS\ (\text{Å})---$ whole	ring	tail
A	3.4	$*CO^4\text{-}NH^4$	(ji)	2.106	1.36	1.59	0.75
		$CO^5\text{-}NH^2$	(qe)	2.181			
		$CO^2\text{-}NH^2$	(fe)	2.351			
B	9.5	$*CO^4\text{-}NH^5$	(jn)	2.352	2.41	2.12	1.29
C	6.9	$CO^6\text{-}NH^8$	(su)	2.026	2.68	1.85	0.95
D	1.3				2.59	2.44	1.99
E	3.6	$*CO^4\text{-}NH^4$	(ji)	2.101	2.36	1.73	0.89
		$CO^2\text{-}NH^5$	(fn)	2.328			
		$CO^4\text{-}\#NH^9$	(my)	2.363			
		$CO^2\text{-}NH^6$	(fr)	2.374			
F	8.0	$CO^7\text{-}\#NH^9$	(tz)	2.333	1.86	1.91	0.98
		$CO^4\text{-}*NH^4$	(ml)	2.375			
G	1.5	$CO^6\text{-}\#NH^9$	(sy)	2.283	2.38	2.19	0.93
H	0.0	$CO^3\text{-}NH^1$	(hc)	2.198	2.28	2.40	0.70
		$CO^9\text{-}NH^1$	(xa)	2.310			
		$CO^8\text{-}NH^9$	(vz)	2.311			
		$CO^2\text{-}*NH^4$	(fk)	2.344			
		$CO^1\text{-}NH^1$	(db)	2.374			
I	14.0	$CO^2\text{-}NH^4$	(fi)	2.182	2.43	2.17	0.98
		$CO^8\text{-}\#NH^9$	(vz)	2.308			
J	4.0	$*CO^4\text{-}NH^4$	(ji)	2.136	2.15	1.55	0.95
		$CO^8\text{-}NH^1$	(vb)	2.211			
		$CO^3\text{-}NH^1$	(ha)	2.213			
		$*CO^5\text{-}NH^5$	(on)	2.273			
		$CO^4\text{-}NH^6$	(mr)	2.314			
		$CO^6\text{-}\#NH^9$	(sz)	2.332			
		$CO^6\text{-}NH^6$	(sr)	2.345			
K	6.8	$CO^4\text{-}NH^6$	(mr)	2.137	2.60	2.39	1.13
		$*CO^4\text{-}NH^4$	(ji)	2.300			
L	2.8	$*CO^5\text{-}NH^5$	(on)	2.175	2.30	2.08	1.14
		$CO^4\text{-}NH^6$	(mr)	2.277			
		$CO^5\text{-}*NH^5$	(qp)	2.286			
		$CO^7\text{-}NH^1$	(tc)	2.300			
M	7.7	$*CO^5\text{-}NH^5$	(on)	2.203	2.51	1.96	1.13
		$CO^1\text{-}NH^1$	(dc)	2.244			
		$CO^5\text{-}*NH^5$	(qp)	2.326			
N	11.4	$CO^3\text{-}NH^6$	(hr)	2.184	2.52	2.27	1.13
		$*CO^4\text{-}NH^5$	(jn)	2.236			
		$CO^1\text{-}NH^1$	(db)	2.264			

Table 7 (cont'd)

Conformer	Relative energy in kcal/mol	Hydrogen bonds with residue number superscripted. Distance between the groups is in Å. The italicized letters are shown in Fig. 1.			$---$ RMS (Å) $---$ whole	ring	tail
O	10.9	‡CO^9-NH^1	(xa)	2.188	1.83	1.39	0.81
		*CO^5-NH^6	(or)	2.199			
		CO^5-NH^2	(qe)	2.201			
		CO^6-NH^6	(sr)	2.311			
		CO^8-‡NH^9	(vz)	2.336			
P	8.9	*CO^4-NH^5	(jn)	2.141	2.01	1.55	0.59
		CO^3-‡NH^9	(hy)	2.200			
		CO^7-NH^1	(tb)	2.226			
		CO^6-NH^1	(sc)	2.285			
		CO^4-NH^4	(mi)	2.250			
		‡CO^9-NH^9	(xw)	2.331			
		CO^8-NH^1	(va)	2.363			
Q	4.3	CO^5-NH^2	(qe)	2.200	2.09	1.57	0.85
		CO^5-NH^5	(qn)	2.234			
R	9.5				2.44	2.34	1.45
S	11.1	CO^1-NH^4	(di)	2.079	1.97	1.83	0.89
		CO^1-NH^5	(dn)	2.306			
		CO^6-NH^6	(sr)	2.367			
T	3.5	CO^4-‡NH^9	(my)	2.189	2.49	2.70	0.54
		‡CO^9-NH^1	(xb)	2.261			
		CO^4-NH^4	(mi)	2.290			
		CO^2-NH^4	(fi)	2.336			
U	4.7	*CO^4-NH^2	(je)	2.264	2.74	2.49	2.70
		*CO^5-NH^5	(on)	2.364			
		CO^5-NH^4	(ol)	2.368			
V	4.7	CO^2-NH^5	(fn)	2.168	2.00	1.80	0.72
		CO^5-NH^2	(qe)	2.243			
		CO^1-NH^1	(db)	2.345			
W	1.9	*CO^4-NH^4	(ji)	2.127	2.63	2.45	0.74
		CO^6-NH^9	(sw)	2.136			
		*CO^4-NH^3	(jg)	2.268			
		*CO^5-NH^4	(on)	2.301			
		‡CO^9-NH^2	(xe)	2.371			
X	10.0	CO^4-NH^1	(mb)	2.224	2.13	2.09	0.93
		CO^2-NH^4	(fi)	2.233			
		‡CO^9-NH^6	(xr)	2.287			
		CO^6-*NH^5	(sp)	2.296			
		CO^5-*NH^4	(ql)	2.344			
Y	9.2	CO^5-NH^2	(qe)	2.099	2.15	1.68	0.63
		CO^6-NH^9	(sw)	2.111			
		CO^1-NH^3	(dg)	2.203			
		CO^2-NH^5	(fn)	2.293			
		CO^6-NH^6	(sr)	2.373			

Table 7 (cont'd)

Conformer	Relative energy in kcal/mol	Hydrogen bonds with residue number superscripted. Distance between the groups is in Å. The italicized letters are shown in Fig. 1.			− − − RMS (Å) − − − whole	ring	tail
Z	7.8	CO^3-NH^1	(ha)	2.271	2.27	1.86	0.84
		*CO^5-NH^5	(on)	2.314			
		CO^5-NH^1	(qb)	2.334			
Z1	1.9	CO^3-$\ddagger NH^9$	(hz)	2.203	2.53	2.20	0.82
		*CO^4-NH^4	(ji)	2.284			
		$\ddagger CO^9$-*NH^4	(xl)	2.352			
Z2	3.0	CO^3-NH^1	(hb)	2.242	2.43	2.25	0.94
		CO^1-NH^1	(dc)	2.324			

neously and, at least for the rigid geometry model, the molecule becomes trapped in a relatively high-energy local minimum. Successful formation could be very dependent on starting conformation, but at least two different points were chosen for each investigation of hydrogen bond formation (Table 4).

In the present study there is little difference in the energy of conformers 'annealed' by MD from energy-minimized conformers and those started, *de novo*, in extended or 2_7 ribbon[5] conformations. It would appear that the choice of starting conformation for a peptide is not important in molecular dynamics simulations, at least for a small constrained molecule. Equally, it suggests that the rigid geometry search, much faster than molecular dynamics because only torsion angles are varied, has not been excluded from regions of the potential surface because it cannot open valence angles to lower barriers to rotation.

A metastable conformer has been identified which has conformational features in common with a proposed active conformation,[22] and its relative energy rises appreciably upon substituting penicillamine at position 1 and energy-minimizing. It is suggested that the active form for Pen^1–oxytocin is too high in energy to be induced by receptor interaction.

5. Summary

This chapter has described the technique of molecular dynamics as well as provide details of our investigations into a better understanding of phase space. This *underpinning* research serves to increase our knowledge of the dynamical processes in the simulations on peptides and proteins; additional applications and advances are discussed in Chapter 8. It also provides guidance in the development of protocols for molecular design.

As a first step in the rational design of peptide hormones, where *rational* is taken to mean using the endogenous peptide *and* its predicted conformational characteristics as the 'leader' substance, it is necessary to reproduce, through simulations, characteristics of peptides prevalent in experimental analyses.[65,81] This is with the provisos that peptides are inherently flexible and influenced by crystal packing forces, close vicinity of other molecules and solvent, especially non-biological solvents. Thus, although many features of, for example, the crystal structure of deamino-oxytocin have been reproduced in the lowest-energy conformer of oxytocin predicted using molecular mechanics, it is only one of a number of candidates, whose relative energies have been predicted, for that of the bioactive form.

Until precise three-dimensional structures of neuropeptide receptors have been determined, active- and receptor-bound forms will have to be deduced. An approach described in this chapter has been to use data on predicted analogue conformers and their experimental potencies, to deduce the active form in different regions, for TRH.[66] The importance of relative energies in the calculations justifies the importance placed on locating the global minimum.[65,69] For the work on TRH, a model for solvent effects based on Onsager's reaction field has been used,[65,66] whilst in other studies a water site model has been included.[97,102,103] It is recognized that hydrogen bonding between the peptide and water might be replaced by similar interactions with the receptor.

A model for the pharmacophore orientation is described here for TRH in terms of inter-ring distances, which is taken to correspond to the distance between essential functional groups. Some of the data from conformational analysis of the analogues, the relative energy and coordinates, are used so that comparison can be drawn between all the conformers from all the analogues. As seen for TRH, there is a range of assays, with analogue potency variable depending on the test. Nonetheless, the correct *trend* in activity is reproduced using the computational model.[66]

One approach taken here, is that the predicted active conformer of the native peptide should be able to accommodate residue/group substitutions which increase potency,[81] where potency is distinguished from increased activity due to stability to enzyme-induced degradation. This procedure has been demonstrated in this chapter for the enkephalins and oxytocin, by reference to agonists and an antagonist, respectively. For agonists, the testing of both stable and metastable conformers acknowledges that interaction with a receptor can induce conformations that might never be found in solution. For antagonists, a range of stable and metastable conformers is required, to notice when a change is incorporated only at high energy cost. This technique is an alternative to *template forcing*,[104] which requires a decision on which features of a native peptide and its analogue(s) should be overlaid. The equivalent type of analysis for proteins is a kind of computer-aided site-directed mutagenesis.[105]

In contrast to most peptides, which have linear sequences, the conformational freedom of oxytocin is restricted by the disulphide bond, and it is possible that alternative searching procedures are required. In a preliminary study along those lines, a side-chain optimization procedure has been invoked, which has been advantageous. This procedure complements the energy minimization and molecular dynamics algorithms, the latter serving as both a probe of conformational space and a test of the ability of SIMPLEX-type procedures to escape from local minima, and of the value of starting from a variety of points on the potential surface.

The modelled structure of oxytocin by molecular mechanics with the lowest RMS deviation from the crystal structure of deamino-oxytocin is also the one of lowest potential energy and has the same hydrogen bonds as in the crystal. 'Annealing' from the coordinates of the crystal structure and from the lowest-energy conformer from energy minimization gives almost identical RMS deviations from the crystal structure of deamino-oxytocin. Use of residue substitutions into predicted stable and metastable conformers, and of other experimental data, has permitted a detailed structure to be predicted for the active form. Although QSAR methods have been applied to oxytocin and its analogues,[106] they do not permit a detailed atomic model of the active conformation to be proposed.

The next stage is to test, on the computer, whether the predicted active conformation can withstand chemical modifications that we may wish to make, including non-peptide mimetics, before chemical synthesis. This can be performed by the techniques outlined in this chapter, and provides testable hypotheses.[102]

In conclusion, all experimental conformational data point towards peptides having an ensemble of interconverting conformers at room temperature. Judicious choice of starting conformation, incorporation of experimental structure–activity data, and a deeper understanding of simulation procedures, may turn out to be the most important approach in deducing the most populated conformations and the active and bound forms for peptides.

Acknowledgements

David Ward and Jin Li were supported in part by SERC grants GR/D 98754 and GR/E 38177 respectively. Andy Brass, Yuan Chen and Eric Platt were supported by EEC Biotechnology Action Programme (BAP) Contract 0149–UK. The computer calculations were performed on the Control Data Corporation Cyber 7600, 176, 205, and Amdahl 5890–300E and VP1200 computers at Manchester Computing Centre, University of Manchester, UK, and on the Silicon Graphics *Powerseries* Iris 4D/240GTX Workstation at University of Manchester Medical School.

References

1. 'Design of potent, orally effective, nonpeptidal antagonists of the peptide hormone cholecystokinin' B. E. Evans, M. G. Bock, K. E. Rittle, R. M. DiPardo, W. L. Whitter, D. F. Veber, P. S. Anderson and R. M. Freidinger (1986) *Proc. Nat. Acad. Sci. USA*, **83**, 4918–4922.

2. 'Neuropeptides and their processing: Targets for drug design' J. W. van Nispen and R. M. Pinder (1987) *Ann. Reports Med. Chem.*, **22**, 51–62.

3. 'The design of biologically active polypeptides' B. Robson (1983) *CRC Crit. Rev. Biochem.*, **14**, (4), 273–296.

4. 'Recent progress in the rational design of peptide hormones and neurotransmitters' V. J. Hruby, J. L. Krstenansky and W. L. Cody (1984) *Ann. Reports Med. Chem.*, **19**, 303–312.

5. *Introduction to Proteins and Protein Engineering* B. Robson and J. Garnier (1986) Elsevier, Amsterdam.

6. 'ρ–σ–π analysis. A method for the correlation of biological activity and chemical structure' C. Hansch and T. Fujita (1964) *J. Am. Chem. Soc.*, **86**, 1616–1626.

7. *QSAR: Quantitative Structure–Activity Relationships in Drug Design* (J. L. Fauchère, ed.) (1989) Alan R. Liss, New York.

8. 'Structure–activity relationships of enkephalin-like peptides' J. S. Morley (1980) *Ann. Rev. Pharmacol. Toxicol.*, **20**, 81–110.

9. 'Peptide quantitative structure–activity relationships, a multivariate approach' S. Hellberg, M. Sjöström, B. Skagerberg and S. Wold (1987) *J. Med. Chem.*, **30**, 1126–1135.

10. 'Secondary structure prediction: combination of three different methods' V. Biou, J. F. Gibrat, J. M. Levin, B. Robson and J. Garnier (1988) *Protein Eng.*, **2** (3) 185–191.

11. 'Protein folding by restrained energy minimization and molecular dynamics' M. Levitt (1983) *J. Mol. Biol.*, **170**, 723–764.

12. 'Dynamics of folded proteins' J. A. McCammon, B. R. Gelin and M. Karplus (1977) *Nature*, **267**, 585–590.

13. *Dynamics of Proteins and Nucleic Acids* J. A. McCammon and S. C. Harvey (1987) Cambridge University Press.

14. *Computer Simulation of Biomolecular Systems: Theoretical and Experimental Applications* (W. F. van Gunsteren and P. K. Weiner, eds.) (1989) Escom, Leiden.

15. 'Refined models for computer calculations in protein engineering: Calibration and testing of atomic potential functions compatible with more efficient calculations' B. Robson and E. Platt (1986) *J. Mol. Biol.*, **188**, 259–281.

16. 'Conformation of amino acid side-chains in proteins' J. Janin, S. Wodak, M. Levitt and B. Maigret (1978) *J. Mol. Biol.*, **125**, 357–386.

17. 'Statistical and energetic analysis of side-chain conformations in oligopeptides' E. Benedetti, G. Morelli, G. Nemethy and H. A. Scheraga (1983) *Int. J. Peptide Protein Res.*, **22**, 1–15.

18. 'An analysis of side-chain conformation in proteins' T. N. Bhat, V. Sasisekharan and M. Vijayan (1979) *Int. J. Peptide Protein Res.*, **3**, 170–184.

19. 'Protein secondary structure prediction with a neural network' L. H. Holley and M. Karplus (1989) *Proc. Nat. Acad. Sci. USA*, **86**, 152–156.

20. 'Prediction of β-turns in proteins using neural networks' M. J. McGregor, T. P. Flores and M. J. E. Sternberg (1989) *Protein Eng.*, **2** (7) 521–526.

21. *Drug Design: Fact or Fantasy* (G. Jolles and K. R. H. Wooldridge, eds.) (1984) Academic Press, New York.

22. 'Implications of the X-ray structure of deamino-oxytocin to agonist/antagonist-receptor interactions' V. J. Hruby (1987) *Trends Pharmacol. Sci.*, **8** (9) 336–339.

23. 'A simplex method for function minimization' J. A. Nelder and R. Mead (1965) *Computer J.*, **7**, 308–313.

24. 'An all atom force field for simulation of proteins and nucleic acids' S. J. Weiner, P. A. Kollman, D. T. Nguyen and D. A. Case (1986) *J. Comp. Chem.*, **7** (2) 230–252.

25. 'Studies in molecular dynamics I. General method' B. J. Alder and T. E. Wainwright (1960) *J. Chem. Phys.*, **31**, 459–466.

26. 'Correlations in the motion of atoms in liquid argon' A. Rahman (1960) *Phys. Rev.*, **136A**, 405–411.
27. 'Structural and energetic effects of truncating long ranged interactions in ionic and polar fluids' C. L. Brooks, B. M. Pettitt and M. Karplus (1985) *J. Chem. Phys.*, **83**, 5897–5908.
28. 'Conformational dynamics detected by nuclear magnetic resonance NOE values and *J* coupling constants' H. Kessler, C. Griesinger, J. Lautz, A. Müller, W. F. van Gunsteren and H. J. C. Berendsen (1988) *J. Am. Chem. Soc.*, **110**, 3393–3396.
29. *Computer Simulations using Particles* R. W. Hockney and J. W. Eastwood (1981) McGraw-Hill, New York.
30. 'Some multistep methods for use in molecular dynamics calculations' D. Beeman (1976) *J. Comput. Phys.*, **20**, 30–139.
31. 'Glassy models of protein folding' J. D. Bryngelson (1988), Ph.D. Thesis, University of Illinois.
32. 'Spin glasses and the statistical mechanics of protein folding' J. D. Bryngelson and P. G. Wolynes (1987) *Proc. Nat. Acad. Sci. USA*, **84**, 7524–7528.
33. 'Temperature-dependent molecular dynamics and restrained X-ray refinement simulations of a Z-DNA hexamer' E. Westhof, B. Chevrier, S. L. Gallion, P. K. Weiner and R. M. Levy (1986) *J. Mol. Biol.*, **190**, 699–712.
34. 'Expert system for protein engineering: its application in the study of chloramphenicol acetyltransferase and avian pancreatic polypeptide' B. Robson, E. Platt, R. V. Fishleigh, A. Marsden and P. Millard (1987) *J. Mol. Graphics*, **5** (1) 8–17.
35. 'Calculating three-dimensional molecular structure from atom–atom distance information: cyclosporin A' J. Lautz, H. Kessler, J. M. Blaney, R. M. Scheek and W. F. van Gunsteren (1989) *Int. J. Peptide Protein Res.*, **33**, 281–288.
36. 'A unified formulation of the critical temperature molecular dynamics methods' S. Nosé and M. L. Klein (1984) *J. Chem. Phys.*, **81**, 511–519.
37. 'Polymorphic transitions in single crystals. A new molecular dynamics method' M. Parinello and A. Rahman (1981) *J. Appl. Phys.*, **52**, 7182–7190.
38. 'Molecular dynamics simulations at constant pressure and/or temperature' H. C. Anderson (1981) *J. Chem. Phys.*, **72**, 2384–2393.
39. 'Molecular dynamics simulations of fluorite structure crystals' A. M. Brass (1987) Ph.D. Thesis, University of Edinburgh.
40. 'Interionic potentials in alkali halides and their use in simulation of molten salts' M. J. L. Sangster and M. Dixon (1976) *Adv. Phys.*, **25**, 247–342.
41. *Theory of Thermal Reaction Scattering. The Use of Reactions for the Investigation of Condensed Matter*, W. Marshall and S. W. Lovesy (1971) Clarendon Press, Oxford.
42. 'Regular and irregular motion' M. V. Berry (1978) in *Topics in Non-Linear Dynamics* (S. Jorma, ed.) AIP Conference Proceedings (No. 46) pp. 16–120.
43. J. G. Powles, Personal communication.
44. *Dynamical Systems IV* V. I. Arnold (1988) Springer-Verlag, Germany.
45. *Dynamical Systems, Theory and Applications* J. Moser (1975) Lecture Notes in Physics, **38**, Springer-Verlag, New York.
46. *Advanced Mathematical Methods for Scientists and Engineers* C. M. Bender and S. A. Orszag (1978) McGraw-Hill, Singapore.
47. *Chaos* A. V. Holden (1986) Manchester University Press.
48. *Nonlinear Dynamics and Chaos* J. M. T. Thompson (1986) Wiley, Chichester.
49. 'Self-generated chaotic behaviour in non-linear mechanics' R. H. G. Helleman (1980). In: *Fundamental Problems in Statistical Mechanics* V (E. G. Cohen, ed.) pp. 165–234. North-Holland, Amsterdam.
50. 'Simple mathematical models with very complicated dynamics' R. M. May (1976) *Nature*, **261**, 459–466.
51. 'Quantitative universality for a class of nonlinear transformations' M. J. Feigenbaum (1978) *J. Stat. Phys.*, **19**, 25–52.

52. 'Lyapunov characteristic exponents for smooth dynamical systems and for Hamiltonian systems: a method of computing all of them' G. L. Benettin, L. Galgani, A. Giorgilli and J. M. Strelcyn (1980) *Meccanica*, **15**, 11–30.

53. 'Characteristic Lyapunov exponents and ergodic theory' Y. B. Pesin (1977) *Russ. Math. Surveys*, **32** (4), 55–114.

54. 'A spectral analysis method of obtaining molecular spectra from classical trajectories' D. W. Noid, M. L. Koszykowski and R. A. Marcus (1977) *J. Chem. Phys.*, **67**, 404–408.

55. *Regular and Stochastic Motion* A. J. Lichtenberg and M. A. Lieberman (1983) Springer-Verlag, New York, pp. 213–258.

56. 'A universal instability of many dimensional oscillator system' B. Chirikov (1979) *Phys. Reports*, **52**, 263–265.

57. 'A method for determining a stochastic transition' J. M. Green (1979) *J. Math. Phys.*, **20**, 1183–1201.

58. 'The relationship of conformation to biological action of oxytocin and its analogues' V. J. Hruby (1985) In: *Oxytocin: Clinical and Laboratory Studies* (J. A. Amico and A. G. Robinson, eds.), pp. 405–414, Elsevier, Amsterdam.

59. *Thyrotropin releasing hormone: biomedical significance* (G. Metcalf and I. M. D. Jackson, eds.) (1989) *Ann. N.Y. Acad. of Sci.*, **553**.

60. 'Diverse roles of thyrotropin-releasing hormone in brain, pituitary and spinal function' N. A. Sharif (1985) *Trends Pharmacol. Sci.*, **6** (3), 119–122.

61. 'Synthesis of thyrotropin-releasing hormone analogues. 1. Complete dissociation of central nervous system effects from thyrotropin-releasing activity' T. Szirtes, L. Kisfaludy, E. Pálosi and L. Szporny (1984) *J. Med. Chem.*, **27**, 741–745.

62. 'Regulatory peptides as a source of new drugs – the clinical prospects for analogues of TRH which are resistant to metabolic degradation' G. Metcalf (1982) *Brain Res. Rev.*, **4**, 389–408.

63. 'Controlled acute trial of a thyrotrophin releasing hormone analogue (RX77368) in motor neuron disease' R. J. Guiloff, D. J. A. Eckland, C. Demaine, R. C. Hoare, K. D. Macrae and S. J. Lightman (1987) *J. Neurology, Neurosurgery, and Psychiatry*, **50**, 1359–1370.

64. 'Treatment with the thyrotropin-releasing hormone analog CG3703 restores magnesium homeostasis following traumatic brain injury in rats' R. Vink, T. K. McIntosh and A. I. Faden (1988) *Brain Res.*, **460**, 184–188.

65. 'Conformational study of thyrotrophin-releasing hormone: I. Aspects of importance in the design of novel TRH analogues' D. J. Ward, E. C. Griffiths and B. Robson (1986) *Int. J. Peptide Protein Res.*, **27**, 461–471.

66. 'Comparative conformation–activity relationships for hormonally- and centrally-acting TRH analogues' D. J. Ward, P. W. Finn, E. C. Griffiths and B. Robson (1987) *Int. J. Peptide Protein Res.*, **30**, 263–274.

67. 'X-ray conformational analysis of the potent thyroliberin analogue L-pyroglutamyl-β-(2-thienyl)-L-alanyl-L-prolinamide' B. Stensland and S. Castensson (1982) *J. Mol. Biol.*, **161**, 257–268.

68. 'Electric moments of molecules in liquids' L. Onsager (1936) *J. Am. Chem. Soc.*, **58**, 1486–1493.

69. 'Computer-aided drug design' G. R. Marshall (1987) *Ann. Rev. Pharmacol. Toxicol.*, **27**, 193–213.

70. 'Distance geometry approach to rationalizing binding data' G. M. Crippen (1979) *J. Med. Chem.*, **22** (8), 988–997.

71. 'Identification of two related pentapeptides from the brain with potent opiate agonist activity' J. Hughes, T. W. Smith, H. W. Kosterlitz, L. A. Fothergill, B. A. Morgan and H. R. Morris (1975) *Nature*, **258**, 577–579.

72. 'The Wellcome Foundation Lecture 1982: Opioid peptides and their receptors' H. W. Kosterlitz (1985) *Proc. R. Soc. Lond.*, B **225**, 27–40.

73. 'Crystal structure of leucine-enkephalin' A. Camerman, D. Mastropaolo, I. Karle, J. Karle and N. Camerman (1983) *Nature*, **306**, 447–450.
74. 'Crystal structure of methionine-enkephalin' D. Mastropaolo, A. Camerman and N. Camerman (1986) *Biochem. Biophys. Res. Comm.*, **134** (2), 698–703.
75. 'The crystal structure of [Met^5]enkephalin and a third form of [Leu^5]enkephalin: Observations of a novel pleated β-sheet' J. F. Griffin, D. A. Langs, G. D. Smith, T. L. Blundell, I. J. Tickle and S. Bedarkar (1986) *Proc. Nat. Acad. Sci. USA*, **83**, 3272–3276.
76. 'Calculations on crystal packing of a flexible molecule, Leu-enkephalin' L. Glasser and H. A. Scheraga (1988) *J. Mol. Biol.*, **199**, 513–524.
77. 'Fluorescence study on the solution conformation of dynorphin in comparison to enkephalin' P. W. Schiller (1983) *Int. J. Peptide Protein Res.*, **21**, 307–312.
78. 'Preferred conformation, orientation and accumulation of dynorphin A-(1–13)-tridecapeptide on the surface of neutral lipid membranes' D. Erne, D. F. Sargent and R. Schwyzer (1985) *Biochemistry*, **24** (16), 4261–4263.
79. 'Conformational analysis of dynorphins [1–17] and [1–8]' E. C. Griffiths, B. Robson and D. J. Ward (1986) *Brit. J. Pharmacol.*, **88**, Suppl. 361.
80. 'Evidence for a folded conformation of methionine- and leucine-enkephalin in a membrane environment' B. A. Behnam and C. M. Deber (1984) *J. Biol. Chem.*, **259** (23), 14935–14940.
81. 'A conformational study of Met- and Leu-enkephalins' E. C. Griffiths, B. Robson and D. J. Ward (1986) *Brit. J. Pharmacol.*, **87**, Suppl. 177.
82. 'The conformational properties of the delta opioid peptide [D-Pen^2,D-Pen^5]enkephalin in aqueous solution determined by NMR and energy minimization calculations' V. J. Hruby, L.-F. Kao, B. M. Pettitt and M. Karplus (1988) *J. Am. Chem. Soc.*, **110**, 3351–3359.
83. 'Cloning and expression of cDNA for human diazepam binding inhibitor, a natural ligand for an allosteric site of the γ-aminobutyric acid type a receptor' P. W. Gray, D. Glaister, P. H. Seeburg, A. Guidotti and E. Costa (1986) *Proc. Nat. Acad. Sci. USA*, **83**, 7547–7551.
84. 'Oxytocin is a precursor of potent behaviourally active neuropeptides' J. P. H. Burbach, P. Bohus, G. L. Kovacs, J. W. van Nispen, H. M. Greven and D. De Wied (1983) *Eur. J. Pharmacol.*, **94**, 125–131.
85. 'Neurohypophyseal principles and memory processes' J. M. van Ree, B. Bohus, D. H. G. Versteeg and D. De Wied (1978) *Biochem. Pharmacol.*, **27**, 1793–1800.
86. 'Induction of maternal behavior in virgin rats after intracerebroventricular administration of oxytocin' C. A. Pederson and A. J. Prange Jr (1979) *Proc. Nat. Acad. Sci. USA*, **76**, 6661–6665.
87. 'Proposed conformation of oxytocin in solution' D. W. Urry and R. Walter (1971) *Proc. Nat. Acad. Sci. USA*, **68** (5), 956–958.
88. 'Proposed conformations of oxytocin and selected analogs in dimethyl sulfoxide as deduced from proton magnetic resonance studies' A. I. R. Brewster, V. J. Hruby, J. A. Glasel and A. E. Tonelli (1973) *Biochemistry*, **12**, 5294–5304.
89. '300-MHz nuclear magnetic resonance study of oxytocin in aqueous solution: Conformational implications' A. I. R. Brewster and V. J. Hruby (1973) *Proc. Nat. Acad. Sci. USA*, **70** (12), 3806–3809.
90. 'Amide hydrogen exchange rates of peptides in H_2O solution by ^1H nuclear magnetic resonance transfer of solvent saturation method – Conformations of oxytocin and lysine vasopressin in aqueous solution' N. R. Krishna, D. H. Huang, J. D. Glickson, R. Rowan and R. Walter (1979) *Biophys. J.*, **26**, 345–366.
91. 'A Raman spectroscopic investigation of the disulfide conformation in oxytocin and lysine vasopressin' F. R. Maxfield and H. A. Scheraga (1977) *Biochemistry*, **16**, 4443–4449.
92. 'Deuteron magnetic resonance studies on the microdynamical behaviour of partially

deuterated oxytocin with neurophysin' J. A. Glasel, V. J. Hruby, J. F. McKelvy and A. F. Spatola (1973) *J. Mol. Biol.*, **79**, 555–575.

93. 'Pharmacological, conformational and dynamic properties of cycloleucine-2 analogues of oxytocin and [1-penicillamine]oxytocin' V. J. Hruby, T. W. Rockway, V. Viswanatha and W. Y. Chan (1983) *Int. J. Peptide Protein Res.*, **21**, 24–34.

94. 'Crystal structure analysis of deamino-oxytocin: Conformational flexibility and receptor binding' S. P. Wood, I. J. Tickle, A. M. Treharne, J. E. Pitts, Y. Mascarenhas, J. Y. Li, J. Husain, S. Cooper, T. L. Blundell, V. J. Hruby, A. Buku, A. J. Fischman and H. R. Wyssbrod (1986) *Science*, **232**, 633–636.

95. 'β-hairpin families in globular proteins' B. L. Sibanda and J. M. Thornton (1985) *Nature*, **316**, 170–174.

96. 'Turns in peptides and proteins' G. D. Rose, L. M. Gierasch and J. A. Smith (1985). In: *Advances in Protein Chemistry*, **37** (J. Edsall, F. M. Richards and C. B. Anfinsen, eds.) pp. 1–109. Academic Press, New York.

97. 'Development and testing of protocols for computer-aided design of peptide drugs, using oxytocin' D. J. Ward, Y. Chen, E. Platt and B. Robson (1990) *J. Theoret. Biol.*, (in press).

98. 'Use of buildup and energy-minimization procedures to compute low-energy structures of the backbone of enkephalin' M. Vásquez and H. A. Scheraga (1985) *Biopolymers*, **24** (8) 1437–1447.

99. 'Conformational energy analysis of oxytocin molecule' V. Krchňák (1983) In: *Peptides 1982; Proceedings of the European Peptide Symposium, 17th meeting* (K. Blaha and P. Malon, eds.) pp. 713–716. Walter de Gruyter, Berlin.

100. 'Conformational energy studies of oxytocin and its cyclic moiety' D. Kotelchuck, H. A. Scheraga and R. Walter (1972) *Proc. Nat. Acad. Sci. USA*, **69** (12) 3629–3633.

101. 'Identification of sites in oxytocin involved in uterine receptor recognition and activation' R. Walter (1977) *Fed. Proc.*, **36**, 1872–1876.

102. 'Computer-aided design and physiological testing of a luteinising hormone-releasing hormone analogue for 'adjuvant-free' immunocastration' C. A. Morrison, R. V. Fishleigh, D. J. Ward and B. Robson (1987) *FEBS Lett.*, **214** (1), 65–70.

103. 'Prediction of preferred solution conformers of analogues and fragments of neurotensin' D. J. Ward, R. V. Fishleigh, E. Platt, E. C. Griffiths and B. Robson (1986) *Regul. Peptides*, **15**, 197.

104. 'Design of peptide analogs: Theoretical simulation of conformation, energetics, and dynamics' R. S. Struthers, A. T. Hagler and J. Rivier (1984) In: *Conformationally Directed Drug Design: Peptides and Nucleic Acids as Templates or Targets* (J. A. Vida and M. Gordon, eds.) pp. 239–261, American Chemical Society.

105. 'Modelling of α-lactalbumin from the known structure of hen egg white lysozyme using molecular dynamics' B. Robson and E. Platt (1987) *J. Computer-Aided Mol. Design*, **1**, 17–22.

106. 'Principal property values for six non-natural amino acids and their application to a structure–activity relationship for oxytocin peptide analogues' S. Wold, L. Eriksson, S. Hellberg, J. Jonsson, M. Sjöström, B. Skagerberg and C. Wikström (1987) *Can. J. Chem.*, **65**, 1814–1820.

107. Cambridge Structural Database (1988) Cambridge Crystallographic Data Centre, University Chemical Laboratory, Lensfield Road, Cambridge CB2 1EW, UK.

—— *Chapter 5* ——————————————————

Synthetic chemistry and the design of peptide-based drugs

V. J. Hruby, W. Kazmierski, A. M. Kawasaki and T. O. Matsunaga

I. Introduction

Peptide hormones and neurotransmitters are the key regulators of the vast majority of cellular, intercellular, organ, and other physiological processes. New structures and putative pharmacological roles are being discovered in ever-increasing numbers, and use of molecular biological methods suggest that this explosive growth is set to continue. These peptides possess enormous potential for the treatment and cure of many diseases which plague humans and other animals. Despite this enormous potential, peptide hormones and neurotransmitters have received relatively little attention for their development as drugs. The causes for this are numerous, but there appear to be two major reasons. First and foremost, most organic chemists, medicinal chemists, and even biochemists are relatively uninformed about principles related to the design, synthesis, conformational analysis, and biochemical/biological functions of peptides. Textbooks in these areas are surprisingly poor, reflecting little of the excitement and exponential growth in this field during the past few decades, and often even what does exist is overlooked in undergraduate and graduate courses. The second and closely related reason is the relative complexity of peptide structures compared to the steroids, heterocyclics, terpenes, alkaloids, and macrolides more familiar to most organic and medicinal chemists. Thus, when dealing with a peptide of molecular weight about 1500 (rather small for many peptides currently under study), many workers tend to avoid serious consideration of how it may be examined for relationships between structure and chemical, physical, and biological properties. Alternatively, peptides are seen simply as a collection of peptide (i.e. amide) bonds, which is the least interesting aspect of peptide structural and biological chemistry.

1.1. AIMS

Our aim in this chapter, therefore, is to provide a brief outline of peptide chemistry, in relation to peptide structure and biological activities, and how chemical principles can be applied to the design and synthesis of peptides with designed physical chemical and biological properties. Space limitations do not allow us to provide an extensive historical perspective. For those interested in the development of various chemical aspects of peptide chemistry, some starting points for synthesis would be the books of Bodanszky et al.[1] and Schröder and Lübke,[2] and for conformational analysis, of Hruby.[3,4] However, it should be pointed out that interest in peptides dates back to the earliest days of organic chemistry, and Emil Fisher, one of the great nineteenth-century chemists, was intensely interested in amino acid and peptide chemistry. Despite outstanding contributions from many chemists, including Vincent du Vigneaud who won the Nobel Prize in 1955 for his work on the isolation, structure determination, and total synthesis of the peptide hormone oxytocin, it is only since about 1980 that a more extensive interest in developing peptides as drugs and pharmaceuticals has emerged. In this chapter we outline some of the approaches which are being developed in this area.

1.2. AN INTERDISCIPLINARY APPROACH: CHEMICAL, PHYSICAL AND BIOLOGICAL CONSIDERATIONS

Peptides, depsipeptides and pseudopeptides make up a very diverse and large class of compounds and include, in addition to the hormones, neurotransmitters, enzyme inhibitors, endocrines, pericrines and the other peptides of higher animals, the antibiotics, ionophores, etc. found in bacteria, yeasts and other micro-organisms. Our major focus in this chapter will be on peptides found in higher animals, since these are the major natural products which are being used as leads for the development of peptide drugs. A major reason for the current interest in peptide ligands as drugs is that they are usually non-toxic and potent, but perhaps most importantly, nature has selected them as the major modulators of cellular function, and of intercellular and perhaps intracellular communication. In complex multicellular systems such as animals, specific, rapid, and complex intercellular communication is essential for survival as the organism responds to its environment. Current drugs which interact with these systems are generally much more toxic than peptides, especially if they were initially derived from plants and single-cell organisms. Apparently in the course of evolution, natural products from these lower life forms, such as terpenes, alkaloids, various heterocyclics, etc., became toxic to eukaryotic cells being developed by the complex multicellular species. These cells therefore turned to the translation products of genes for their control, peptides. However, peptides, which are generally formed by synthesis from much larger proteins, possess a number of problems which need to be resolved. For example, many peptide hormones and neurotransmitters (though not all) are readily biodegraded by proteolytic enzymes, and are designed by nature to be so, since they often serve as biological switches. The ease of degradation of many peptide hormones and neurotransmitters has led many scientists to suggest that peptides are not viable candidates as drugs. In our view, this is not correct thinking. Though clearly peptides in general are biodegradable, many are stable in the body for very long periods of time. If this were not the case, complex life, as we know it, would not be possible. True, our understanding of which structural, conformational, and topographical properties lead to relative peptide stability and which do not, is still primitive. However, our experience suggests that easily biodegradable peptide hormones and neurotransmitters can be quite readily redesigned to give analogues with virtually the same potency but with little tendency to biodegradation. Use of conformational constraints, amide bond replacements, carefully considered configurational changes, etc. can all be used to accomplish this goal. Similar, though not identical, arguments can be made regarding other putative deficiencies of peptides as drugs, such as problems of penetration of membranes in the gut or brain, of biodistribution, etc. Basically, very little has been done to understand these problems and find solutions. Undoubtedly that will change too.

From the above discussion, it is clear that solutions to problems of developing peptide ligands as drugs will require a highly interdisciplinary scientific approach. Elements of biology, chemistry, and physics must all be used together to obtain success. This need can perhaps be most readily illustrated by a brief consideration of the requirements for design of peptide ligands for a specific biological activity. First, this requires a sophisticated appreciation of structure–biological activity relationships. Having said this, one immediately recognizes a need for evaluating the conformational and structural properties of the peptide on the one hand, and assessing multiple biological activities and properties on the other. Then, these two must be integrated so as to develop an understanding of the possible relationships between them. Structural and conformational studies alone, no matter how sophisticated, will not provide the kind of insights that lead to a coherent approach to design better ligands. On the other hand, neither will sophisticated biological studies necessarily provide clues as to how to improve the potency, selectivity, biostability, etc. of any given peptide.

Clearly, what is needed is a highly integrated utilization of both biological and physical/chemical studies, and such an integrated approach is now possible. Modern biology has provided more specific and comprehensive *in vitro* and *in vivo* bioassays and binding assays which can provide detailed descriptions of the biochemical and biophysical events related to biological activity. These include characterization of partial agonists or antagonists, which in turn can provide leads for the development of receptor antagonists. Furthermore, it is important to recognize that most peptide hormones and neurotransmitters interact with complex multiple receptor subtypes, and thus it is necessary to utilize multiple bioassays to obtain insights into the differential structural, conformational, and dynamic requirements for each individual receptor subtype. Similar statements can be made regarding enzymes and receptors for peptides and proteins. Fortunately, such assays have been developed for many peptide hormones and neuropeptides, and continued rapid development can be expected in this area. Furthermore, it is important to realize that most receptors are exquisitely sensitive to changes in structure, conformation and topography for a peptide ligand. Thus it is no wonder that a 'simple assay' can often assess changes of these physical/chemical properties in a much more sensitive way than even the most sophisticated biophysical tools. On the physical/chemical side, it is no longer adequate to simply replace one or more amino acid residues with other eukaryotic amino acid residues, though such studies are still needed to provide information about the specific amino acid residues, functional groups, and their spatial relationships which provide a particular 'biological' activity. However, because of the inherent flexibility of small peptides (and even larger ones) it is also important to develop critical insights into the conformational and topographical requirements for biological activity. Thus, in addition to conformational analysis using NMR spectroscopy, circular dichroism spectroscopy, X-ray crystallography when possible, etc., it will generally be necessary to develop more constrained analogues. This will require

- development of a better understanding of the relationships between structure and conformational preferences in peptides;
- development of methods for asymmetric synthesis and for conformational restriction to provide the necessary structures which can fix or bias particular conformational preferences in a manner compatible with bioactive conformations of peptides and proteins.

In this regard, it has become apparent that the use of conformational constraints[5] can provide a powerful approach to the development of peptide ligands which can interact more specifically with a particular receptor, and thereby provide a working hypothesis of the 'biologically active conformation' for a particular ligand. Constrained analogs can then be further tested by applying additional structural or conformational modifications. Obviously, only when such an approach is carefully tested and refined by appropriate biological assays will it be possible to determine the validity of the chemical and physical principles that are being applied.

In summary, therefore, it is important to emphasize that any rational and realistic approach to the design of peptide ligands requires an integrated multidisciplinary approach involving at least:

- organic synthetic and structural chemistry;
- spectroscopic and other physical and biophysical methods;
- use of modern computer assisted molecular modeling and molecular mechanics and molecular dynamics calculations;
- detailed quantitative multiple biological assay systems.

2. Theory

2.1. SYNTHETIC METHODS

2.1.1. *Amino acids and amino acid analogues*

The design of suitable peptide analogues depends on the availability and purity of appropriate amino acids suitable for peptide synthesis. It is critical to emphasize that the purity of a synthetic peptide cannot generally exceed the purity of the amino acid derivatives used in its synthesis. In the case of the 20 or so eukaryotic amino acids, and a number of other amino acids which are available from natural sources, amino acids of high purity and stereochemical integrity are available from commercial sources. However, many D-amino acids have a small (1–3%) contamination with the L-enantiomer. This is important to realize, since there are now numerous cases where the biological activity of diastereoisomeric peptides (in which just one of the amino acid residues is D rather than L or vice versa) differ in potency by factors of 100 or more, or where partial antagonist or agonist bioactivity is seen for only one isomer. In these cases, a small (1%) impurity can be responsible for all the observed biological activity. Thus it is critical that the protected amino acids used in peptide synthesis be not only analytically pure, but also stereochemically pure.

Increasingly, there is a need for special amino acids with particular side-chain moieties, multiple asymmetric centres, etc., in the design of peptide analogues. The synthesis of amino acids has been of interest to organic chemists since the earliest days of synthetic chemistry, and numerous methods for their syntheses were developed. These syntheses generally provided racemic products, and chemical and enzymatic methods were developed for separation of the enantiomers. Many of these methods are very useful even today, and the work in this area to 1960 has been excellently summarized by Greenstein and Winitz.[6] To some extent this information has been updated in a recent book edited by Barrett.[7]

As the complexity of peptide design increases, an increased need arises for more complex amino acid derivatives. Organic chemists have been rather slow in responding to this need. Nonetheless, more rapid progress is now being made, and some recent approaches can be seen by the recent appearance of a Tetrahedron Symposium[8] and a book by Williams.[9] In addition, many new methods are under development in this area suggesting that progress will accelerate in the future, particularly in new methods and successes in the asymmetric synthesis of amino acids in high stereochemical purity.

2.1.2. *Solution methods of peptide synthesis*

Until the invention of the solid-phase methods of peptide synthesis 25 years ago, peptides were made in solution utilizing the batch method of synthesis generally used by synthetic chemists. In this approach, the growing peptide chain, properly protected, is used and a properly protected amino acid or peptide is then attached

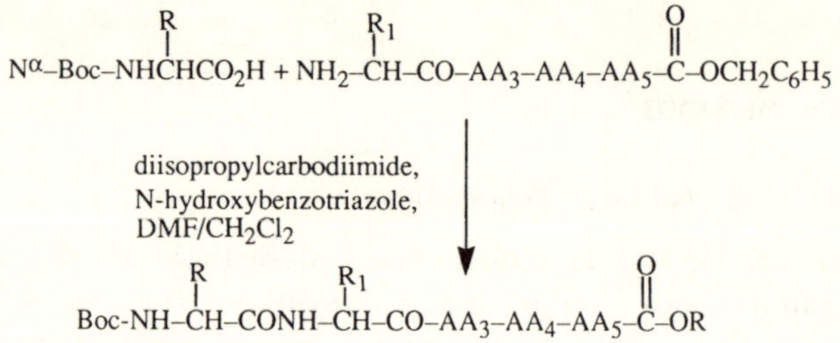

$$N^{\alpha}\text{-Boc-NHCHCO}_2H + NH_2\text{-CH-CO-AA}_3\text{-AA}_4\text{-AA}_5\text{-C-OCH}_2C_6H_5$$

with R on the first residue, R_1 on the second, and O on the carbonyl.

diisopropylcarbodiimide,
N-hydroxybenzotriazole,
DMF/CH$_2$Cl$_2$

$$\text{Boc-NH-CH-CONH-CH-CO-AA}_3\text{-AA}_4\text{-AA}_5\text{-C-OR}$$

with R and R_1 substituents and O carbonyl.

Figure 1   Coupling of a protected amino acid residue to a polypeptide.

to the other peptide (Fig. 1). Generally the amino acid or peptide is added to the growing peptide chain from the N-terminus to minimize problems of racemization. Numerous N^{α}-protecting groups have been developed, and an even larger number of methods to form the peptide bond have been suggested. Many of these have been discussed and reviewed in the nine-volume treatise edited first by Gross and Meienhofer and then by Udenfriend and Meienhofer,[10] including strategies for functional group protection. These will not be discussed here, though it should be emphasized that there is still a need for further orthogonal methods for peptide synthesis. The methodology shown in Fig. 1 is shown to emphasize three important considerations (and limitations) of solution synthesis of peptides:

- *proper regiochemistry* – this is maintained by appropriate protection of amino and carboxy termini;
- *quantitative coupling* – coupling reagents which provide quantitative coupling and few side reactions;
- *suppression of racemization* – choice of urethane amino protection if amino acid is added, or use of Gly or Pro as carboxy terminal residue in a peptide to be coupled.

Solution methods of peptide synthesis are still widely applicable:

- for the synthesis of small peptides (<4–6 residues) where it is much more efficient than solid-phase synthesis;
- for many industrial applications;
- for difficult coupling requiring elevated temperatures or other special conditions;
- for synthesis of certain proteins of higher molecular weight;
- for developing new synthetic methodologies or protecting groups.

2.1.3. *Solid-phase methods of peptide synthesis and their automation*

The development of the solid-phase method by Merrifield[11] represents one of the most important developments in organic chemistry in the twentieth century. Despite the obvious importance and fundamental significance of this advance, the method met with surprising opposition from many prominent peptide chemists, many of whom have never, or only recently, incorporated this elegant

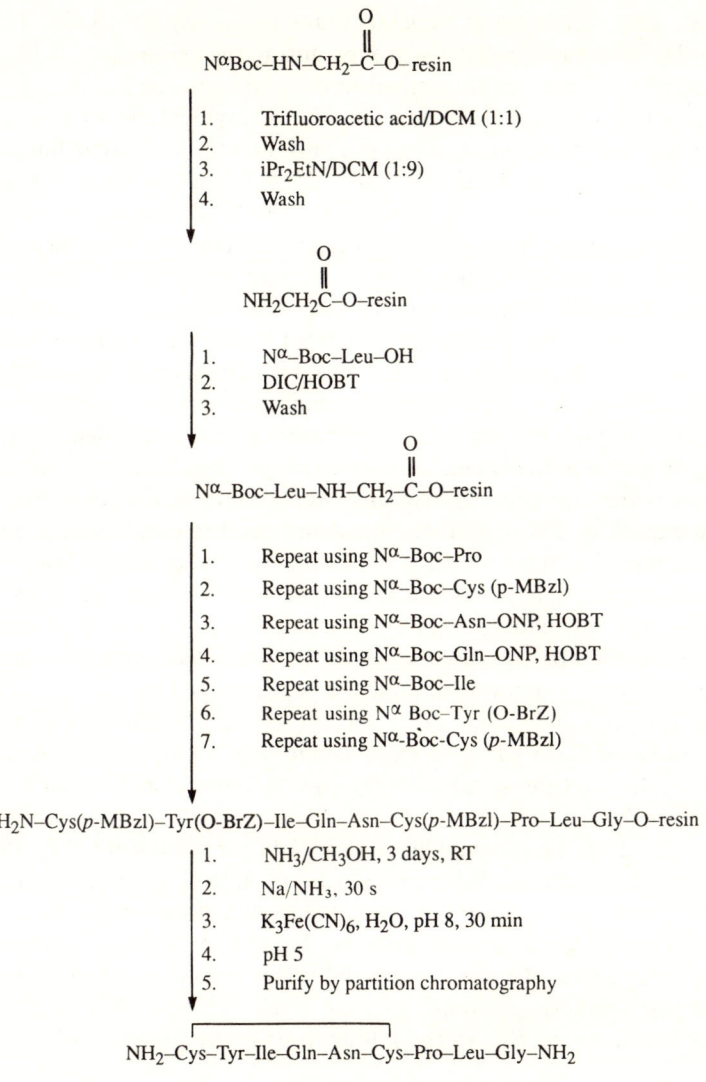

Figure 2 A 'typical' example of solid-phase peptide synthesis. The synthesis of oxytocin.

method into their laboratories. The reasons for this seem to have more to do with human behavior than scientific rationality, since the solid-phase method is indeed beautiful in its scientific elegance and simplicity. Obviously, of course, its proper application for the synthesis of peptides and proteins requires the most sophisticated thinking and attention to detail. The outline of a 'typical' solid-phase synthesis is shown in Fig. 2. This simple synthesis of oxytocin is given using the N^α-Boc strategy which is the most widely used in peptide synthesis.[12,13] Also widely used is the N^α-Fmoc strategy[14,15] which is currently somewhat more expensive, but is growing in popularity due to the gentler conditions used to remove the N^α-amino protecting groups during the synthesis.

As outlined in Fig. 2, the C-terminal amino acid is attached to the solid support (usually polystyrene cross-linked with 1% divinylbenzene to give a porous polymer which expands in most organic solvents) through some functionalized moiety (the substitution level is usually 0.25–0.70 mmol/g resin, though higher and lower substitutions can be used). Then the N^α-Boc group is removed, the resin washed, neutralized and washed again to remove excess base. The next amino acid is attached to the growing peptide chain by a suitable coupling procedure, generally using a two- to three-fold excess of the amino acid to ensure $> 99\%$ coupling. Coupling is repeated if less than quantitative coupling has occurred, or the unreacted amino group is permanently acylated (generally with acetic anhydride) to ensure that the chain does not continue to grow during the subsequent synthetic steps. This minimizes the presence of deletion sequences missing only one amino acid, and tends to give small amounts of truncated sequences which are generally easier to separate from the final product than deletion sequences. The peptide chain is then extended, usually one amino acid at a time (though fragment coupling is possible and is sometimes used) until the peptide chain, properly protected, has been assembled. Cleavage from the resin depends on the resin chosen and the end-groups needed. Here liquid ammonia in anhydrous methanol was used to give the protected carboxamide terminal peptide. If a carboxylate were desired, the peptide could be cleaved in liquid HF $(0°)$, etc. The disulphide bond is then formed, though this can sometimes be done on the resin as well. In addition, cyclic lactam-containing peptides can be made directly on the solid-phase resin in very high yield (greater than can be obtained in solution).[16]

The method can be automated, and several instruments are currently on the market for this purpose. With modern minicomputers, careful control of all synthetic procedures is possible, though the manufacturers have the unfortunate tendency to provide instruments with only limited possibilities for adaptation to different chemistries. In any case, solid-phase peptide synthesis has revolutionized the synthesis of simple peptides, and perhaps 80% of all synthetic peptides are made via this methodology in its many forms.

2.2. CHEMICAL MODIFICATIONS – DESIGN PRINCIPLES

As already mentioned, preparation of analogues of natural peptides provides a powerful and, in principle, rational approach to the design of analogues with unique biological and physicochemical properties. For biologically active peptides, the common assumption is that their biological activities depend on their three-dimensional structures (the function code). However, in view of their flexibility, most peptide hormones and neurotransmitters can assume a variety of conformations, and the question arises as to which of these are of biological significance. For this reason, much effort has been made to develop an approach using conformational constraints to explore the conformational requirement at a receptor.[5,17,18] The working hypothesis in this approach is that the imposed constraints will be maintained at the receptor, allowing a true evaluation of the validity of this constraint for ligand–receptor interaction. Proof for this assumption must await isolation of receptors and determination of the bound conformation of the ligands in the peptide–receptor complex. However, the validity of this approach seems reasonable, since many successful applications have now appeared. Alternatively, it should be remembered that conformational flexibility can be functionally important in biological systems using peptides as ligands:

- the ability of a peptide ligand to adapt different conformations may enhance its ability to recognize and interact with several receptor sub-types or even different receptors;
- it may enhance the ability of a peptide to 'adjust' to evolutionary changes of receptors;
- it can more readily be recognized by various degrading enzymes;
- it may allow it to more readily adopt the different conformations needed for recognition and transduction.[19,20]

The fact that small peptides form different conformations in 'bound' states than in 'free' solution has now been demonstrated in several cases. For example, Milon *et al.*[21] have utilized transferred nuclear Overhauser effects (TRNOE) to study the conformation of α-mating factor, Leu-enkephalin, and mastoparun to phospholipid membranes. We[22] have used the same method to demonstrate in the oxytocin/neurophysin I complex that the conformation of oxytocin bound to neurophysin is quite different from that found in the X-ray structure of deamino-oxytocin.[23] In fact, such dynamic conformational effects are not limited to the peptide area. For example, Glasel and Borer[24] have studied the conformations of nalorphine and levorphanol bound to binding sites of two anti-opiate antibodies and demonstrated significant differences in the free and bound forms of these ligands.

These examples and others indicate that successful delineation of the relationships of the 'structure code' to the 'function code' in peptides will require special techniques to obtain direct or circumstantial evidence regarding the 'biologically active conformations' of the molecule.

Therefore, we now turn to a brief discussion of the most powerful approach found to date for helping solve this problem by the use of conformational constraints.

2.2.1. *Amino acids for secondary structure*

Due to their structural properties, certain amino acids can impose local conformational constraints in peptides. For example, α-aminoisobutyric acid, $(CH_3)_2C(NH_2)CO_2H$, is found in a number of peptide antibiotics and has been used for some time to locally constrain a peptide residue to the α-helical region, though it can also be found in a 3_{10} helix and possibly in other regions of conformational space such as β-turns. It and other α-substituted amino acids, though clearly constrained, can thus adapt somewhat to their environment. However, because of the energy penalty for such an adaptation, they constitute a valuable approach to local conformational constraints. Reviews on the conformational properties of α,α'-substituted amino acids have recently appeared.[25,26] The peptide bond can in principle be replaced by a carbon–carbon double bond with a *trans* conformation. Though this has been done in several cases, generally the results have been disappointing. For example, cholecystokinin-4 (CCK-4) analogues with a *trans* carbon–carbon double bond are generally inactive.[27] The *cis* conformational constraint could utilize the *cis* olefin bond, but this has been synthetically unsuccessful. However, the tetrazole appears to be useful for this purpose,[28] though more work is needed to determine if its stereostructural properties are compatible with peptide and protein structures.[11] Another approach has been that of Van der Elst *et al.*[29] who utilized O-(aminomethyl)phenyl acetic (O-AMPA) derivatives as a *cis* amide bond surrogate in somatostatin analogues. Unfortunately, the resulting analogues did not display any inhibition of growth hormone release characteristic of somatostatin analogues, perhaps in part due to the C_8 δ-turn conformation which was seen in the modified structure.

Cyclopropyl amino acids, however, provide the only known residue to strongly appear to prefer the 'bridge' region of conformational space for a residue ($\varphi = -90°$, $\psi = 0°$).

Pipecolic acid (Pip, piperidine-2-carboxylic acid, or homoproline) appears to have properties somewhat different from those of proline. Whereas proline favours neither the *trans* or *cis* peptide bond, Pip favours the *trans* bond slightly. Whereas the proline ring forms either an envelope or half-chair (twist) conformation, the chair forms of the six-membered ring are more stable than the boat forms in Pip, and the peptide –CONH– must assume an axial orientation. These general observations are true for internal residues but not for N-terminal residues.[32] For Pip, $\varphi = -115 \pm 15°$ (though φ can be $-60°$ for the boat conformation) and $\psi \approx 40°$. For Pro, however, $\varphi = -65 \pm 15°$ and $\psi = 155 \pm 10°$ (*trans*) or $-35 \pm 10°$ (*cis'*) with others also accessible, especially $125 \pm 10°$ when Pro is in the $i+1$ position of a type-II β-bend.

Other related amino acid derivatives have been reviewed.[30]

2.2.2. *Amino acids for topographical bias.* (*Side-chain constraints* – χ_1, χ_2, *etc.*)

On the surface where contact between ligand and receptor (acceptor) occurs, the relative spatial arrangements of the side-chain groups on the peptide pharmacophore are critical for receptor recognition and probably determine, in many cases, the affinity and selectivity of the peptide for a particular receptor protein. While virtually all discussions of peptide and protein conformation provide detailed examination of conformational properties, these discussions usually are confined to the secondary structure (α-helix, β-turn, β-sheet, etc.). However, in considering what is needed in problems of molecular recognition, whether one is

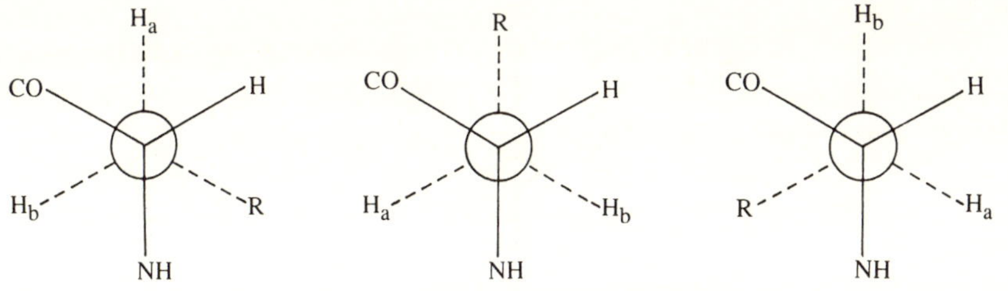

$$\chi^1 = -60° \qquad\qquad \chi^1 = \pm180° \qquad\qquad \chi^1 = +60°$$

Figure 3 Side-chain conformations *gauche*$(-)$ $(-60°)$, *trans* $(\pm 180°)$ and *gauche*$(+)$ $(+60°)$ for a typical α-amino acid residue about the C_α–C_β bond.

talking about the protein folding problem, the 'docking' problem, receptor–ligand interaction, etc., little thought has been given to the topographical problem. For discussion purposes, we define the topography as the 'relative, co-operative three-dimensional arrangements of the side-chain groups in a polypeptide'. Thus it is possible to have a different topography with the same secondary structure conformation (e.g. α-helix, β-turn, etc.), and alternatively to have the same topography but a different secondary structure conformation. Unfortunately, very little has been done to examine this problem especially as it applies to the host–guest biological problem mentioned above. Therefore, we have begun to develop approaches to study this problem. In these approaches, we seek to constrain, bias, or fix side-chain groups to one of the conformers generally assumed by amino acid side-chain groups: *gauche*$(-)$ $(\chi^1 = -60°)$. *trans* $(\chi^1 = \pm 180°)$; and *gauche*$(+)$ $(\chi^1 = +60°)$ (Fig. 3). These classical staggered rotamer conformations are those generally found in proteins. By biasing or fixing side-chain groups of critical amino acid residues it should be possible to modify the surface architecture of already constrained analogues, and thereby modify in very specific ways the ligand–receptor interaction. We will discuss some of the approaches we believe will be useful in this approach. The discussion here will be very short, but the potential of this approach is enormous.

β-Substituted amino acids in peptides have the potential to bias side-chain conformations by virtue of non-bonded interactions between vicinal substituents. For example, β-methylphenylalanine has four different stereochemical structures, and, as seen in Fig. 4 for the S,S and S,R structures different side-chain conformations are favoured in each case, with *trans* favoured for the S,S and *gauche*$(-)$ for the S,R isomers. Similar arguments can be made for the R,R and R,S diastereoisomers, but, of course in this case, the change in stereochemistry at the α-carbon will affect the topographical relationships of the side-chain groups relative to the rest of the peptide side-chain groups within the same overall conformational family. We have shown[31] that for analogues of [D-Pen², D-Pen⁵]enkephalin in which the Phe⁴ was replaced by each of the isomers of β-methyl-p-nitrophenylalanine (β-Me-pNO$_2$Phe) compounds of widely different potencies and specificities for δ and μ receptors occurred, though the overall

(a) 2S, 3S isomer

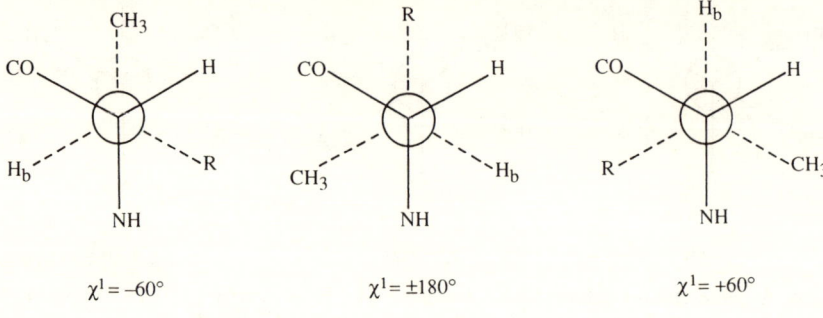

$$\chi^1 = -60° \qquad\qquad \chi^1 = \pm180° \qquad\qquad \chi^1 = +60°$$

(b) 2S, 3R isomer

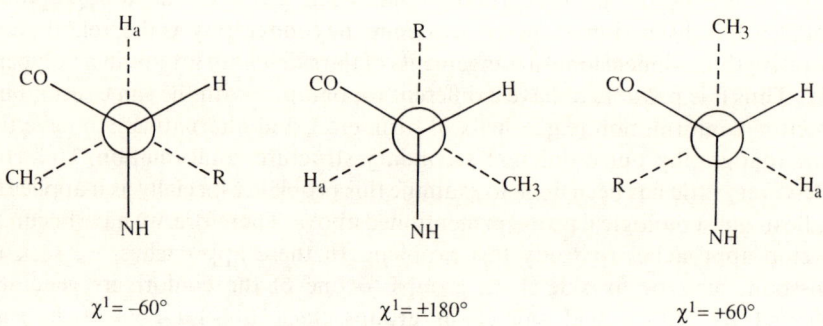

$$\chi^1 = -60° \qquad\qquad \chi^1 = \pm180° \qquad\qquad \chi^1 = +60°$$

Figure 4 Low-energy side-chain conformations for the 2S, 3S (a) and 2S, 3R (b) isomers of a β-substituted α-amino acid about the $C_\alpha-C_\beta$ bond. Note that the low-energy conformer is different for the isomers.

conformation remained the same in all cases. Thus, though the same conformation was retained, topographical changes *only* led to large differences in opioid receptor potency and selectivity.

Topographical bias of side-chain moieties may be more stringently enforced on aromatic (or aliphatic) amino acid residues by connecting their α-amino group to the 2'-aromatic position with a methylene group to give 1,2,3,4-tetrahydroisoquinoline-2-carboxylic acid (Tic, Fig. 5). This led to restriction of χ_1 to either $-60°$ or $+60°$ (the *trans* side-chain rotamer is excluded). It has been shown[32] that the *gauche*($-$) conformation is characteristic for this amino acid when it is the N-terminal residue, whereas acylated derivatives (amino acids or carboxylic acid derivatives) exhibit a strong conformational bias to a *gauche*($+$) side-chain conformation (Hruby and Kazmierski, unpublished results). Utilizing these principles, the topography of the potent and receptor specific μ opioid receptor antagonists of the general formula D-Phe–Cys–Tyr–D–Trp–Xxx–Thr–

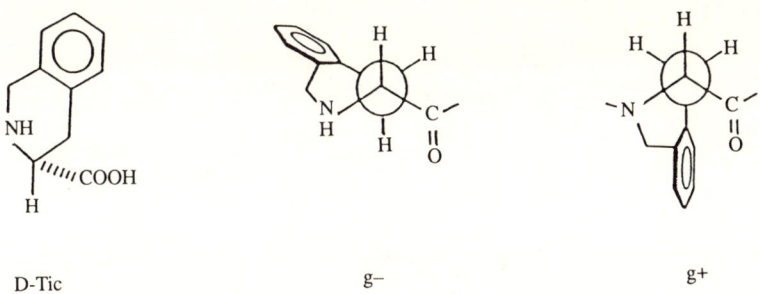

D-Tic   g–   g+

Figure 5 The structure of D-1,2,3,4-tetrahydroisoquinoline-2-carboxylic acid (D-Tic) and its two low-energy chair-like conformations.

Pen–Thr-NH$_2$ (where Xxx = Lys, Orn, Arg) has been modified to constrain side-chain residues on opposite sides of the molecule in Gly–D-Tic–Cys–Tyr–D-Trp–Orn–Thr–Pen–Thr-NH$_2$ (Fig. 6)[30] and D-Phe–Cys–Tic–D-Trp–Orn–Thr–Pen–Thr-NH$_2$ (Fig. 7, Hruby and Kazmierski, unpublished) to give analogues with very weak potencies. On the contrary, restricting these aromatic amino acid side-chains to *gauche*(−) in D-Tic–Cys–Tyr–D-Trp–Lys–Thr–Pen–Thr-NH$_2$ (Fig. 8) resulted in a highly potent and selective μ opioid receptor antagonist.[32,33] This and other possible related amino acids (see Toniolo[30] for a few additional amino acid derivatives) can clearly be used to explore important topographical requirements for different receptors and acceptors.

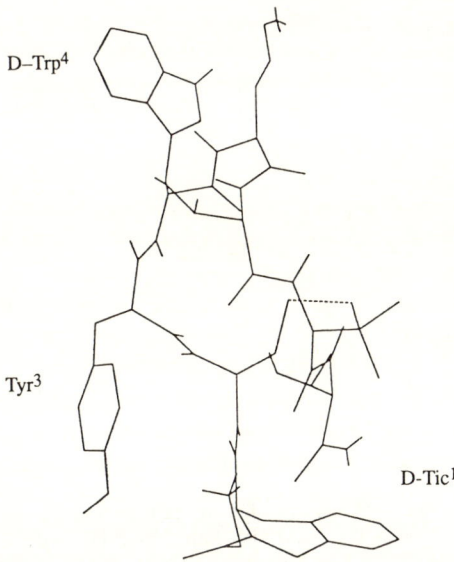

Figure 6 The low-energy conformation for the weak μ opioid receptor antagonist Gly–D-Tic–Cys–D-Trp–Orn–Thr–Pen–Thr-NH$_2$.

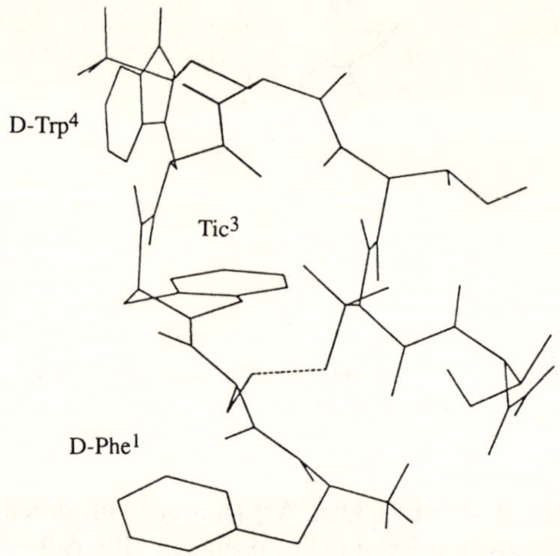

Figure 7 The low-energy conformation for the weak μ opioid receptor antagonist D-Phe–Cys–Tic–D-Trp–Orn–Thr–Pen–Thr-NH$_2$. In both this analogue and the analogue in Fig. 6, the conformation remains the same, but the topographical relationships of the side-chain groups in positions 1 and 3 are not compatible with potent interactions with the μ opioid receptor in these cases.

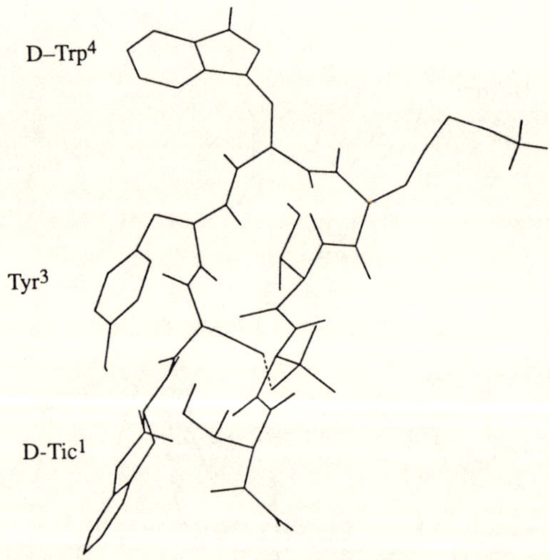

Figure 8 The low-energy conformation for the potent and selective μ opioid receptor antagonist D-Tic–Cys–Tyr–D-Trp–Lys–Thr–Pen–Thr-NH$_2$. Again, the backbone conformation is the same as in Figs 6 and 7, but now the topographical relationships between the side-chain groups in positions 1 and 3 are highly compatible with potent interactions with the μ opioid receptor.

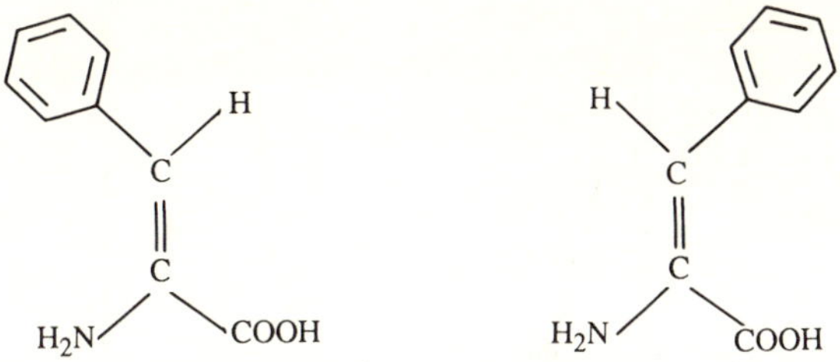

Figure 9 Structures of Δ^Z- and Δ^E-dehydrophenylalanine.

α,β-Unsaturated amino acids (Fig. 9) offer a special opportunity in that they have only two side-chain conformations with the Δ^E conformation similar to the *trans* conformation (Fig. 8), while the more readily obtained Δ^Z conformation places the aromatic rings approximately halfway between the *gauche*($-$) and *gauche*($+$) conformations (compare Fig. 3 and Fig. 9). Also, it should be noted that the α-carbon loses its chirality and this may provide significant populations of semi-extended conformations for peptides containing this amino acid residue.[34]

Similar thinking, of course, can be applied to other diastereotopic positions in a peptide.

2.2.3. *Side-chain to side-chain cyclizations*

Since natural peptides and proteins often choose this approach via the Cys–Cys disulphide bridge, this is an obvious approach to stabilizing peptide conformational structure. Even within this context, however, further constraint is possible,[5] in fact necessary, in order to stabilize particular cyclic conformations. The penicillamine residue (S–(CH$_3$)$_2$CCH(NH$_2$)CO$_2$H), β,β-dimethyl cysteine) is particularly valuable for constraining the disulphide dihedral angle to greater than 100°, making it possible to utilize CD and the quadrant rule to determine whether the disulphide is right-handed or left-handed. This is very useful information that can provide important insights into receptor requirements for the cystine moiety. A dynamic interconversion may be involved in favourable cases allowing conformational transitions necessary for going from the binding state to the transduction state in a ligand–receptor interaction. For example, in deamino-oxytocin, the disulphide bridge exists in both a right-handed and a left-handed chirality.[23] However, in Pen[1]-containing oxytocin antagonists, only the right-handed chirality is found. Thus, we have suggested[23,35] that the transition of the disulphide conformation may be of critical importance to agonist *vs.* antagonist biological activity. In addition, it is possible to replace the S–S bridge with C–S, S–C, Se–C, C–C, S, etc. Essentially all of these have been tried, sometimes with quite exciting results such as in oxytocin, where superagonists were obtained in some cases.[20]

Conversion of linear peptide into cyclic disulphide-containing conformationally constrained peptides has met with considerable success, especially when

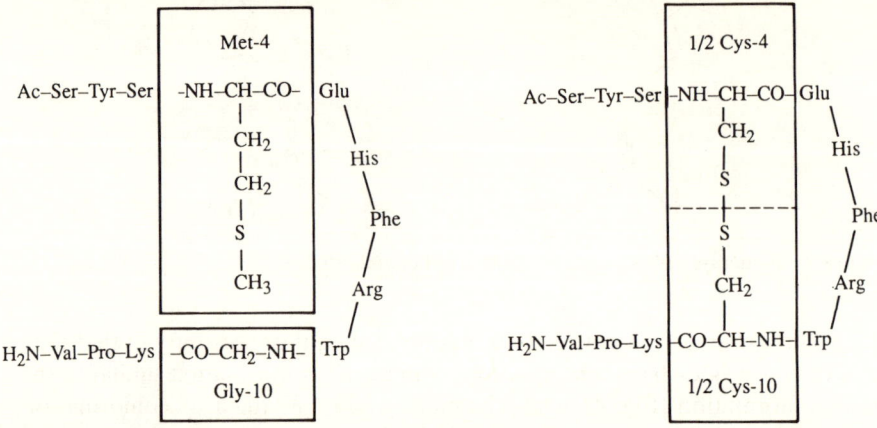

Figure 10 Structures of α-melanotropin (left) and Ac-[Cys^4,Cys^{10}]α-MSH showing the pseudoisosteric relationship between the linear hormone and the superpotent cyclic analogue.

pseudoisosteric cyclization was used.[5] For example, based on conformational considerations, in conjunction with structure–function relationships, it was postulated that α-melanotropin (Ac–Ser–Tyr–Ser–Met–Glu–His–Phe–Arg–Trp–Gly–Lys–Pro–Val–NH$_2$) in its 'biologically active' conformation possessed a reverse turn conformation about the Phe^7 moiety. This led Hruby et al. to stabilize this putative reverse turn conformation by isosteric replacement (Fig. 10) of the Met^4 and Gly^{10} residues by cysteine residues followed by oxidative cyclization to the cyclic disulphide analogue [Cys^4, Cys^{10}]α-MSH resulting in a superagonist^{36} (Fig. 11). Perhaps as interesting was the conversion of methionine enkephalin, a linear pentapeptide of the sequence H–Tyr–Gly–Gly–Phe–Met–OH, to a highly potent and δ opioid receptor selective cyclic, conformationally constrained D-penicillamine-containing analogue H–Tyr–D-Pen–Gly–Phe–D-Pen–OH (DPDPE, Fig. 12).^{37}

A somewhat different approach in which the ring size was severely contracted, by preparing a disulphide ring of progressively reduced size, eventually led to a cyclic hexapeptide derived from somatostatin-14 (Ala–Gly–Cys–Lys–Asn–Phe–Phe–Trp–Lys–Thr–Phe–Thr–Ser–Cys–OH) which has the formula Pro–Phe–D-Trp–Lys–Thr–Phe. This was equipotent to somatostatin and orally active (for a review see reference 38). This was achieved by a systematic reduction in ring size until the primary structural and conformational elements necessary for interaction with the somatostatin receptors resulted in inhibition of growth hormone release from the pituitary, but glucagon and insulin release from the pancreas. Interestingly, other studies using differently constrained somatostatin analogues by the Sandoz group^{39} has led to the development of cyclic disulphide-containing octapeptide analogues of somatostatin which are superactive analogues in inhibition of growth hormone release with considerable selectivity for

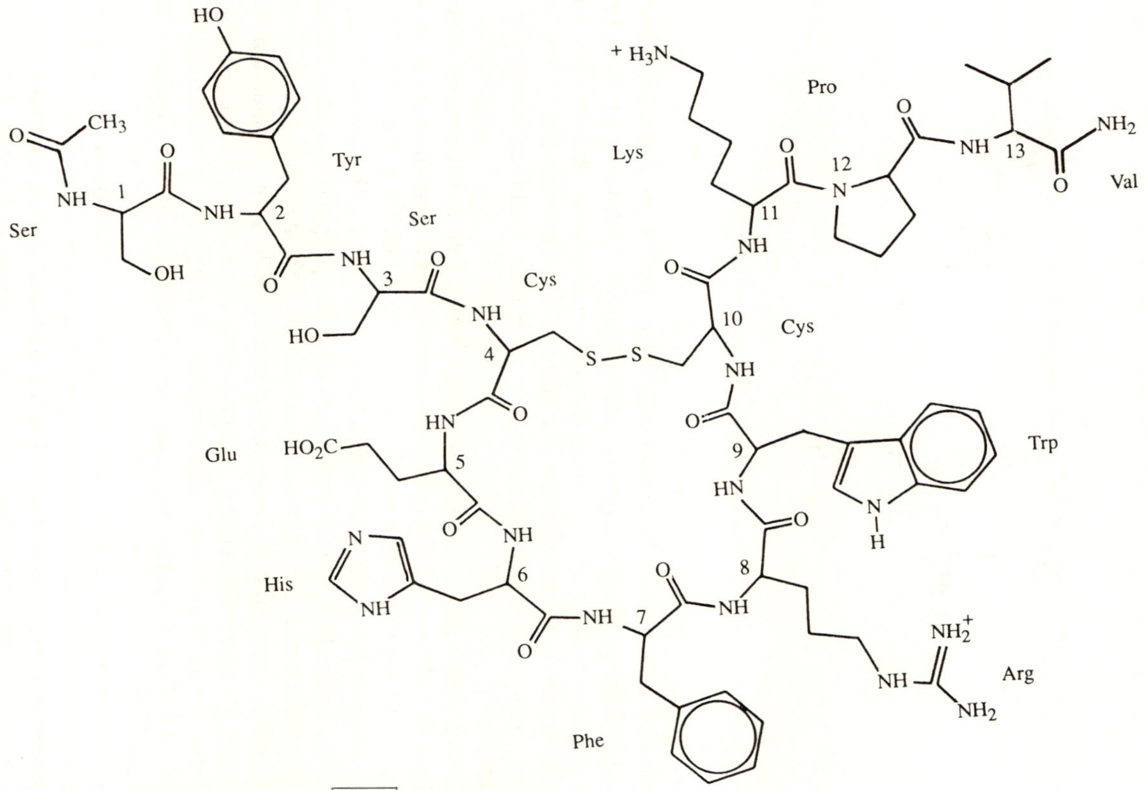

Figure 11 Structure of Ac-[Cys^4, Cys^{10}]αMSH.

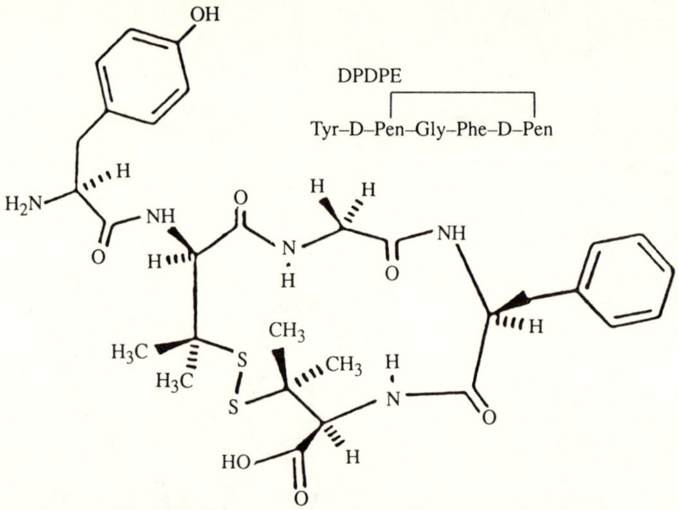

Figure 12 Structure of [D-Pen², D-Pen⁵]enkephalin (DPDPE).

the pituitary receptors *vs.* the pancreatic receptors.[39] In a somewhat different approach, Hruby and co-workers have taken the somatostatin structure and by a series of conformational and topographical constraints converted it from a compound active at somatostatin receptors to a compound with a very high potency and exceptional selectivity for μ opioid receptors, acting as an antagonist.[32,33,40,41]

Another important side-chain to side-chain cyclization involves lactam formation via the carboxylate side-group on Asp or Glu and the ε-amino group on Lys. This has not yet been used as extensively. Schiller *et al.* have used it in a cyclic lactam series of enkephalin with some success, with H–Tyr–D-Orn–Phe–Asp–NH$_2$ being quite potent and selective for the μ opioid receptor[42] though similar studies with dynorphins were less successful.[43] On the other hand, using a combination of computer-aided molecular modelling in conjunction with molecular dynamics simulations and molecular mechanics calculations, a side-chain group interconversion and subsequent transannular cyclic lactam formation has led to a superpotent and super-prolonged-acting α-melanotropin analogue.[18,44]

An exciting example of conformational space restriction on biological activity via cyclic lactam formation is the recent preliminary report following the conversion of a very weak oxytocin agonist analogue to a highly potent oxytocin antagonist by conversion of the monocyclic compound [β-Mpa, Glu⁴, Cys⁶, Lys⁸] oxytocin to the bicyclic analogue [β-Mpa¹, Glu⁴, Cys⁶, Lys⁸] oxytocin.[45] Such an example clearly has much insight to offer on the relationship of topography and relative conformational rigidity on agonist and antagonist biological activity.

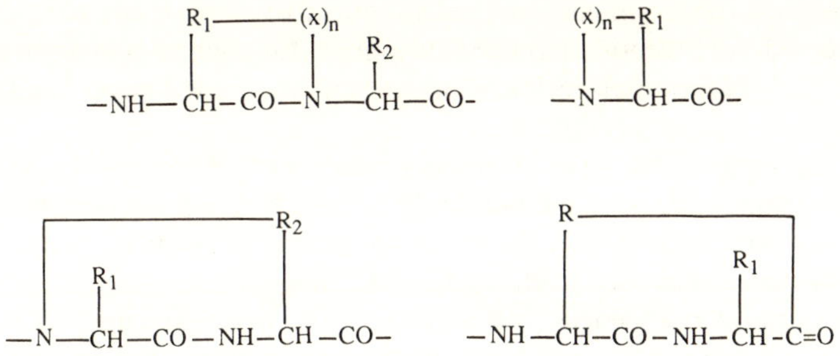

Figure 13 Structures of some possible cyclic side-chain to backbone lactam analogues that might be considered in peptide structures.

2.2.4. *Side-chain to backbone cyclizations*

Cyclization of peptides from a side-chain functional group to the peptide backbone has been widely used as a mode of conformational constraint. Perhaps the most widely used method is cyclic lactam formation from a side-chain carboxylate to a backbone amino or alternatively of a side-chain amino to a backbone carboxyl group. This approach has been used successfully by Schiller *et al.* to develop highly μ opioid receptor selective peptide ligands (reviewed in reference 46). In this case, ε-amino groups from D-amino acid residues in position 2 are reacted with the carboxylate terminal group. However, Chipens and co-workers[47] have utilized this kind of cyclization for a variety of peptides including bradykinin, kallidin, angiotensin, neurotensin, substance P, and others, and many of these peptides retain considerable biological activity.

Other types of side-chain lactams can also be envisioned (Fig. 13). In the first class, the side-chain group of the n residue forms a lactam bridge to the N of the $n+1$ residue. This has been used for example in the synthesis of a number of CCK_4 analogues,[48] and in the synthesis of LHRH and other peptide hormone analogues.[49] In these compounds with five- or six-membered rings, the χ_i angle is restricted to the range $-125° \pm 10°$, and thus these residues can occupy the $i+1$ residue of a type-II' β-bend ($\varphi = 60°, \psi = -120°$), a position usually occupied by R(D)-amino acid residues. When larger rings are formed (seven- to nine-membered), it appears that these help stabilize a type-II' turn instead.

There are many other possibilities that can be imagined here, and a number of them have been tried (reviewed by Toniolo[30]).

2.2.5. *Backbone to backbone cyclizations*

In principle, covalent attachments between any two positions along a peptide backbone are possible, and obviously they would restrict the intervening sequence to a cyclic structure. Such conformational constraints have not been developed except for the trivial N-terminal to C-terminal cyclization. In small peptide hormones and neurotransmitters where this has been tried, for example, in enkephalins, oxytocin, etc., biologically inactive compounds have generally been observed. An exception has been for somatostatin where bridges between C-terminal and N-terminal residues in several analogues gave highly active compounds.[38]

C_α to C_α cyclizations have essentially been covered already by disulphide bridges, lactam bridges, thioether bridges, etc., but, of course, many other possibilities exist waiting only for the development of synthetic methodologies which will make their preparations feasible, and for an examination of their conformational consequences to determine if the three-dimensional structural properties are compatible with protein structure and peptide–protein interactions. Many possibilities exist for C_α–N (already partially discussed), C_α–C', C'–N, N–N, and C'–C' type backbone to backbone conformational restrictions. Some effort has been made to obtain such compounds, though at present most of the structures thus synthesized do not accommodate chiral side-chain functionalities, and thus it would be anticipated that such peptide derivatives and analogues, especially if modified at residues which were important for interaction of the peptide with its receptor(s), would be inactive. What has been done thus far has been reviewed by Toniolo,[30] but clearly we are only at the beginning of exploring the enormous number of possibilities for such modifications. Hence, there is great opportunity for innovation and discovery which hopefully will appeal to organic chemists.

2.2.6. Use of peptide mimetics

The replacement of some or all of the peptide moiety with other organic functionalities, and with maintenance of the biological potency and specificity of the peptides, and their general lack of toxicity, has intrigued organic chemists for years. Though numerous successes have been made in the area of enzyme inhibitors such as angiotensin converting enzyme (ACE) inhibitors, for peptide hormones and neurotransmitters, the field is still in its infancy. The discovery[50] that benzodiazepine derivatives could act as competitive antagonists of CCK A receptors (and perhaps other peptide receptors as well) is of great interest, but as with the putative mimicry of the enkephalins and endorphins for interaction at opioid receptors by morphine and the numerous related and not so related synthetic alkaloids, the discovery relied on screening of natural products from non-mammalian sources. De novo design of peptide mimics is in its infancy. Most of the successful applications have been in the area of amide bond replacements. An excellent review exists in this area[51] and will not be discussed here. As for peptide mimetics with secondary structure, a number of mimetics for β-turns, γ-turns, α-helical segments, and even β-sheets have been proposed and synthesized, and in several cases (see for example the work of Kemp and co-workers[52–54]) these mimetics have been incorporated into peptides and demonstrated to have the proposed conformations (for a review see Toniolo[30]). However, most of the mimetics to date have been synthesized without the appropriate side-chain functional groups (e.g. phenyl, indole, isopropyl, etc.) found in peptides. Thus, even if these are good mimics for the backbone conformation of a peptide, because the side-chain groups of the replaced amino acid residues are missing, the topography of the molecules is much different from that of the original peptide ligand. Thus, if the replaced structural element is important for interaction of the peptide with its receptor, the resulting 'peptide mimetic' will have little or no biological activity. So far, this has been the fate of most peptide mimetics of biologically active peptides. Some hope can be found in the recent report of Hirschmann and colleagues,[55] who used a simple sugar to design a somatostatin mimic, but clearly much work remains to be done. Some very careful thinking regarding the correspondence between peptide structures and other possible peptide 'mimics' (sugars, alkaloids, heterocycles, etc.) will be needed before the success suggested by the morphine/enkephalin and cholecystokinin/ benzodiazepine relationships is realized. However, the possible rewards are great and increasing studies in this area are likely.

3. Practical methodology

3.1. TIME AND COST CONSIDERATIONS FOR SYNTHETIC METHODOLOGY

The development of increasingly efficient methods of peptide synthesis is clearly of great concern to those interested in developing peptides as drugs. Considerable progress has been made. Though inflation has been very high for most chemicals, solvents, and equipment during the past 10–15 years, the cost in real terms for many amino acids and protected amino acids has increased less during that time. Improvements in the technologies for the preparation of starting protected amino acids, in coupling efficiencies, in suppression of racemization and other side reactions, and in purification techniques, have all decreased the time and material costs for peptide synthesis. At the present, it appears that solid-phase synthesis is the method of choice for most peptide syntheses for research purposes and even for most industrial syntheses of peptides greater than 10–20 residues. For smaller peptides prepared for industrial purposes, batchwise solution syntheses may still be the method of choice, and solution techniques will undoubtedly be needed for the synthesis of some of the more constrained or otherwise modified peptides and pseudopeptides. In some of the latter cases, mixed solution and solid-phase methods will be needed as fragment condensation on solid-phase synthesis becomes more widely used. Increasingly, the cost of solvents becomes an issue in terms both of raw material cost and of disposal problems. Solvents such as dichloromethane, benzene, etc. need to be used and disposed of with great care and sometimes at considerable expense.

3.2. ANALYTICAL CONSIDERATIONS

3.2.1. *Choice of purification methods*

In view of their special properties (generally soluble in water, not soluble in most organic solvents) purification of unprotected peptides generally is done in aqueous media or in mixed organic/aqueous solvents such as acetonitrile/0.1% aqueous trifluoroacetic acid. Often it is necessary to use buffers or salts to maintain pH and/or to prevent self-association and other chemical/physical changes that lead to precipitation. The most widely used purification methods for peptides include:

- gel permeation chromatography;
- ion exchange chromatography;
- partition chromatography;
- reverse-phase high-pressure liquid chromatography (HPLC);
- other forms of high- and medium-pressure liquid chromatography;
- counter-current distribution;
- recrystallization or reprecipitation.

The choice of methodology is very dependent on the properties of the peptide and on the anticipated or known side-products. For example, if racemization is expected in some steps of the synthesis, separation of diastereoisomers will be necessary. This is usually best accomplished by reverse-phase HPLC or by partition chromatography. Very careful gel permeation chromatography on a resin of the appropriate pore size (depending on the molecular weight of the peptide) is generally recommended especially to get rid of small molecules and ions which peptides tend to sequester. The use of salt gradients, solvent gradients, pH gradients, etc. is generally very valuable and takes advantage of the fact that peptides are usually polyions. The development of these and other purification methods for biopolymers especially peptides and proteins has been enormous in recent years, and provide synthetic peptide chemistry with a critical ally. The continuing developments in this area will greatly aid the development of peptides as drugs and research tools.

3.2.2. Criteria for purity

How pure is pure enough for a peptide? And what are adequate criteria for purity? The arguments are endless, and because peptides generally are not crystalline, the old organic criteria of a 'constant melting point' are not relevant. Traditional elemental analysis is not of much use as a criteria of purity because most peptides contain water and other solvents either sequestered or as salts with functional groups. 'Corrections' for these constituents, often present in unknown or non-stoichiometric quantities, must be made which are essentially arbitrary. Thus, in general, other criteria for purity are needed, and fortunately again developments of the last 10–20 years have greatly aided the determination of purity and structural integrity.

The simple answer to the first question, of course, is as pure as possible! This is very important for biologically active peptides, since as discussed earlier it is not unknown for a minor impurity (even less than 1%) to account for the observed biological activity. This can be especially critical in the development of antagonists where often weak partial antagonists can serve as an initial lead.

However, even a 1% impurity of potent full agonist could completely mask the activity of the principal product. Thus, care in selection and use of analytical methods to establish purity is essential.

For most new peptides, the following analytical methods would appear to be the minimum required:

- amino acid analysis;
- thin-layer chromatography in at least three different solvent systems;
- HPLC in a system different from that used for purification;
- fast atom bombardment mass spectrometry–generally much superior to elemental analysis in determining the presence of the desired structure;
- peptide mapping for analogues of larger peptides (>20 residues) and microsequencing of new structural moieties;
- comparison with authentic peptide if available.

Generally optical rotation is also provided as a physical constant.

Impurities of 1% or greater will generally show up in one or more of the above methods. Many other analytical methods are available and often provide very valuable criteria for purity. These include:

- paper or gel electrophoresis;
- melting points;
- ultracentrifugation;
- elemental analysis;
- optical rotation;
- paper chromatography;
- nuclear magnetic resonance spectroscopy;
- circular dichroism spectroscopy;
- fluorescence spectroscopy; and a variety of other specialized methods.

In many cases these methods should appear in the first group as essential.

3.3. USE OF STRUCTURE–ACTIVITY RELATIONSHIPS

As already emphasized, the development of peptides for medicinal or pharmaceutical purposes requires a close collaboration of chemists and biologists *both* utilizing up-to-date methods in drug design and development. It does little good for development of rational drug design if sophisticated chemical synthetic methods are utilized and the only biological follow-up is a simple *in vivo* screen that provides virtually no information except whether the compound is active or not under the conditions of the screen.

3.3.1. *Multiple bioassay systems*

Clearly the requirement on the biological side is for multiple assay systems. Unfortunately, these are often not used, at least as far as can be judged from patents and the primary literature. A particularly interesting example is the vast majority of studies done in the field of opioid peptides. Though thousands of analogues appear to have been made by numerous drug companies, in many cases only a simple *in vivo* assay (analgesia) or *in vitro* assay (guinea pig ileum) was used, and often only one or the other. This seems very strange to us, since it was well known almost from the time Leu-enkephalin was discovered that multiple opioid receptors exist. With a single assay as feedback, there was little hope that a rational approach to ligand design would be possible. Thus, it is not surprising that the vast majority of peptides made by drug companies working on opioid peptides show little in the way of innovative, rational approaches to peptide design. In fact, most of the innovations which have led to highly selective and potent ligands for the μ and δ receptors have come from academic institutions and research foundations.

As discussed previously, there is ample evidence that most if not all peptide hormones and neurotransmitters have multiple sites of biological activity. Thus, multiple bioassays are immediately needed if one wishes to develop rational approaches to the design of receptor selective ligands. In addition, questions of potency and efficacy are an increased concern. In some cases, compounds that bind very well have poor bioactivity due to low efficacy, while others that bind only poorly can be very potent *in vivo* due to their high efficacy. For example, there are cases (e.g. glucagon) where a 2% occupancy of the glucagon receptor can give a full *in vivo* response (for glycogenolysis). Thus, in addition to an *in vivo* assay, there generally is a need to have biochemical assays including binding assays and biochemical assays (e.g. second-messenger assays) to supplement and/or complement *in vitro* and *in vivo* bioassays. With these tools in hand, rapid development of receptor-selective analogues is often possible, which can then be used as a clinical lead for design and synthesis of a clinical candidate.

3.3.2. *Use of classical structural modifications*

While the preceding discussions have emphasized the use of conformational constraints to develop a rational approach to peptide ligand design, it is important to recognize, nonetheless, the need for classical structural modifications in the development of a rational approach. Obviously, if one can build in constraints, especially cyclic or pseudocyclic constraints, early in the process, development will be greatly aided. When confronted with a new peptide, however, it is usually necessary to establish the residues critical to biological action by more conventional methods. A few of the most important will be mentioned here.

Often the N- and/or C-terminal residue in a biologically active peptide is not critical for biological activity. This often can be established by removing one amino acid residue at a time from each end, and then if the evidence warrants it, from both ends simultaneously. Partial agonist and/or antagonist activities have often occurred in such studies and should be looked for from the very beginning.

D-Amino acids, if placed in the proper place in the sequence, can often have very salutary (or disastrous) consequences for bioactivity of peptide hormones and neurotransmitters, and hence can provide important insights in conformational and topographical requirements of a particular receptor and/or insights into important sites of biodegradation. For use in stabilizing secondary structures, it often is useful to examine the probabilities for different secondary structures in the peptide of interest using Chou–Fasman or other similar methods. In any case, a systematic replacement of L-amino acids by D-amino acids, one at a time, can provide important insights.

Glycine residues provide special challenges and opportunities in peptide design. The conformational space available to glycine far exceeds that of other α-amino acids and thus it is perhaps not surprising that it often serves the function of a critical 'hinge' for some secondary structural feature. Furthermore, in most cases, the diastereotopic α-hydrogens are distinctly different from the view of both the peptide *and* the receptor. Hence, substitution of an R group (e.g. CH_3- or $C_6H_5-CH_2-$) on the pro-R or pro-S face can have much different consequences.

As for the importance of various R groups to a particular peptide–receptor interaction, the systematic replacement of each with a $-CH_3$ group can provide some important insights, though one must always be aware of the possible conformational consequences which may be associated with such structural changes. This is especially true for aromatic residues and charged residues since perturbations caused by their removal (or addition) are often large.

3.3.3. *Developing 'leads'*

Once a suitable, hopefully conformationally constrained, lead compound has been obtained based on its potency, receptor selectivity, efficacy, antagonist activity or whatever, the most rapid development can best occur within the context of a conformational model. Both intuition and the use of modern methods for conformational analysis, computer-assisted molecular modelling and molecular mechanics and molecular dynamic calculations should be incorporated into the design process at this point in the development. It is very important not just to make as many compounds as seem 'easy' to make, but rather to develop a *model* for structure–activity relationships that *can be tested* by systematic rational design. Simultaneous structural changes which will stabilize or enhance conformational or topographical features believed to be important for biological activity, which will protect the molecule from biodegradation, and which will enhance biophysical studies as well, are needed. It may also be useful to re-evaluate the choice of biological assays and expand into areas of critical concern related to possible side reactions, etc. Close collaboration between biologists, chemists, biophysicists, and theoretical investigators is most important for rapid development of the lead compound. This is the juncture at which greatest diversity of talent and knowledge is needed in the development process.

4. Results of approaches

Despite its recent development, there has been considerable success in the development of peptide ligands with greater promise as potential drugs and pharmaceuticals, and many more peptides are on the market than may be generally appreciated. These include analogues of (occasionally the natural product itself) oxytocin, vasopressin, calcitonin, ACTH, LHRH, somatostatin, insulin, and many others. In previous sections, we have already mentioned several successful uses of various design principles which have led to unique peptide analogues. Here we provide a summary on the prospects of these approaches to provide peptides with desirable biological and pharmacological properties. We conclude with a brief discussion of four specific examples.

4.1. POTENCY

Naturally occurring peptide hormones and neurotransmitters tend to be quite potent substances, with EC_{50} values for biological activity generally in the nanomolar ± 100 range. Thus, the major task in ligand design is to maintain this level of potency. In general, this has been quite easy. In fact, for many peptide hormones and neurotransmitters, it has been possible to modify 50% or more of the residues (sometimes virtually every residue) and still remain within a factor of 10 of the potency of the native peptide. It would appear that there are many ways in which the host–guest problem can be solved for a particular peptide and its receptor. Though we still do not understand the basic principles which would allow us to design these ligands more quickly and with greater fidelity, past experience suggests that high potency is accessible.

4.2. RECEPTOR SELECTIVITY

As with potency, development of selectivity can often be achieved to a fairly high degree *provided one has the multiple bioassays needed to examine the various receptor systems.* With these in hand, it is possible to rapidly develop receptor selective analogues once one has a lead, that is a compound with 10-fold or greater selectivity for a particular receptors. Generally such leads come from a structural, conformational, or topographical modification that clearly establishes differential effects at the different receptors. For economical reasons, appropriate binding assays are the usual way to go, but in some cases bioassays that can be done with large n numbers per day can be as useful. It is critical to have rapid feedback between chemist and pharmacologist to ensure the most rapid development of a highly selective ligand.

4.3. DEVELOPMENT OF ANTAGONISTS

Antagonists of peptide hormones and neurotransmitters have great value not only for their use in understanding the biology of a particular ligand and its receptor(s), but also because there is a great need in clinical medicine and in diagnostics for peptides. However, the development of good antagonists for peptide hormones and neurotransmitters is difficult, and requires the most stringent attention to detail by both the chemists and the biologists involved. There are numerous reasons for these difficulties. On the biological side, competitive antagonists must bind to the receptor in question, must readily displace the natural endogenous ligand for the receptor, have absolutely no ability to transduce the message (partial agonists and antagonists are often fully active *in vivo*), and have this full activity in humans (often compounds which act as antagonists in one species are agonists or partial agonists in another). Thus, assays must be developed which not only establish binding, competitive binding and efficacy, but also examine second messengers that may be influenced differently for agonists or antagonists and where, as a result, other endogenous events can override the antagonist effects. On the chemical side, from the standpoint of design, the problem is equally difficult. This is largely due to the fact that competitive antagonists bind differently to the receptor than agonists, a conclusion which is not widely appreciated yet. The consequences are that all or most of what has been learned from structure–function analysis in the agonist series must be re-examined in the antagonist series. Since the antagonist can compete with the agonist for a particular binding site, it is quite clear they utilize different kinds of interaction. In the case of an agonist, the binding interaction must eventually lead to the transduced, biologically active state of the peptide–receptor complex. In the case of the antagonist, the binding interaction must be such as to preclude the peptide–receptor complex ever attaining the conformation required for transduction. In the end then, one must have a lead, that is an analogue which is an inhibitor, or at least a partial antagonist or agonist. Biochemical assays are generally the only way to obtain the latter and hence can be extremely useful for developing leads toward receptor antagonists.

4.4. METABOLIC STABILITY

Though many peptide hormones and neurotransmitters were designed by nature to be metabolically unstable, this is not necessarily an intrinsic property of peptides. By the use of conformational constraints, of appropriately placed D-amino acid residues, of non-eukaryotic amino acids, and of other methods, peptides with very high stability to the proteolytic enzymes found in the serum, kidney, brain, and liver can be obtained. There is still a tremendous amount of research that needs to be done in this area of 'biosynthesis', but it appears that in general, agonists and antagonist peptides which are not readily biodegraded can be designed. We have recently reviewed some aspects of this research.[56]

4.5. PROLONGED BIOACTIVITY

Prolonging the biological activity once it has been administered often is a goal in developing a peptide drug. Especially if antagonist action is needed, prolonged biological activity is usually highly desirable. This can be accomplished in several ways including:

- reducing biodegradation;
- reducing elimination through the kidney or other routes;
- increasing bioavailability;
- decreasing the off rate from the peptide-receptor complex;
- increasing the time in the 'receptor-compartment';
- a combination of these.

In the development of several peptide ligands, we have been quite successful in this regard. We have developed α-melanotropin agonist analogues which are active *in vitro* for hours to days, and *in vivo* for days to weeks, and with little apparent down regulation,[44,57-59] oxytocin antagonist analogues with half-lives for inhibitory activity of over 4 h,[60] and μ opioid receptor selective antagonists with half-lives for inhibitory activity of over 3 h.[61] The systematic way to do this is not clear, though conformational constraints can be helpful. Much more effort is needed in this area.

5. Discussion: use of conformationally constrained ligands

In this section, we provide brief outlines of examples of the approach outlined in this chapter. It is meant to be instructive, not exhaustive. Interested readers should examine the original literature.

5.1. DEVELOPMENT OF HIGHLY RECEPTOR-SELECTIVE POTENT CONSTRAINED PEPTIDE AGONISTS AND ANTAGONISTS FOR δ AND μ RECEPTORS

Since the discovery of the endogenous enkephalins, an enormous amount of research has been done in the opioid peptides and non-peptides area.[62] It is now accepted that the enkephalins are highly flexible molecules that can assume an ensemble of energetically preferred conformations.[63] It has been shown that the opioid receptors are heterogeneous and consist of at least three subtypes, μ, δ, κ, and possibly ε, etc. It is thought that the complexity of the pharmacological responses (analgesia, respiratory depression, physiological dependence and tolerance, gut motility, etc.) to the opioids is due in part to their non-selective binding to opioid receptor subtypes. Before the physiological role of opioids can be understood, highly receptor-selective ligands (agonist and antagonists) for the opioid receptor subtypes must be developed.

The use of modern pharmacological methods (binding assays, *in vitro* bioassays, etc.) in conjunction with spectroscopic methods (2-D NMR, laser spectroscopy, etc.) of conformational analysis has resulted in the design and synthesis of conformationally constrained opioid peptides that are highly receptor-selective. This section will encompass recent significant advances made in this area.

As discussed previously, the most selective and potent μ opioid receptor antagonists reported have been the somatostatin-derived octapeptides.[33,40,41] Somatostatin, a cyclic tetradecapeptide, has been found to weakly interact with opioid receptors[64,65] and the somatostatin fragment analogue D-Phe–Cys–Phe–D-Trp–Lys–Thr–Cys–Thr(ol) (SMS 201–995), a superpotent growth hormone release inhibitor, has antagonistic properties against enkephalins in bioassays.[66] The prospect of maximizing the opioid receptor interaction while minimizing that of somatostatin was investigated. Some of the more notable analogues which were synthesized are shown in Table 1. Substitution of the Cys^7 in an octapeptide analogue of somatostatin with penicillamine (β,β-dimethylcysteine) gave the analogue 1 which exhibited the first significant somatostatin/opioid (μ) receptor selectivity (0.18), although the μ potency was low and cross-reactivity with the δ receptor was observed. A 30-fold increase in somatostatin/μ selectivity was found when Phe^3 of analogue 1 was replaced with Tyr to yield 2. Also a three-fold increase in μ potency and a twofold increase in μ vs. δ selectivity was observed (Table 1) with analogue 2. The first highly potent and selective μ opioid receptor analogue CTP (3) resulted when the C-terminal carboxylate was replaced with the carboxamide. This substitution led to the expected increase in μ vs. δ selectivity (21-fold) and furthermore resulted in a 72-fold increase in μ vs. somatostatin selectivity and an 83-fold increase in μ potency.[40]

Substitution of the basic residue Lys^5 of CTP with Arg or Orn resulted in analogues CTAP (4)[41] and CTOP (5).[41] Relative to CTP, CTAP showed a 10-fold increase in μ vs. somatostatin and an eightfold increase in μ vs. δ selectivities while retaining high μ potency. CTOP exhibited an even more pronounced μ vs. somatostatin selectivity (28-fold) and a fourfold increase in μ vs. δ selectivity while also maintaining a high μ potency relative to CTP.

Conformational analysis of CTP utilizing high field ^1H- and ^{13}C-NMR in aqueous[67] and DMSO[68] solutions established that the core tetrapeptide sequence of CTP (–Tyr–D-Trp–Lys–Thr–) exists in a β-II' conformation. Also, the helicity of the disulphide bond was shown to be negative.[69] The tetrapeptide conformation template was used in conjunction with constrained amino acids to investigate the conformational requirements of the μ opioid receptor for the first position in CTP. The constrained amino acids D-phenylglycine (D-Pgl), D-tetrahydroisoquinoline carboxylate (D-Tic) and D-N-methylphenylalanine (D-N-MePhe) were substituted into the first position of CTP to give the analogues 6,[40] 7,[33] and 8,[33] respectively. The [D-Tic^1]CTP analogue 7 showed striking increases in somatostatin/μ (50-fold) and δ/μ (26-fold) selectivities, and moreover displayed improved μ potency (3.2-fold) relative to CTP.

Furthermore, [D-Tic^1]CTP has been shown to be superior to CTP as a μ selective antagonist peripherally as well as centrally.[33] The CTP analogues 6 and 8 no longer possessed significant μ selectivity and their potencies were drastically decreased.

It was determined by use of NMR that the solution conformations of 6, 7, and

Table 1 Binding potencies and receptor selectivities of somatostatin and somatostatin-octapeptide analogues in opioid and somatostatin binding assays with rat brain homogenates.

Compound	$[^3H]Nal$	$IC_{50}(nM)$ $[^3H]DADLE$	$[^{125}I]-$ $CGP23996$	IC_{50} ratio $\dfrac{[^3H]DADLE}{[^3H]Nal}$	IC_{50} ratio $\dfrac{[^{125}I]CGP}{[^3H]Nal}$
Somatostatin^a	27 400	16 400	6	0.60	0.0002
D-Phe–Cys–Phe–D-Trp–Lys–Thr–Pen–Thr–OH^a (1)	931	5400	170	5.8	0.18
D-Phe–Cys–Tyr–D-Trp–Lys–Thr–Pen–Thr–OH^a (2)	290	3800	1600	13	5.5
D-Phe–Cys–Tyr–D-Trp–Lys–Thr–Pen–Thr–NH$_2$^{a} (CTP) (3)	3.5	950	1462	271	395

	$[^3H]CTOP$	$IC_{50}(nM)$ $[^3H]DPDPE$	$[^{125}I]-$ $CGP23996$	IC_{50} ratio $\dfrac{[^3H]DPDPE}{[^3H]CTOP}$	IC_{50} ratio $\dfrac{[^{125}I]CGP}{[^3H]CTOP}$
CTP (3)	3.7	1153	1462	312	395
[Arg^5]CTP^b(CTAP) (4)	2.10	5314	8452	2530	4025
[Orn^5]CTP^b (CTOP) (5)	4.30	5598	47 704	1301	11 094
[D-Pgl^1]CTP^a (6)	350^d ± 97	2800 ± 570	—	8	
[D-Tic^1]CTP^c (7)	1.15 ± 0.03	9320 ± 546	23 000	8080	20 000
[D-N-MePhe^1]CTP^c (8)	284 ± 36	10% at 10^{-5}M	—	> 35	

Data from ^{a} reference 40, ^{b} reference 41, ^{c} reference 33. ^{d} [^{3}H]Naloxone used.

8 possess the same β-II' conformation, centered about the core tetrapeptide sequence, as does CTP.[32] Furthermore, the analogues **6, 7,** and **8** exhibit a negative disulphide helicity as with CTP.[21] The marked difference between [D-Tic[1]]CTP and the analogues **6** and **8** was the spatial relationship of the aromatic groups in the first and third positions. Analogue **7** exists in more extended topology in which the aromatic group of D-Tic is at a relatively maximum distance from the Tyr[3] residue. The D-Pgl analogue adopts a folded conformation which positions the aromatic groups closest together while the N-MePhe analogue assumes a conformation which results in an intermediate interaromatic distance. It was concluded that the rotamer populations of the first residue determine the interaromatic distances in these analogues. If the first residue assumes mainly a *gauche*$(-)$ (*g*-) conformation, the interaromatic distance will be maximized while the $g+$ or *trans* (*t*) conformations will minimize the parameter. In fact, based on the coupling constants ($J_{\alpha,\beta}$), it was concluded that the D-Tic residue of **7** exists exclusively in the $g-$ conformation while the N-MePhe residue exists mainly in the *t* (but also in $g-$ and $g+$) conformations.[32] Thus it appears that the interaromatic distance and consequently the bioactivity of the CTP analogues **6, 7,** and **8** can be manipulated by use of conformationally biased amino acids at the N-terminal position of the somatostatin-derived octapeptides.

The most potent and selective peptide μ agonists have been those derived from the linear peptides, dermorphins, and the linear tetra- or pentapeptides, the β-casomorphins. These two classes of opioid peptides have the Phe residue in the third position as opposed to the four position with the enkephalins. It had been observed that if the 15-membered ring size of the analogue Tyr–D-Lys–Phe–Glu–NH_2 was decreased to a 13-membered ring present in the cyclic lactam **9** (Table 2), μ receptor affinity (10 nM) was decreased; however δ/μ selectivity was improved.[69] Transposition of the Orn and Asp residues of **9** to give analogue **10** resulted in a very similar μ and δ binding profile[70] (Table 3); however, the peripheral bioassays showed a decrease (14-fold) of potency in the guinea pig ileum (GPI) and an apparent loss of δ/μ selectivity (sevenfold) as indicated by the MVD/GPI IC_{50} ratio[71] (Table 2). The ring size of **10** (13-membered) was

Table 2 Potencies of cyclic enkephalin and cyclic dermorphin analogues in the guinea pig ileum (GPI) and mouse vas deferens (MVD) assays.

Compound	IC_{50} (nM)		IC_{50} ratio
	GPI	*MVD*	*MVD/GPI*
H-Tyr–D-Orn–Phe–Asp-NH_2^{a} (**9**)	36.2 ± 3.7	3880 ± 840	107
H-Tyr–D-Asp–Phe–Orn-NH_2^{a} (**10**)	522 ± 102	8570 ± 3500	16.4
H-Tyr–D-Asp–Phe–A_2bu-NH_2^{a} (**11**)	279 ± 16	2030 ± 690	7.3
H-Tyr–c[-N^ε-D-Lys–Gly–Phe–Leu-]^b (**12**)	4.8 ± 1.8	141 ± 28	29.4
H-Tyr–c[-N^ε-D-Lys–Gly–Phe(p-NO_2)–Leu-]^c (**13**)	0.50 ± 0.14	13.4 ± 6.3	26.6

Data from ^{a} reference 71, ^{b} reference 72, and ^{c} reference 73.

Table 3 Binding potencies and receptor selectivities of cyclic dermorphin analogues in binding assays with rat brain homogenates.

Compound	K_i (nM)		K_i ratio
	$[^3H]DAGO$	$[^3H]DSLET$	$[^3H]DSLET/[^3H]DAGO$
H-Tyr–D-Orn–Phe–Asp-NH_2^a (**9**)	10.4 ± 3.7	2220 ± 65	213
H-Tyr–D-Asp–Phe–Orn-NH_2^b (**10**)	9.55 ± 2.5	1320 ± 150	138
H-Tyr–D-Asp–Phe–A_2bu-NH_2^c (**11**)	24.8 ± 1.1	4170 ± 430	168
DAGO	3.90 ± 0.80		

Data from [a] reference 69, [b] reference 70, and [c] reference 71.

decreased to a 12-membered ring by substitution of Orn by α,γ-diaminobutyric acid (A_2bu) to give analogue **11**[24] (Table 3). The μ binding affinity of **11** decreased 2.5-fold relative to **10** but the δ/μ selectivity improved slightly.[71] As in the case of **10** in the peripheral bioassays the analogue **11** showed large decreases in μ potency and selectivity relative to **9** (Table 2). The analogues **9, 10, 11** showed K_e values (1–2 nM) for naloxone as antagonist in the GPI which precluded interaction with the κ opioid receptor. Various other modified analogues of **9** were synthesized, but none displayed increased bioactivities.[24]

Of note are several side-chains to C-terminal carboxylate cyclized enkephalins, e.g., the analogue **12**[72] (Table 2) which have shown high μ potency and moderate μ vs. δ selectivity in peripheral bioassays. The substitution of Phe in **12** with para-nitrophenylalanine (p-NO_2Phe) to give **13**[26] led to the expected increase in μ potency[73] (10-fold) but the μ vs. δ selectivity remained unchanged relative to **12** in peripheral bioassays (Table 2).

Conformational constraint has also been utilized to develop highly selective and potent agonists for the δ opioid receptor. The cyclic enkephalin amides (Enk-NH_2) **14** and **15** (Table 4) were the first constrained Met-enkephalin disulphide

Table 4 Potencies of enkephalinamide and enkephalin analogues in the guinea pig ileum (GPI) and mouse vas deferens (MVD) assays.

Compound	IC_{50} (nM)		IC_{50} ratio
	GPI	MVD	GPI/MVD
[D-Cys²,Cys⁵]Enk-NH_2^a (**14**)	1.51 ± 0.03	ND	ND
[D-Cys²,D-Cys⁵]Enk-NH_2^a (**15**)	0.78 ± 0.01	ND	ND
[D-Pen²,Cys⁵]Enk^b (**16**)	213 ± 63	0.32 ± 0.03	666
[D-Pen²,D-Cys⁵]Enk^b (**17**)	1350 ± 340	6.27 ± 1.2	215
[D-Cys²,D-Pen⁵]Enk^c (**18**)	67 ± 1.3	0.13 ± 0.06	515
[D-Pen²,Pen⁵]Enk^c (**19**)	2720 ± 50	2.50 ± 0.03	1088
[D-Pen²,D-Pen⁵]Enk^c (**20**)	6930 ± 124	2.19 ± 0.30	3164

Data from [a] reference 75, [b] reference 77, and [c] reference 37. ND, not determined.

Table 5 Binding potencies and receptor selectivities for cyclic enkephalin analogues in rat brain homogenate.

Compound	IC_{50} (nM)		IC_{50} ratio
	$[^3H]Nal$	$[^3H]DADLE$	$[^3H]Nal/[^3H]DADLE$
[D-Pen^2,Cys^5]Enk^a (16)	178 ± 16	11.7 ± 1.2	15.2
[D-Pen^2,D-Cys^5]Enk^a (17)	157 ± 74	26.0 ± 0.5	6.0
[D-Cys^2,D-Pen^5]Enk^a (18)	22.0 ± 2.8	3.5 ± 0.8	6.3
[D-Pen^2,Pen^5]Enk^{a,b} (19)	3710 ± 740	10.0 ± 2.2	371
[DPen^2,D-Pen^5]Enk^{a,b} (20)	2840 ± 670	16.2 ± 0.9	175
		2.3 ± 0.3^d	1235^d

	IC_{50} (nM)		IC_{50} ratio
	$[^3H]DAGO^c$	$[^3H]DPDPE^c$	$[^3H]DAGO/[^3H]DPDPE$
[D-Pen^2,D-Pen^5]Enk (20)	2230 ± 131	18.4 ± 3.9	121
[D-Cys^2,D-Pen^5]Enk (18)	22.2 ± 2.3	7.6 ± 3.0	2.9
[(3S)Me-D-Cys^2,D-Pen^5]Enk (21)	28.7 ± 1.4	13.7 ± 6.1	2.1
[(3R)Me-D-Cys^2,D-Pen^5]Enk (22)	88.8 ± 22.2	13.3 ± 3.5	6.7

Data from ^{a} reference 37, ^{b} reference 80, and ^{c} reference 81. ^{d} [^{3}H]DPDPE was used.

analogues reported.[74,75] Both of the cyclic analogues 14 and 15 were found to be slightly μ receptor selective.[75] However, substitution of the second residue of 14 or 15 with D-Pen afforded enkephalin amide analogues which were δ selective.[76] Transformation of the enkephalin amides to carboxylate analogues gave 16[77] and 17,[78] and in the GPI and MVD bioassays the expected increases in δ selectivity (666 and 215) and binding potency (0.3 and 6 nM) were found. However, the D-Pen^2 analogues 16 and 17 exhibited relatively low μ vs. δ selectivity in binding studies[37] with rat brain homogenate[5] (Table 5) in light of the peripheral bioassay results. Also the D-Cys^2,D-Pen^5 analogue 18 was prepared, and it showed similar pharmacological profiles in the GPI/MVD bioassays and the binding studies.[37]

After analysis of conformational models of D-Pen^2 and D-Pen^5 enkephalin analogues, it appeared that the β-hydrogens of the half-cystine residue were not sterically hindered.[79] Thus, it was reasoned that these analogues might be able to accommodate a second Pen residue in place of the Cys. In fact, the bis-penicillamine analogues [D-Pen^2, L-Pen^5]enkephalin (DPLPE, 19)[37] and [D-Pen^2,D-Pen^5]- enkephalin (DPDPE, 20)[37] were synthesized in good yields. Both DPLPE and DPDPE displayed very high potencies (2.5 and 2.2 nM) in the MVD bioassay and possessed extremely high GPI/MVD IC_{50} ratios[37] (1088 and 3164 respectively) (Table 4). Furthermore, in binding studies[37] DPLPE showed good δ affinity (1–10 nM)[37,80] and very high μ/δ selectivity (371) as did DPDPE, with δ receptor affinity of 16 nM and μ/δ selectivity of 175 (Table 5). Other workers have confirmed the high δ selectivity and good δ affinity of DPLPE and

DPDPE[81] and have determined that neither interact with the κ opioid receptor.[81]

The very high δ opioid receptor selectivity and good δ potency of DPDPE have prompted investigations into the steric and conformational factors involved in this high selectivity. Previous work had indicated that the geminal dimethyl groups of the Pen[2] residue imparted high δ selectivity to the D-Pen enkephalin analogues.[77] Recently the steric interactions of the gem dimethyl groups of DPDPE have been investigated in depth,[82] and of the stereochemical and topographical requirements of the Phe[4] residue in DPDPE for interaction with the δ receptor selectivity.[83,84] Diastereoisomeric D-cysteine derivatives, (2S,3S)methylcysteine [(3S)Me–D-Cys] and (2S,3R)methylcysteine [(3R)Me–D-Cys] were prepared and were used to synthesize the DPDPE analogues 21 and 22 (Table 5). The D-Cys[2] analogue 18 was also included in this study. By use of ^{1}H-NMR it was shown that the D-Cys[2] and (3S)Me–D-Cys[2] analogues (18 and 21) possessed conformations similar to DPDPE in aqueous solutions, while the (3R)Me-D-Cys[2] analogue 22 appeared to adopt a conformation which differed from DPDPE and the D-Cys[2] analogue. It was reasoned that since DPDPE and analogues 18 and 21 were conformationally similar, their differing pharmacological profiles could be explained on the basis of the steric interactions of the 3-methyl groups with the δ and μ receptors. The μ affinity of the (3S)Me–D-Cys analogue was slightly decreased to that of the D-Cys analogue but much greater than that of DPDPE. Since the presence of the (3S)Me group in 21 permitted a high μ affinity, it was reasoned that the pro-R methyl group of the D-Pen[2] residue in DPDPE was responsible for the extremely low μ affinity of DPDPE. The δ affinity of the (3S)Me–D-Cys analogue was observed to be intermediate between DPDPE and the D-Cys[2] analogue. This evidence led to the proposal that both the pro-R and pro-S methyl groups of the D-Pen[2] residue of DPDPE contributes to a minor adverse interaction with the δ receptor.

Recently NMR methods, in conjunction with energy minimization calculations, were utilized to propose several low-energy conformations for DPDPE.[85,86] The conformation favoured by the authors adopts a bent boat-like shape with the Tyr[1] and Phe[4] aromatic rings in close topographical proximity and with both functions positioned above the 14-membered ring. There appear to be no strong intramolecular hydrogen bonds stabilizing the ring system, which is characterized by a type IV β-like turn. Stabilization of the conformation appears to result from the interactions of the aromatic groups of Tyr[1] and Phe[4] with the β-methyl groups and sulphur of the D-Pen[2] residue. It has been suggested that the latter-mentioned lipophilic topographical arrangement may result in the high δ selectivity of DPDPE. The overall conformation of DPDPE imparts amphiphilic properties of DPDPE due to a lipophilic side and a hydrophilic side.[84,86]

The first reported δ selective opioid peptides were the linear hexapeptide analogues of Leu-enkephalin, Tyr–D-Ser–Gly–Phe–Leu–Thr (DSLET) and Tyr–D-Thr–Gly–Phe–Leu–Thr (DTLET).[87] Although the δ binding affinities of DSLET and DTLET are high, cross-reactivity with the μ opioid receptor is observed. Conformational analysis of DTLET and DPLPE by use of NMR[88]

Table 6 Inhibitory potencies of enkephalin analogues on the specific binding of 2 nM [^{3}H]DSBULET at δ-site and 1 nM [^{3}H]DAGO at μ-site in rat brain tissue at 35°C.

| Compound | K_i (nM) | | K_i ratio |
	[^{3}H]DSBULET	[^{3}H]DAGO	[^{3}H]DAGO/[^{3}H]DSBULET
Tyr–D-Ser–Gly–Phe–Leu–Thr^a (DSLET)	3.80 ± 0.63	31.0 ± 5.0	8.2
Tyr–D-Thr–Gly–Phe–Leu–Thr^a (DTLET)	1.61 ± 0.22	25.3 ± 2.5	16
Tyr–D-Ser(O^tBu)–Gly–Phe–Leu–Thr^b (DSBULET) (23)	2.81 ± 0.64	374 ± 35	130
Tyr–D-Ser(O^tBu)–Gly–Phe–Leu–Thr(O^tBu)^b (BUBU) (24)	1.69 ± 0.45	480 ± 44	280
Tyr–D-Thr(O^tBu)–Gly–Phe–Leu–Thr^c (DTBULET) (25)	1150 ± 150	4500 ± 920	3.9
Tyr–D-Thr–Gly–Phe–Leu–Thr–(O^tBu)^c (DTLETBU) (26)	2.57 ± 2.70^d	66.3 ± 5.6	25.8
DPDPE	8.85 ± 1.69	993 ± 151	110
DPLPE	7.08 ± 1.17	873 ± 210	120
DAGO	629 ± 13	3.90 ± 0.80	0.0062

Data from ^{a} reference 87, ^{b} reference 90, ^{c} reference 91, ^{d} [^{3}H]DTLET used (1 nM, 35°C, 40 min).

Table 7 Potencies of E-cyclopropylphenylalanine (cpPhe) enkephalin analogues in the mouse vas deferens (MVD) and guinea pig ileum (GPI) assays.[a]

Compound	$IC_{50}\,(nM)$	
	MVD	GPI
[D-Ala^2,E-(2R,3S)–cpPhe^4,Leu^5]Enk (**27**)	2000	3700
[D-Ala^2,E-(2S,3R)–cpPhe^4,Leu^5]Enk (**28**)	5800	14 000
DADLE	0.51	46.1
DPDPE	3.78	6930[b]

Data from [a] reference 93 and [b] reference 37.

and theoretical calculations[89] suggested that introduction of steric bulk at the 2 and/or 6 position of DSLET or DTLET could decrease their μ affinity. Thus the hydroxyl functions of DSLET were etherized with the O-tertbutyl (O^tBu) group to give the D-Ser^2(O^tBu) analogue (DSBULET)[90] **23** and the D-Ser^2(O^tBu),– Thr^6(O^tBu) analogue (BUBU)[38] **24** (Table 6). The analogues DSBULET exhibits δ vs. μ selectivity and affinity which is comparable to that of DPDPE. The specific binding of [^{3}H]DSBULET was reported to be high (61% at $K_D = 2.2$ nM) in the rat brain membranes. Also, no significant interaction with the guinea pig cerebellum κ receptors was observed for DSBULET. The bis-protected analogue BUBU displayed slightly higher δ affinity and about a twofold increase in δ vs. μ selectivity relative to DSBULET. High δ affinity but only moderate δ vs. μ selectivity resulted from the Thr^6 protection of DTLET (DTLETBU,[39] **26**).[91] Of note was the D-Thr^2(O^tBu) analogue (DTBULET, **25**) which showed a drastic loss of both δ affinity and selectivity.[91]

Substitution of a highly constrained amino acid, E-cyclopropyl-phenylalanine (E-cpPhe), led to the diastereomeric enkephalin (Enk) analogues [D-Ala^2,E-(2R,3S) cpPhe^4,Leu^5]EA[92] (**27**) and [D-Ala^2, E-(2S,3R)cpPhe^4,Leu^5]Enk (**28**). Both of the constrained analogues **27** and **28** were found to be completely inactive in the GPI and MVD bioassays[92] (Table 7). However, binding studies with rat brain showed the E-(2R,3S)cpPhe^4 analogue **27** to possess high δ affinity against [^{3}H]DADLE or [^{3}H]DPDPE (Table 8).[93] Furthermore, analogue **27** exhibited δ vs. μ selectivity comparable to that of DPDPE. Of note was that the

Table 8 Receptor binding affinities of E-cyclopropylphenylalanine (cpPhe) enkephalin analogues in rat brain homogenates.[a]

Compound	$K_i\,(nM)$		
	[^{3}H]DADLE	[^{3}H]DAGO	[^{3}H]DPDPE
[D-Ala^2,E-(2R,3S)–cpPhe^4,Leu^5]Enk (**27**)	13.0	3290	16.8
[D-Ala^2,E-(2S,3R)–cpPhe^4,Leu^5]Enk (**28**)	2560	1960	ND
DADLE	1.37	12.9	ND
DPDPE	10.7	ND	12.6

[a] Data from reference 93. ND, not determined.

diastereomeric E-$(2S,3R)$cpPhe^4 analogue (28) showed no δ affinity or selectivity. These results have led the investigators to propose that the δ receptors in the rat brain differ from those in the MVD and that the E-$(2R,3S)$cpPhe^4 analogue 27 is able to distinguish between the two, while the E-$(2S,3R)$-cpPhe^4 analogue 28 or DPDPE cannot.^{94} Finally, the E-$(2R,3S)$cpPhe^4 analogue showed no antagonistic activity in the MVD or GPI bioassays.^{93} Recently, it has been reported that the δ-selective analogue 27 modulates μ receptor-mediated thermal analgesia by morphine.^{94}

In comparison to the μ and δ opioid receptor areas, structure–function or conformational studies for κ receptor ligands have been limited. Most of the structure–function studies of the κ-selective ligands have been based on the putative endogenous ligands Dyn A, Dyn B, and α-neoendorphin:

H-Tyr–Gly–Gly–Phe–Leu–Arg–Arg–Ile–Arg–Pro–Lys–Leu–Lys–Trp–Asp–
Asn–Gln-OH
Dyn A

H-Tyr–Gly–Gly–Phe–Leu–Arg–Arg–Gln–Phe–Lys–Val–Val–Thr–OH
Dyn B

H-Tyr–Gly–Gly–Phe–Leu–Arg–Lys–Tyr–Pro–Lys-OH
α-Neoendorphin

Recently the structure–function relationships for the Dyn peptides have been reviewed.^{95} Some of the more important points will be discussed. Sequential removal of the C-terminal amino acids of Dyn A-(1–13) established that the basic residues Arg-7, Lys-11, and Lys-13 were required for high κ selectivity and/or potency. Also the deletion of residues 14–17 does not affect the bioactivity of Dyn A. Substitution of lipophilic residues and certain D-amino acids at position 8 of Dyn A increases κ receptor selectivity. These observations may suggest a reverse turn at this position. Similarly the substitution with D-Pro in position 10 of Dyn A-(1–11) or (1–13) effects greater κ selectivity and is compatible with high κ receptor potency. Again this may be suggestive of a reverse turn and/or the need for an N-substituted amino acid in this position for high κ selectivity and/or potency.^{95}

The first conformationally constrained Dyn A analogue reported was the cyclic disulphide tridecapeptide 29^{96} (Table 9). In the GPI and MVD bioassays, analogue 29 exhibited high potency in the GPI and possessed a high MVD/GPI IC_{50} ratio. However, in binding studies with the rat brain analogue 29 displayed a high μ affinity (Table 10).^{96} Recently, various cyclic lactams of Dyn A were prepared (analogues 30–33, Tables 9 and 10).^{97} The analogue 30 showed high μ potency in the GPI, but displayed a naloxone K_e value (1.5 nM) which precluded interaction with the κ receptor. Analogues 31–33 exhibited low potencies in both the GPI and MVD bioassays and also showed non-κ naloxone K_e values (Table 9). In the binding studies analogues 31 and 33 possessed high μ affinities and moderate δ/μ selectivities.^{97} Binding affinities against a κ ligand were not reported.

Table 9 Potencies of cyclic dynorphin A (Dyn A) analogues in the guinea pig ileum (GPI) and mouse vas deferens (MVD) assays.[a]

Compound	GPI $IC_{50}(nM)$	$K_e[Nal(nM)]$[c]	MVD $IC_{50}(nM)$	MVD/GPI IC_{50} ratio
[D-Cys^2,Cys^5]Dyn A$_{1-13}$[b] (**29**)	0.032 ± 0.008	ND	7.65 ± 1.34	241
H-Tyr–D-Orn–Gly–Phe–Asp–Arg–Arg–Ile-NH$_2$ (**30**)	2.93 ± 0.57	1.49 ± 0.17	8.84 ± 1.85	3.02
H-Tyr–Gly–Gly–Phe–Orn–Arg–Arg–Asp–Arg–Pro–Lys–Leu–Lys-NH$_2$ (**31**)	1970 ± 490	1.84 ± 0.26	2590 ± 360	1.31
H-Tyr–Gly–Gly–Phe–Orn–Arg–Arg–Ile–Arg–Asp–Lys–Leu–Lys-NH$_2$ (**32**)	667 ± 63	4.13 ± 0.56	35100 ± 5400	52.6
H-Tyr–Gly–Gly–Phe–Orn–Arg–Arg–Ile–Arg–Pro–Lys–Leu–Asp-NH$_2$ (**33**)	687 ± 73	1.72 ± 0.15	18700 ± 3300	27.2
Dyn A$_{1-8}$	131 ± 19	13.4 ± 2.6	13.4 ± 3.4	0.102
Dyn A$_{1-13}$	1.46 ± 0.41	29.2 ± 9.2	12.7 ± 5.6	8.70
[Leu^5]enkephalin	246 ± 39	1.53 ± 0.43	11.4 ± 1.1	0.0463

Data from [a] reference 97 and [b] reference 96. [c] K_e-values determined with naloxone as antagonist. ND, not determined.

Table 10 Binding potencies and receptor selectivities for cyclic dynorphin A (Dyn A) analogues.[a]

Compound	$K_i(nM)$ [^{3}H]DAGO	[^{3}H]DSLET	K_i ratio [^{3}H]DSLET
[D-Cys^2,Cys^5]Dyn A$_{1-13}$[b] (**29**)	0.94 ± 0.29[c]	3.35[d]	3.57
H-Tyr–D-Orn–Gly–Phe–Asp–Arg–Arg–Ile-NH$_2$ (**30**)	0.0555 ± 0.0036	0.893 ± 0.196	16.1
H-Tyr–Gly–Gly–Phe–Orn–Arg–Arg–Asp–Arg–Pro–Leu–Lys-NH$_2$ (**31**)	6.84 ± 1.16	253 ± 51	37.0
H-Tyr–Gly–Gly–Phe–Orn–Arg–Arg–Ile–Arg–Asp–Lys–Leu–Lys-NH$_2$ (**32**)	38.6 ± 1.3	323 ± 34	8.37
H-Tyr–Gly–Gly–Phe–Orn–Arg–Arg–Ile–Arg–Pro–Lys–Leu–Asp-NH$_2$ (**33**)	3.39 ± 2.03	163 ± 11	48.1
Dyn A$_{1-8}$	1.63 ± 0.25	4.51 ± 0.61	2.77
Dyn A$_{1-13}$	3.95 ± 0.36	3.74 ± 0.40	0.947
[Leu^5]enkephalin	9.43 ± 2.07	2.53 ± 0.35	0.268

Data from [a] reference 97 and [b] reference 96. [c] [^{3}H]Naloxone used. [d] [^{3}H]DADLE used.

5.2. DEVELOPMENT OF AGONISTS AND ANTAGONISTS FOR LHRH RECEPTORS

Luteinizing hormone–releasing hormone (LHRH) is a decapeptide that is found in the anterior pituitary of various mammalian species. The sequence is as follows:

$$\text{pyro-Glu–His–Trp–Ser–Tyr–Gly–Leu–Arg–Pro–Gly–NH}_2$$

Its role as a neurohormone was found to be the regulation of the release of both luteinizing and follicle stimulating hormones from the hypothalamus.

Interestingly, both the design and synthesis of agonists and antagonists have at one time or another proved to be desirable. Potent and long-term agonists were at one time desired because of their desensitizing effects from chronic long-term administration. Their clinical importance stemmed from cases where control of sex steroid secretion was desired. For the last two decades, the focus of research has also been directed towards the design of potent LHRH antagonists as contraceptive agents. Antagonizing LHRH will in turn prevent release of luteinizing hormone (LH) and follicle stimulating hormone (FSH), hence acting as an anti-ovulatory agent.

Initial work utilizing structure–activity relationships have pointed to the importance of the substituent in position 6.[98] One of the initial structural modifications that led to increased potency was the substitution of a D-Ala for the Gly at the 6 position.[99] This led to an analogue 3.7 times more active than native LHRH. The substitution was good as well for increasing the potency of the antagonist des-His^2–[D-Ala^6] LHRH. Addition of the N-methyl leucine yielded a further increase in activity.[99] The fact that an L-Ala^6 analogue had lower potency prompted Monahan and co-workers to eliminate the presence of a α-helix (D-amino acids would disrupt a helix) but rather, to suggest a β-turn in the sequence Ser^4–Tyr^5–Gly^6–Leu^7 to be possibly necessary for receptor binding. This turn would strategically place the Gly or D-Ala in the $i+2$ position which, according to work by Venkatachalam, would be optimal for a β-bend centred around these four residues. This rationale led Donzel and co-workers[100] to replace Ser^4 and Leu^7 with Glu and Orn respectively. A putative β-turn was then generated through amide linkage cyclization. This yielded an analogue that had no activity but, nonetheless, addressed the first attempts to define the conformational requirements for LHRH hormone receptors by the use of constrained analogues. Other modifications based on this same scheme, utilizing components as charge transfer anmplexes, led to potencies of 0.1 % or less.[101] N to C terminal cyclization yielded agonists only slightly more potent.[102]

A different approach to constraining the putative β-turn led to more successful analogues. Freidinger and co-workers,[103] using computer superposition of the postulated β-turn segment localized about residues 5–8, proposed that a single five-membered γ-lactam constraint may mimic this turn. Their simulations revealed that the pro-S hydrogen of Gly^6 was in close proximity to the N$^{\alpha}$-hydrogen of Leu^7. Replacement of these hydrogens with a methylene bridge would adequately constrain this turn segment in the molecule. This led to the synthesis of a unique γ-lactam. Using an *in vitro* assay of a pituitary cell culture system, they found that their analogue was 8–9 times more potent than LHRH.

5.2.1. *Antagonists*

The realization that LHRH antagonists could be of benefit as potent contraceptive agents has prompted the design over the past 16 years of over 3000 analogues of variable potency and duration of action. Preliminary data from clinical investigators has been promising.[104]

Generally speaking, compounds of the structure N-acetyl L- or D- $(AA)^1$, D-4-$(HAL)Phe^2$, D-Trp^3, D-Trp^6 (or D-Nal^6)–LHRH where $(AA)^1$ can be residues such as Ala, Phe, Trp, etc., and HAL are either Cl or F halogens, are good antagonists and can prevent ovulation in rats at low doses (μg quantities).[104] Based upon information about the existence of Type II β-turns at positions 5–8, Freidinger *et al.*[105] synthesized a cyclic hexapeptide with the purpose of stabilizing this turn. To test this hypothesis in the antagonist series, they synthesized cyclo(Tyr–D-Trp–Leu–Arg–Trp–Pro) (D-Trp was replaced for Gly because the substitution was thought to favour β-turns as well as enhance potency). The compound was found to reduce LH secretory rates and appeared to fit the model for competitive antagonism.

Approaches to design by utilizing theoretical simulations (i.e. molecular dynamics, energy minimization, and template forcing) were adopted by Roeske and co-workers[106] and Struthers *et al.*[107] Of interest were the analogues by Rivier and co-workers who synthesized an analogue with substitutions at positions 1 (Ac-D-3-(2′-naphthyl)alanine), 2 [D-(4-chlorophenyl)alanine], 3-[D-3-(3′-pyridyl)]alanine, 5-(arginine or tyrosine) and 6-[D-3-(3′-pyridyl)-alanine] or D-arginine(A). In addition, conformational constraints were imposed by substitution in positions 4 and 10 with Cys and subsequent cyclization. The second class of peptides involved end-to-end cyclization of cyclo(1–10)-[D-Pro^3, 4-Cl-D-Phe^2, D-$Trp^{3,6}$, N-Me-Leu^7, β-Ala^{10}]GnRH. The latter compound was also used to propose a putative receptor binding conformation for the antagonist.[108] These led to the first evidence that conformational restrictions *per se* could be applied fruitfully to the antagonist series of peptides. Structural studies of these and similarly constrained molecules by NMR methods (i.e. NOESY) have confirmed the presence of a β-type turn between residues 5 and 8.[108]

Although active but marginally potent, observations by NMR analysis and subsequent computer analyses suggested that the Ser^4 and Pro^9 were of close proximity (2.5 Å) to each other and pointed towards the centre of the ring. This suggested the possibility of stabilization via bridging with two or three methylene groups. The fact that the cyclic 4–9 analogues were similar in potency to A suggested that the desired molecules had similar conformational properties. This prompted an examination of 4 to 10 bridging as well. A series of peptides based on [Ac-D-Nal^1, D-Cpa^2, D-Pal^3, Arg^5, D-Pal^6, D-Ala^{10}]GnRH were chosen because of their relatively high potency in the antiovulatory assay (AOA) combined with low potency in the rat mast cell histamine release assay. Substitution of Asp in position 4 and Dpr in position 10 followed by subsequent cyclization, led to a compound that was similar in potency to the parent compound in the AOA, but unfortunately also became more potent as to histamine release. Nonetheless, the synthesis of this constrained cyclic analogue that continues to display biological activity and potency within a factor of 2 of the linear analogue has allowed closer examination of the conformational requirements by the ligand during binding to the receptor. Work with this important lead continues to be undertaken.

5.3. DEVELOPMENT OF CYCLIC DISULPHIDE AND LACTAM ANALOGUES OF α-MELANOTROPIN WHICH ARE SUPERPOTENT AND LONG-ACTING

When discussing work on conformational restriction of melanotropins, one generally refers to the linear tridecapeptide α-MSH since this compound probably bestows the most physically relevant features. It is a linear molecule of the following sequence:

$$\text{Ac-Ser-Tyr-Ser-Met-Glu-His-Phe-Arg-Trp-Gly-Lys-Pro-Val-NH}_2$$

It comprises the initial 13 amino acid sequence of ACTH_{1-39}.

Prior to the time that potent cyclic analogues were developed, initial structure–activity relationships involving the linear peptide were undertaken. Most interesting of these were the experiments by Bool and co-workers[109] who revealed that heat–alkali treatment of α-MSH led to a response that was slower but greatly prolonged. Gas chromatographic methods revealed that the greatest extent of racemization was in the 6–8 sequence (His–Phe–Arg–Trp). After extensive model building and Chou–Fasman calculations, as well as conformational considerations, this led to the synthesis of [Nle[4], D-Phe[7]]α-MSH[58,110] that was 60- and 5-fold more potent than native α-MSH in the *in vitro* frog and lizard skin bioassays respectively and had very prolonged biological activities *in vitro* and *in vivo*.[57,58]

Efforts to further define the topographical (side-chain) and conformational (backbone) features of this analogue were directed towards applying conformational restrictions. One clear goal in this approach was to maintain the putative reverse turn stability that is imposed by D-amino acids in the 6–8 region. Hruby and co-workers, noting that replacement of Gly[10] and Met[4] with cysteines followed by oxidation to the disulphide, would lead to a pseudoisosteric substitution, synthesized [Cys[4], Cys[10]]α-MSH.[34,35] This peptide was found to be more than tenfold more potent than α-MSH in the *in vitro* frog skin bioassay and at least twofold more potent in the *in vitro* lizard skin bioassay.

Removal of the residues 1–3 from the analogue, resulting in Ac-[Cys[4], Cys[10]] α-MSH$_{4-13}$-NH$_2$ led to an analogue slightly more potent in the frog skin but less potent in the lizard skin *in vitro* assays. An interesting note that underlies the results was the fact that these cyclic peptides were more potent than their linear sequence counterparts which implies that a more favourable conformation for the receptor is being assumed in these restricted cyclic peptides. This fact supports the basis for the applications of conformational restrictions in the synthesis of highly potent and prolonged-acting melanotropins. From this point, a variety of different analogues were synthesized to determine if

- ring expansion or contraction;
- stereochemical changes;
- N-terminal acyl deletions and desamino compounds; or
- addition of more conformationally constraining residues (i.e. D-penicillamine), would significantly affect potency and transduction.[111–114]

Alteration of the 23-membered ring by substitution of Cys^4 with mercapto-acetic acid, or L-homocysteine^4 (Hcy^4) yielded 22- and 24-membered rings. Both analogues were poor (500-fold less potent) in the frog skin assay. However, somewhat surprisingly, the 24-membered ring was slightly enhanced (~ 1.2 fold) in potency on the lizard skin.[112] This suggested the preference for the 23-membered ring but, in addition, importantly suggested that conformational and topographical requirements for the frog skin receptor vs. the lizard skin receptor may indeed be different, as had been previously suggested. Stereochemical changes by the substitution of $D\text{-}Cys^4$ for $L\text{-}Cys^4$ led to losses in potency (tenfold) in the frog skin assay but full potency in the lizard skin, thus supporting the different receptor requirements for frog and lizard. Removal of the N-terminal acyl group or the amine terminus by the synthesis of $[\overline{Mpa^4, Cys^{10}}]\alpha\text{-}MSH_{4-13}NH_2$ led to little or no change in potency,[112] thus suggesting that these were not essential for potency.

The use of D-penicillamine in lieu of cysteine in α-MSH analogues has been tried, but did not improve potency.[113] Suffice it to say that substitution of penicillamine in position 4 did not affect potency dramatically while Cys^{10} substitution by penicillamine led to decreases of about tenfold relative to the $\overline{Cys^4, Cys^{10}}$ analogues. Further discussion can be found in a recent review.[114]

The adaptations of computer modelling techniques and molecular dynamic simulations as a rational method for the design of α-MSH analogues was recently explored by Al-Obeidi and co-workers.[16,115–117] Using quenched dynamics simulations,[16,117] which allow one to explore conformational space by some-times permitting a molecule to traverse rotational barriers, Al Obeidi and co-workers observed that a modification of glycine in position 10 for lysine would permit the formation of a salt bridge with the Glu (or Asp) in position 5. This had not previously been observed with lysine in its usual 11 position and glycine in position 10. In addition, the dynamic simulations of this analogue obligated the formation of a reverse turn in the His–Phe–Arg–Trp part of the molecule, as well as a segregation of lipophilic and hydrophilic side-chains to opposite faces of the molecule relative to the ring. This observation prompted the synthesis of a cyclic lactam bridged analogue $Ac\text{-}[Nle^4,\overline{Asp^5,D\text{-}Phe^7,Lys^{10}}]\alpha\text{-}MSH_{4-10}\text{-}NH_2$. The analogue was found to be superpotent in both the lizard skin (90-fold more potent) and melanoma tyrosinase (100-fold more potent) assays relative to α-MSH. The frog skin assay, paradoxically, exhibited only half as much potency for the cyclic analogue relative to α-MSH. Clearly, this is an example of a case whereby conformationally constrained analogues can have very species-dependent affects for the same receptor.

Acknowledgements

The work in our laboratories was supported in part by grants from the U.S. Public Health Service and the National Service Foundation. We thank Natasha Johnson and Cheryl McKinley for help in typing and editing this chapter.

References

1. *Peptide Synthesis*, 2nd edn, M. Bodanszky, Y.S. Klausner and M. Ondetti (1976) Wiley Interscience, New York.
2. *The Peptides*, Vols. I and II E. Schröder and K. Lübke (1966) Academic Press, New York.
3. 'A perspective on the application of physical methods to peptides conformational – biological activity studies' V. J. Hruby (1985). In: *The Peptides: Analysis, Synthesis, Biology. Vol. 7. Conformation in Biology and Drug Design* (C. W. Smith, ed.) pp. 1–14. Academic Press, New York and accompanying chapters.
4. 'Conformations of peptides in solution as determined by NMR spectroscopy and other physical methods' V. J. Hruby (1974). In: *Chemistry and Biochemistry of Amino Acids, Peptides and Proteins*, Vol. 3 (B. Weinstein, ed.) pp. 1–188. Marcel Dekker, New York.
5. 'Conformational restrictions of biologically active peptides via amino acid side chain groups' V. J. Hruby (1982) *Life Sciences*, **31**, 189–199.
6. *Chemistry of the Amino Acids*, Vol. I, II and III. J. P. Greenstein and M. Winitz (1961) John Wiley, New York. Now published by R. E. Krieger Publishing Co., Malabar, FL.
7. *Chemistry and Biochemistry of Amino Acids*, (G. C. Barrett, ed.) (1985) Chapman and Hall, London.
8. 'α-Amino acid synthesis', *Tetrahedron Symposia in Print*, **33** (M. J. O'Donnell, Guest ed.) (1988) **49** (17) pp. 5253–5614.
9. *Synthesis of Optically Active α-Amino Acids* R. M. Williams (1989) Pergamon Press, Oxford.
10. *The Peptides: Analysis, Synthesis, Biology*, Vols. I–IX (E. Gross, J. Meienhofer and S. Udenfriend, eds.) (1979–1987) Academic Press, New York.
11. 'Solid phase peptide synthesis. I. The synthesis of a tetrapeptide' R. B. Merrifield (1963) *J. Amer. Chem. Soc.*, **85**, 2149–2154.
12. 'Solid phase peptide synthesis' B. W. Erickson and R. B. Merrifield (1976). In *The Proteins* (H. Neurath and R. L. Hill, eds.) 3rd edn., Vol. II, pp. 255–527. Academic Press, New York.
13. 'Solid phase peptide synthesis' G. Barany and R. B. Merrifield (1981). In: *The Peptides: Analysis, Synthesis, Biology* (E. Gross and J. Meienhofer, eds.) Vol. II, pp. 1–284. Academic Press, New York.
14. 'The fluorenylmethoxycarbonyl amino protecting group' E. Atherton and R. C. Sheppard (1987). In: *The Peptides: Analysis, Synthesis, Biology* (S. Udenfriend and J. Meienhofer, eds.), Vol. 9, pp. 1–38. Academic Press, New York.

15. 'Solid phase peptide synthesis utilizing 9-fluorenylmethoxycarbonyl amino acids' G.B. Field and R.C. Noble (1989) *Int. J. Peptide Protein Res.*, **35**, 161–214.
16. 'Potent and prolonged acting cyclic lactam analogues of α-melanotropin: Design based on molecular dynamics' F. Al-Obeidi, A.M.L. Castrucci, M.E. Hadley and V.J. Hruby (1989) *J. Med. Chem.*, **32**, 2555–2561.
17. 'Design of peptide superagonists and antagonists: Conformational and dynamic considerations' V.J. Hruby (1984). In: *Conformationally Directed Drug Design* (J.A. Vida and M. Gordon, eds.) pp. 9–29. ACS Symposium Series **251**, Washington, D.C.
18. 'Binding and information transfer in conformationally restricted peptides' V.J. Hruby and M.E. Hadley (1986). In: *Design and Synthesis of Organic Molecules Based on Molecular Recognition* (G. Van Binst, ed.), pp. 269–289. Springer-Verlag, Heidelberg.
19. 'Structure and conformation related to the activity of peptide hormones' V.J. Hruby (1981). In: *Perspectives in Peptide Chemistry* (A. Eberle, R. Geiger and T. Wieland, eds.), pp. 207–220. Karger, Basel.
20. 'Relation of conformation to biological activity in oxytocin, vasopressin and their analogues' V.J. Hruby (1981). In: *Topics in Molecular Pharmacology* (A.S.V. Burgen and G.C.K. Roberts, eds.), pp. 99–126. Elsevier/North-Holland, Amsterdam.
21. 'Conformation and motion of biologically active peptides as bound to phospholipid membrane' A. Milon, K. Wakamatsu, K. Saito, A. Okada, T. Miyazawa and T. Higashijima (1988). In: *Peptides: Chemistry and Biology* (G.R. Marshall, ed.), pp. 71–73. ESCOM, Leiden.
22. 'The conformation of oxytocin bound to neurophysin I' K. Hallenga, N.R. Nirmala, D.D. Smith and V.J. Hruby (1988). In: *Peptides: Chemistry and Biology* (G.R. Marshall, ed.), pp. 39–41. ESCOM, Leiden.
23. 'Crystal structure analysis of deamino-oxytocin: conformational flexibility and receptor binding' S.P. Wood, I.J. Tickel, A.M. Trehorne, J.E. Pitts, Y. Mascarenhas, J.Y. Li, J. Husain, S. Cooper, T.L. Blundell, V.J. Hruby, A. Buku, A.J. Fischman and H.R. Wyssbrod (1986) *Science*, **232**, 633–636.
24. 'NMR studies of flexible opiate conformations at monoclonal antibody binding sites. I. Transferred nuclear Overhauser effects show bound conformations' J.A. Glasel and P.N. Borer (1986) *Biochem. Biophys. Res. Commun.*, **141**, 1267–1273.
25. 'Structure of conformationally constrained peptides: From model compounds to bioactive peptides' C. Toniolo (1989) *Biopolymers*, **28**, 247–257.
26. 'Design of peptides and proteins' W.F. De Grado (1988) *Adv. Protein Chem.*, **39**, 51–124.
27. '*Trans* carbon–carbon double bond isosteres of the peptide bond: General methodology and the synthesis of cholecystokinin (CCK) analogues' Y.K. Shue, G.M. Carrera, Jr., A.M. Nadzan, J.F. Kerwin, H. Kopecka and C.W. Lin (1988). In: *Peptides: Chemistry and Biology* (G.R. Marshall, ed.) pp. 112–114. ESCOM, Leiden.
28. 'Conformational mimicry. 1. 1,5-disubstituted tetrazole ring as a surrogate for the cis amide bond' J. Zabrocki, G.D. Smith, J.B. Dunbar, Jr., H. Iijama and G.R. Marshall (1988) *J. Amer. Chem. Soc.*, **110**, 5875–5880.
29. 'Synthesis and conformational study of a cyclic hexapeptide analogue of somatostatin cyclo(Phe-D-Trp-Lys-Thr-O-AMPA)' P. Van Der Elst, E. Van Der Berg, H. Pepermans, L. Van Der Auwere, R. Zeevus, D. Tourwe and G. Van Binst (1987) *Int. J. Peptide Protein Res.*, **29**, 318–330.
30. 'Conformationally restricted peptides through short-range cyclizations' C. Toniolo (1990) *Int. J. Peptide Protein Res.*
31. 'Cyclic enkephalins which are optically pure isomers of [B-Me-p-NO$_2$-Phe^4]DPDPE possess extraordinary δ-opioid receptor selectivities' V.J. Hruby, G. Toth, O. Prakash, P. Davis and T.F. Burks (1989). In: *Peptides 1988* (G. Jung and E. Bayer, eds.), pp. 616–618. de Gruyter, Berlin.

32. 'A new approach to receptor ligand design: Synthesis and conformation of a new class of potent and highly selective Mu opioid antagonists utilizing tetrahydroisoquinoline carboxylic acid' W. Kazmierski and V. J. Hruby (1988) *Tetrahedron*, **44**, 697–710.
33. 'Design and synthesis of somatostatin analogues with topographical properties that lead to highly potent and specific μ opioid receptor antagonists with greatly reduced binding to somatostatin receptors' W. Kazmierski, W. S. Wire, G. K. Lui, R. J. Knapp, J. E. Shook, T. F. Burks, H. I. Yamamura and V. J. Hruby (1988) *J. Med. Chem.*, **31**, 2170–2177.
34. 'Conformations of dehydrophenylalanine containing peptides' V. S. Chauen, A. K. Sharma, K. Ana, P. K. C. Paul and P. Balaram (1987) *Int. J. Peptide Protein Res.*, **29**, 126–133.
35. 'Implications of the X-ray structure of deamino-oxytocin to agonist/antagonist–receptor interactions' V. J. Hruby (1987) *Trends in Pharmacol. Sci.*, **8**, 336–339.
36. '[4-Half-cystine, 10-Half-cystine]-α-Melanocyte stimulating hormone: A cyclic α-melanotropin showing superagonist biological activity' T. K. Sawyer, V. J. Hruby, P. S. Darman and M. E. Hadley (1982) *Proc. Natl. Acad. Sci. U.S.A.*, **79**, 1751–1755.
37. 'Bis-penicillamine enkephalins possess highly improved specificity toward delta opioid receptors' H. I. Mosberg, R. Hurst, V. J. Hruby, K. Gee, H. I. Yamamura, J. J. Galligan and T. F. Burks (1983) *Proc. Natl. Acad. Sci. U.S.A.*, **80**, 5871–5874.
38. 'Design of novel cyclic hexapeptide somatostatin analogues from a model of the bioactive conformation' R. M. Freidinger and D. F. Veber (1984). In: *Conformationally Directed Drug Design* (J. A. Vida and M. Gordon, eds.) pp. 169–187. ACS Symposium Series, **251**, Washington, D.C.
39. 'Structure–activity relationships of highly potent and specific octapeptide analogues of somatostatin' W. Bauer, U. Briner, W. Doepfner, R. Haller, R. Huguenin, P. Marbach, T. J. Petcher and J. Pless (1983). In: *Peptides 1982* (K. Blaha and P. Malon, eds.), pp. 583–588. de Gruyter, Berlin.
40. 'Conformationally restricted analogues of somatostatin with high μ-opiate receptor specificity' J. T. Pelton, K. Gulya, V. J. Hruby, S. P. Duckles and H. I. Yamamura (1985) *Proc. Natl. Acad. Sci. U.S.A.*, **82**, 236–239.
41. 'Design and synthesis of conformationally constrained somatostatin analogues with high potency and specificity for mu opioid receptors' J. T. Pelton, W. Kazmierski, K. Gulya, H. I. Yamamura and V. J. Hruby (1986) *J. Med. Chem.*, **39**, 2370–2375.
42. 'A novel cyclic opioid peptide analogue showing high preference for μ-receptors' P. W. Schiller, T. M.-D. Nguyen, L. A. Maziak and C. Lemieux (1985) *Biochem. Biophys. Res. Commun.*, **127**, 558–564.
43. 'Synthesis and opioid activity profiles of cyclic dynorphin analogues' P. W. Schiller, T. M.-D. Nguyen and C. Lemieux (1988) *Tetrahedron*, **44**, 733–743.
44. 'Design of a new class of superpotent cyclic α-melanotropins based on quenched dynamic simulations' F. Al Obeidi, M. E. Hadley, B. M. Pettitt and V. J. Hruby (1989) *J. Am. Chem. Soc.*, **111**, 3413–3416.
45. 'Weak oxytocin agonist converted to highly potent oxytocin antagonist through bicyclization' P. S. Hill, J. Slaninova and V. J. Hruby (1989). In: *Peptides: Chemistry and Biology* (G. R. Marshall, ed.) pp. 468–470. ESCOM, Leiden.
46. 'Aspects of conformational restriction in biologically active peptides' P. W. Schiller and J. Di Maio (1983). In: *Peptides: Structure and Function* (V. J. Hruby and D. H. Rich, eds.), pp. 269–278. Pierce Chemical Co., Rockford, IL.
47. 'Principles of active site formation in peptides and proteins' G. Chipens (1983). In: *Peptides: Structure and Function* (V. J. Hruby and D. H. Rich, eds.), pp. 865–868. Pierce Chem. Co., Rockford, IL.
48. 'Synthesis of conformationally constrained CCK-4 analogues containing a substituted gamma lactam ring' D. S. Garvey, A. P. D. May and A. M. Nadzan (1989). In: *Peptides: Chemistry Biology* (G. R. Marshall, ed.), pp. 123–125. ESCOM, Leiden.

49. 'Lactam restriction of peptide conformation in cyclic hexapeptides which alter rumen fermentation' R. M. Freidinger, D. F. Veber, R. Hirschmann and L. M. Paege (1980) *Int. J. Peptide Protein Res.*, **16**, 464–470.

50. 'Design of potent, orally effective, nonpeptide antagonists of the peptide hormone cholecystokinin' B. E. Evans, M. G. Bock, K. E. Rittle, R. M. Di Pardo, W. L. Whitter, D. F. Veber, P. S. Anderson and R. M. Freidinger (1986) *Proc. Natl. Acad. Sci. U.S.A.*, **83**, 4918–4922.

51. 'Peptide backbone modification: A structure–activity analysis of peptides containing amide bond surrogates, conformational constraints and related backbone replacements' A. F. Spatola (1983). In: *Chemistry and Biochemistry of Amino Acids, Peptides and Proteins*, Vol. 7 (B. Weinstein, ed.) pp. 267–357. M. Dekker, New York.

52. '(2S, 5S, 8S, 11S)-1,4-diaza-3-keto-5-carboxy-10-thia-tricyclo-(2.8.0^{4,8})-tridecane, 1 the preferred conformation of 1 (1 ≡ αtemp-OH) and its peptide conjugates αtemp-1-(ala)n-or (n = 1 to 4) and α-temp-L-ala-L-phe-L-lys (εboc)-L-lys(εboc)-L-lys(εboc)-NHMe studies of templates for α-helix formation' D. S. Kemp and T. P. Curran (1988) *Tetrahedron Letters*, 4935–4938.

53. 'Amino acid derivatives that stabilize technology structures of polypeptides–IV. Practical synthesis of L$_1$-alkylamino-3-cyano-G-azabicyclo(3.2.1)oct-3-enes (BEN derivatives) as γ-turn templates' D. S. Kemp and J. S. Carter (1987) *Tetrahedron Letters*, 4645/4648.

54. 'A convenient preparation of derivatives of 3(S)-amino-10(R)-carboxy-1,6-diaza-cyclodeca-2.7-olione. The dilactam of L-2,8-diaminobutyric acid and D-glutamic acid: A β-Turn template' D. S. Kemp and W. E. Stites (1988) *Tetrahedron Letters*, 5057–5060.

55. 'The design and synthesis of a carbohydrate-derived peptidomimetic' K. C. Nicolaou, J. M. Savino, K. Raynor, S. Pietranico, R. M. Freidinger, T. Reisine and R. Hirschmann (1989) Presented at Eleventh American Peptide Symposium, July 9–14, 1989, Abstract LM17.

56. 'Design of peptide hormone and neurotransmitter analogues with high receptor selectivity and biological stability' V. J. Hruby (1990). In: *Peptides, Peptoids and Proteins* (P. Garzone, ed.) Proceedings of the Pittsburgh Pharmacodynamic Conference.

57. 'Calcium-dependent prolonged effects on melanophores of [4-norleucine, 7-D-phenylalanine]α-melanotropin' M. E. Hadley, B. Anderson, C. B. Heward, T. K. Sawyer and V. J. Hruby (1981) *Science*, **213**, 1025–1027.

58. '[Nle^4, DPhe^7] α-Melanocyte stimulating hormone: A highly potent α-melanotropin with ultralong biological activity' T. K. Sawyer, P. J. Sanfilippo, V. J. Hruby, M. H. Engel, C. B. Heward, J. B. Burnett and M. E. Hadley (1981) *Proc. Natl. Acad. Sci. U.S.A.*, **77**, 5754–5758.

59. 'Differentiation of the structural features of melanotropins important for biological potency and prolonged activity *in vitro*' B. C. Wilkes, T. K. Sawyer, V. J. Hruby and M. E. Hadley (1983) *Int. J. Peptide Protein Res.*, **22**, 313–324.

60. 'Design of oxytocin antagonist with prolonged action: Potential tocolytic agents for the treatment of preterm labor' W. Y. Chan, V. J. Hruby, T. W. Rockway and J. Hlavacek (1986) *J. Pharmacol. Exp. Therap.*, **239**, 84–87.

61. 'Mu opioid antagonist properties of a cyclic somatostatin octapeptide in vivo: Identification of mu receptor related functions' J. E. Shook, J. T. Pelton, P. F. Lemicke, F. Porreca, V. J. Hruby and T. F. Burks (1987) *J. Pharmacol. Exp. Therap.*, **242**, 1–7.

62. R. S. Rapaka and R. L. Hawks (1986). In: *Opioid Peptides: Molecular Pharmacology, Biosynthesis, and Analysis*, NIDA Research Monograph **70**, Rockville, MD. *The Peptides: Analysis, Synthesis, Biology, Vol. 6, Opioid Peptides Biology, Chemistry, Genetics* (S. Udenfriend and J. Meienhofer, eds.), Academic Press, New York, 1984.

63. 'Conformational analysis of opioid peptides and the use of conformational restriction in the design of selective analogues' P. W. Schiller (1986). In: *Opioid*

Peptides: Medicinal Chemistry (R. S. Rapaka, G. Barnett and L. Hawks, eds.) pp. 291–311. NIDA Research Monograph **69**, Rockville, MD.

64. 'Somatostatin and ACTH are peptides with partial antagonist-like selectivity for opiate receptors' L. Terenius (1976) *Eur. J. Pharmacol.*, **38**, 211–213.

65. 'Opiate-like naloxone-reversible actions of somatostatin given intra-cerebrally' M. Rezek, V. Havlicek, L. Leybin, F. S. La Bella and H. Friesen (1978) *Can. J. Physiol. Pharmacol.*, **56**, 227–231.

66. 'Opiate antagonistic properties of an octapeptide somatostatin analogue' R. Maurer, B. H. Gaehwiler, H. H. Beuscher, R. C. Hill and D. Roemer (1982) *Proc. Natl. Acad. Sci.*, **79**, 4815–4817.

67. 'Conformation of D-Phe-Cys-Tyr-D-Trp-Lys-Thr-Pen-Thr-NH$_2$ (CTP-NH$_2$), a highly selective mu-opioid antagonist peptide, by ^{1}H-and ^{13}C N.M.R. 'J. T. Pelton, M. Whalon, W. L. Cody and V. J. Hruby (1988) *Int. J. Peptide Protein Res.*, **31**, 109–115.

68. 'Proton n.m.r. investigation of conformational influence of penicillamine residues on the disulfide ring system of opioid receptor selective somatostatin derivatives' E. E. Sugg, D. Tourwe, W. Kazmierski, V. J. Hruby and G. Van Binst (1988) *Int. J. Peptide Protein Res.*, **31**, 192–200.

69. 'A novel cyclic opioid peptide analogue showing high preference for μ-receptors' P. W. Schiller, T. M.-D. Nguyen, L. A. Maziak and C. Lemieux (1985) *Biochem. Biophys. Res. Commun.*, **127**, 558–564.

70. 'Synthesis and activity profiles of novel cyclic opioid peptide monomers and dimers' P. W. Schiller, T. M.-D. Nguyen, C. Lemieux and L. A. Maziak (1985) *J. Med. Chem.*, **28**, 1766–1771.

71. 'Structure–activity relationships of cyclic opioid peptide analogues containing a phenylalanine residue in the 3-position' P. W. Schiller, T. M.-D. Nguyen, L. A. Maziak, B. C. Wilkes and C. Lemieux (1987) *J. Med. Chem.*, **30**, 2094–2099.

72. 'A cyclic enkephalin analog with high in vitro opiate activity' J. DiMaio and P. W. Schiller (1980) *Proc. Natl. Acad. Sci. U.S.A.*, **77**, 7162–7166.
'Synthesis and pharmacological characterization in vitro of cyclic enkephalin analogues: effect of conformational constraints on opiate receptor selectivity' J. DiMaio, T. M.-D. Nguyen, C. Lemieux and P. W. Schiller (1982) *J. Med. Chem.*, **25**, 1432–1438.

73. 'Comparison of μ-, δ-, and κ-receptor binding sites through pharmacologic evaluation of p-nitrophenylalanine analogs of opioid peptides' P. W. Schiller, T. M. -D. Nguyen, J. DiMaio and C. Lemieux (1983) *Life Sci.*, **33**, Supp. 1, 319–322.

74. 'Peptides with morphine-like activity' D. Sarantakis (1979) U.S. Patent 4 098 781.

75. 'Cyclic enkephalin analogs containing a cystine bridge' P. W. Schiller, B. Eggimann, J. DiMaio, C. Lemieux and T. M.-D. Nguyen (1981) *Biochem. Biophys. Res. Commun.*, **101**, 337–343.

76. '[D-Pen^2, L-Cys^5]Enkephalinamide and [D-Pen^2, D-Cys^5]Enkephalinamide, conformationally constrained cyclic enkephalinamide analogs with delta receptor specificity' H. I. Mosberg, R. Hurst, V. J. Hruby, J. J. Galligan, T. F. Burks, K. Gee and H. I. Yamamura (1982) *Biochem. Biophys. Res. Commun.*, **106**, 506–512.

77. 'Conformationally constrained cyclic enkephalin analogs with pronounced delta opioid receptor agonist selectivity' H. I. Mosberg, R. Hurst, V. J. Hruby, J. J. Galligan, T. F. Burks, K. Gee and H. I. Yamamura (1983) *Life Sci.*, **32**, 2565–2569.

78. 'Cyclic penicillamine containing enkephalin analogs display profound delta receptor selectivities' H. I. Mosberg, R. Hurst, V. J. Hruby, K. Gee, K. Akiyama, H. I. Yamamura, J. J. Galligan and T. F. Burks (1983) *Life Sci.*, **33**, Supp. 1, 447–450.

79. 'Design of conformationally constrained cyclic peptides with high delta and mu opioid receptor specificities' V. J. Hruby (1986). In: *Opioid Peptides: Medicinal Chemistry* (R. S. Rapaka, G. Barnett and R. L. Hawks, eds.) pp. 128–147. NIDA Research Monograph **69**, Rockville, MD.

80. 'Characterization of [^{3}H] [2-D-Penicillamine, 5-D-Penicillamine]-enkephalin

($[^3H]$ DPDPE) binding to delta opioid receptor in the rat brain and neuroblastoma-glioma hybrid (NG 108–15) cells' K. Akiyama, K.W. Gee, H.I. Mosberg, V.J. Hruby and H.I. Yamamura (1985) *Proc. Natl. Acad. Sci. U.S.A.*, **82**, 2543–2547.

81. 'Site-directed alkylation of multiple opioid receptors. I. Binding selectivity' I.F. James and A. Goldstein (1984) *Mol. Pharmacol.*, **25**, 337–342.

82. 'Role of steric interactions in the delta opioid receptor selectivity of (D-Pen^2, D-Pen^5) enkephalin' H.I. Mosberg, R.C. Haaseth, K. Ramalingam, A. Mansour, H. Akil and R.W. Woodard (1988) *Int. J. Peptide Protein Res.*, **32**, 1–8.

83. 'Conformation biological activity relationships of conformationally constrained delta specific cyclic enkephalins' V.J. Hruby, L.F. Kao, L.D. Hirning and T.F. Burks (1985). In: *Peptides: Structure and Function* (C.M. Deber, V.J. Hruby and K.D. Kopple, eds.), pp. 487–490. Pierce Chem. Co., Rockford, IL.

84. 'Cyclic Enkephalins which are optically pure isomers of [B-Me-p-NO$_2$Phe^4]-DPDPE possess extraordinary δ-opioid receptor selectivities' V.J. Hruby, G. Toth, O. Prakash, R. Davis and T.F. Burks (1989). In *Peptides 1988* (G. Jung and E. Bayer, eds.), pp. 616–618. de Gruyter, Berlin.

85. 'The conformational properties of delta opioid peptide [D-Pen^2, D-Pen^5]-enkephalin in aqueous solution determined by NMR and energy minimization calculations' V.J. Hruby, L.-F. Kao, B.M. Pettitt and M. Karplus (1988) *J. Amer. Chem. Soc.*, **110**, 3351–3359.

86. 'Conformation–biological activity relationships for receptor selective, conformationally constrained opioid peptides' V.J. Hruby and B.M. Pettitt (1989). In *Computer Aided Drug Design* (T.J. Perun and C. Propst, eds.), pp. 405–460. M. Dekker, New York.

87. 'Deltakephalin, Tyr-D-Thr-Gly-Phe-Leu-Thr: A new highly potent and fully specific agonist for opiate δ-receptors' J.-M. Zajac, G. Gacel, F. Petit, P. Dodey, P. Rossignol and B.P. Roques (1983) *Biochem. Biophys. Res. Commun.*, **111**, 390–397.

88. 'Comparison of conformational properties of linear and cyclic δ selective ligands DTLET (Tyr-D-Thr-Gly-Phe-Leu-Thr) and DPLPE (Tyr-c[D-Pen-Gly-Phe-Pen]) by ^{1}H-n.m.r. spectroscopy' J. Belleney, B.P. Roques and M.C. Fournie-Zaluski (1987) *Int. J. Peptide Protein Res.*, **30**, 356–364.

89. 'Mechanistic structure–activity studies of peptide and nonpeptide flexible opioids: An interdisciplinary approach' G.H. Loew, T. Toll, E. Uyeno, A. Cheng, A. Judd, J. Lawson, C. Keys, P. Amsterdam and W. Polgar (1985). In: *Opioid Peptides: Medicinal Chemistry* (J. Durell, ed.), **69**, pp. 231–265. NIDA, Rockville, MD.

90. '[^{3}H][D-Ser^2(O-tert-butyl), Leu^5] Enkephalyl-Thr^6 and [D-Ser^2 (O-tert-butyl), Leu^5]Enkephalin-Thr^6 (O-tert-butyl). Two new enkephalin analogs with both a good selectivity and a high affinity' P. Delay-Goyet, C. Seguin, G. Gacel and B. Roques (1988) *J. Biol. Chem.*, **263**, 4124–4130.

91. 'Differences in the conformational behavior of the potent and selective Tyr-D-Thr-Gly-Phe-Leu-Thr-(O^tBu) and of the inactive Tyr-D-Thr(O^tBu)Gly-Phe-Leu-Thr. δ-opioid ligands evidenced by ^{1}H.NMR' J. Belleney, G. Gacel, B. Maigret, M.C. Fournie-Zaluski and B. Roques (1988) *Tetrahedron*, **44**, 711–720.

92. 'The synthesis, bioactivity and enzyme stability of D-Ala^2, Δ^EPhe^4, Leu^5-Enkephalins' H. Kimura, C.H. Stammer, Y. Shimohigashi, C. Ren-Lin and J. Stewart (1983) *Biochem. Biophys. Res. Commun.*, **115**, 112–115.

93. 'Δ^E Phe^4-enkephalin analogs' Y. Shimohigashi, T. Costa, A. Pfeiffer, A. Herz, H. Kimura and C.H. Stammer (1987) *FEBS Lett.*, **222**, 71–74.

94. 'A highly selective ligand for brain δ opiate receptors, a Δ^E Phe-enkephalin analog, suppressed μ receptor-mediated thermal analgesia by morphine' Y. Shimohigashi, Y. Takano, H. Kamiya, T. Costa, A. Herz and C.H. Stammer (1988) *FEBS Lett.*, **233**, 289–293.

95. 'Opioid receptors for the Dynorphin peptides' I.F. James (1986) In *NIDA Research Monograph Series*, **70** (R.S. Rapaka, R.L. Hawks, eds.), 192–208. Rockville, MD.

96. 'Comparative structure–function studies with analogs of Dynorphin-(1–13) and [Leu^5]enkephalin' P. W. Schiller, B. Eggiman and T. M.-D. Nguyen (1982) *Life Sci.*, **31**, 1777–1780.

97. 'Synthesis and opioid activity profiles of cyclic Dynorphin analogs' P. W. Schiller, T. M.-D. Nguyen and C. Lemieux (1988) *Tetrahedron*, **44**, 733–743.

98. 'Hypothalamic hypophysiotropic hormones' W. Vale and C. Rivier (1975). In: *Handbook of Psychopharmacology*, Vol. 5 (S. D. Iverson and S. H. Snyder, eds.), pp. 195–238. Plenum, New York.

99. 'The effects of isolated methylated residue on the conformational characteristic of polypeptides' A. E. Tonelli (1976) *Biopolymers*, **15**, 1615–1622.

100. 'Synthesis of a cyclic analog of the luteinizing hormone releasing factor: [Glu^4,D-Ala^6, Orn^7] LRF' B. Donzel, J. Rivier and M. Goodman (1977) *Biopolymers*, **16**, 2587–2590.

101. 'Synthesis of a cyclic charge transfer labeled analogue of the luteinizing hormone-releasing factor' C. Sakarellos, B. Donzel and M. Goodman (1978) *J. Org. Chem.*, **43**, 293–296.

102. 'Cyclic analogues of Luteinizing Hormone-Releasing Hormone with significant biological activities' J. Seprodi, D. H. Coy, J. A. Vilchez-Martinez, E. Pedroza, W-Y. Huang and A. V. Schally (1978) *J. Med. Chem.*, **21** (9), 993–995.

103. 'Bioactive conformation of luteinizing hormone-releasing hormone: evidence from a conformationally constrained analog' R. M. Freidinger, D. F. Veber, D. S. Perlow, J. R. Brooks and R. Saperstein (1980) *Science*, **210**, 656–658.

104. 'Peptide antagonists of LH-RH: Large increases in antiovulatory activities produced by basic D-amino acids in the six position' D. H. Coy, A. Horvath, M. V. Nekola, E. J. Coy, J. Erchogyi and A. V. Schally (1982) *Endocrinology*, **110**, 1445–1447.

105. 'A cyclic hexapeptide LH-RH antagonist' R. M. Freidinger, C. D. Colton, W. C. Randall, S. M. Pitzenberger, D. F. Veber, R. Saperstein, E. J. Brady and B. H. Arison (1985). In: *Peptides: Structure and Function* (C. M. Deber, V. J. Hruby and K. Kopple, eds.), pp. 549–552. Pierce Chem. Co., Rockford, IL.

106. 'LHRH antagonists with restricted conformation: α-methyl residues and disulfides' R. W. Roeske, G. M. Anantharamaiah, F. A. Momany and C. Y. Bowers (1985). In: *Peptides: Structure and Function* (C. M. Deber, V. J. Hruby and K. D. Kopple, eds.), pp. 549–552. Pierce Chem. Co., Rockford, IL.

107. 'Design of peptide analogs: theoretical simulation of conformation, energetics, and dynamics' R. S. Struthers, J. Rivier and A. Hagler (1983). In: *Conformationally Directed Drug Design – Peptides and Nucleic Acids as Templates or Targets* (J. A. Vida and M. Gordon, eds.), pp. 239–261. American Chemical Society, Washington D.C.

108. 'Molecular dynamics and minimum energy conformations of GnRH and analogs: A methodology for computer-aided drug design' R. S. Struthers, J. Rivier and A. Hagler (1984). In: *Macromolecular Structure and Specificity: Computer Assisted Modelling and Applications* (B. Venkataraghavan and R. J. Feldman, eds.), **439**, 81–96. *Ann. N.Y. Acad. Sci.*

109. 'Racemization effects on Melanocyte-stimulating hormones and related compounds' A. M. Bool, G. H. Grey II, M. E. Hadley, C. B. Heward, V. J. Hruby, T. K. Sawyer and Y. C. S. Yang (1981) *J. Endocrinol.*, **88**, 57–65.

110. 'Melanotropins: structural, conformational and biological considerations' V. J. Hruby, B. C. Wilkes, W. L. Cody, T. L. Sawyer and M. E. Hadley (1984). In: *Peptides and Protein Reviews*, Vol. 3 (M. T. W. Hern, ed.), pp. 1–64. Marcel Dekker, New York.

111. 'Structure–activity studies of highly potent cyclic [Cys^4, Cys^{10}]-melanotropin analogs' J. J. Knittel, T. K. Sawyer, V. J. Hruby and M. E. Hadley, (1983) *J. Med. Chem.*, **28**, 125–129.

112. 'Modified ring structures, synthesis and biological activity' M. Lebl, W. L. Cody,

B.C. Wilkes, V.J. Hruby, A.M.L. Castrucci and M.E. Hadley (1984) *Int. J. Peptide Protein Res.*, **24**, 472–478.

113. 'Conformational and biological analysis of α-MSH fragment analogues with sterically constrained amino acid residues' V.J. Hruby, W.L. Cody, A.M.L. Castrucci and M.E. Hadley (1988) *Collect. Czech. Chem. Commun.*, **53**, 2549–2573.

114. 'Cyclic, conformationally constrained melanotropin analogs: structure–function and conformational relationships' W.L. Cody, M.E. Hadley and V.J. Hruby (1988). In: *The Melanotropins* Vol. III. *Mechanisms of Actions and Biomedical Applications* (M.E. Hadley, ed.), pp. 75–92. CRC Press, Boca Raton, FL.

115. 'Design of potent linear α-melanotropin 4–10 analogues modified in positions 5 and 10' F. Al-Obeidi, V.J. Hruby, A.M.L. Castrucci and M.E. Hadley (1989) *J. Med. Chem.*, **32**, 174–179.

116. 'Design of a new class of superpotent cyclic α-melanatropins based on quenched dynamic simulations' F. Al-Obeidi, M.E. Hadley, B.M. Pettitt and V.J. Hruby. *J. Am. Chem. Soc.*, (1989) **111**. 3413–3416.

117. 'Conformational constraint in the design of receptor selective peptides: conformational analysis and molecular dynamics' V.J. Hruby, W. Kazmierski, B.M. Pettitt and F. Al Obeidi (1988). In: *Molecular Biology of Brain and Endocrine Peptidergic Systems* (M. Chretien and K.W. McKerns, eds.), pp. 13–27. Plenum, New York.

—— *Chapter 6* ————————————————————

Genetic approaches to peptide and polypeptide synthesis and design

J. Rosamond

I. Introduction

Although the basis of genetics was established by Mendel over a century ago, it was only in the 1940s that Mendel's rules were given a molecular basis with the demonstration by Avery, MacLeod and McCarthy that genetic information was stored in DNA and not protein as had previously been thought. This was the first major step into the era spanning 15 years that saw the structure of DNA determined, the genetic code deciphered, and the processes of transcription and translation described. It was this work that laid the foundations of molecular research in biology and established what became known as the central dogma of molecular biology, which asserts that genetic information is stored in the cell as DNA where it serves two functions: firstly, to transmit the genetic information from that cell to subsequent generations; and secondly, to allow the genetic information to be expressed in a cell by transcription to RNA and translation to proteins that can perform a variety of functions. They can act as structural components of the cell or as catalysts for reactions involving other proteins or small molecules; they can act as regulators of bodily function or to prevent infection and remove toxic agents. A consequence of the relationship between DNA and protein is that the protein complement of the cell and the amino-acid sequences of individual proteins are determined by the sequences of bases in the DNA of that cell. In theory then, if you change the DNA content, you change the protein composition.

The ability to change or manipulate the DNA content of a cell by transferring DNA from one cell to another has been known for many years. Conjugal transfer in *Escherichia coli* or infection of bacteria with viral transducing particles can lead to inherited changes in the genetic content of the recipient cell, and techniques such as these were used widely to investigate a variety of biological problems, most notably the control circuits of bacterial operons. However, since the early 1970s, we have been able to mimic these natural systems not only to transfer specific DNA molecules into cells, but also to construct *in vitro* new combinations of DNA molecules carrying specific, precisely defined DNA base sequences. Central to this are the techniques of modern molecular biology, frequently referred to as recombinant DNA or genetic engineering techniques, which have revolutionized almost all areas of science concerned with biological systems, from botany to medicine.[1] It is because of these procedures, allowing us to manipulate the genetic composition of an organism, that we can induce cells to produce proteins that either they would not otherwise synthesize or else to make natural proteins in amounts that would not usually be found in that cell type. Such techniques are important not only in basic research, where they have been used to produce proteins to study oncogenesis, hormone–receptor interactions, and cellular signalling pathways, but also for the rapidly expanding biotechnology industry, where genetic manipulation techniques can be used to produce proteins of therapeutic value, such as factor VIII, insulin, interferons, and tissue plasminogen activator. It is this latter application that will be emphasized in this chapter.

Regardless of the overall purpose of the work, whether for pure research or some immediate biotechnological application, there is a fundamental requirement for techniques to isolate specific DNA fragments or perhaps more accurately, DNA fragments carrying specific genes. These gene cloning techniques allow DNA fragments from any organism to be joined to a carrier molecule such that the recombinant or chimaeric molecule can be stably maintained in a host cell and transmitted to all progeny derived from that cell. The basic steps of this process (summarized in Fig. 1) are the joining of the DNA fragment and the carrier molecule *in vitro*, the transfer of this recombinant to an appropriate host cell, and the selection amongst the cell population for those members carrying the required recombinant DNA molecule. The variety of methods available for joining specific DNA fragments and for selecting the required clones are major factors in ensuring the broadest application of gene cloning. However, one of the key components of the whole process is the carrier molecule, the *vector*, which carries a number of functional regions designed amongst other things to facilitate selection of recombinants in the host cell and to ensure the replication of the chimaera during cell division. In addition, some vectors contain sequences designed not only to allow replication but also to permit expression of the cloned DNA so that it synthesizes its protein product. The importance of this is that it allows the expression of foreign DNA (such as plant or animal DNA, or even synthetic genes) in relatively simple cells such as bacteria or yeast. Expression vectors that have this property are an essential component of gene cloning strategies designed to produce pharmaceutically important proteins and peptides, and their use is considered in more detail in Section 2.

Once a gene has been cloned, then a wide array of analytical methods can be

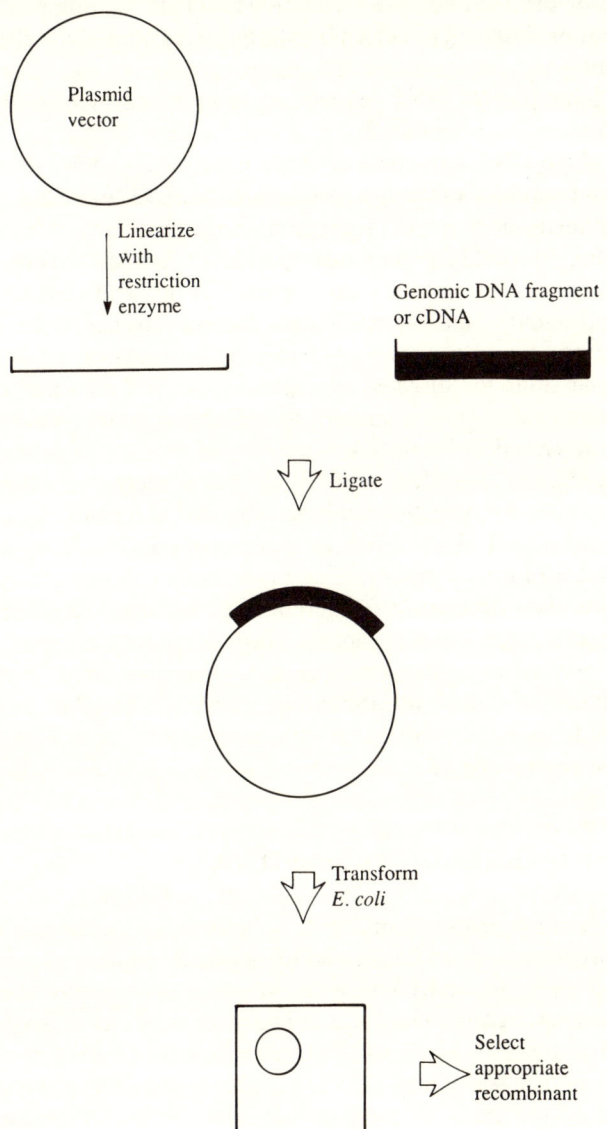

Figure 1 Outline of the basic protocol for constructing recombinant DNA molecules.

used to characterize the gene and its protein product including a number that rely on hybridization or immunological techniques. Of these, some of the most important are those designed to determine the base sequence of the cloned DNA, as this represents the ultimate molecular characterization of a gene as a protein-coding unit. Techniques for DNA sequencing and the information that this can provide are considered in Section 3.

Knowledge of the DNA sequence of a gene immediately allows the prediction of the amino-acid sequence of the gene product in the absence of the protein itself. Analysis of the functionally important regions in the protein can then be carried out by changing or mutagenizing one specific nucleotide within the DNA sequence of the gene, leading to a single, defined amino-acid change within the protein. The particular value of site-directed mutagenesis lies in the fact that it permits precise structure–function relationships to be established for any protein sequence. Applications for directed mutagenesis in the functional characterization of proteins and its potential for generating novel pharmaceutical compounds is described in Section 4.

Finally, this chapter considers some of the more recent and therefore less exploited techniques for genetic manipulation and discusses their potential application in the design of novel bioactive oligopeptides. Chief amongst these techniques are those for the construction of conditional, dominant lethal alleles of cloned genes as these offer powerful methods for identifying regions of proteins that interact with other cellular factors such as activators and substrates. Mutations of this nature offer enormous, as yet unrealized, potential for therapeutic strategies aimed at disrupting unwanted cellular processes by preventing specific essential protein–protein interactions from occurring. This potential is discussed in Section 5.

2. Isolation and expression of cloned DNA

All procedures to clone genes rely on the same basic components: generating and joining DNA fragments, introducing the recombinant molecules into a cell in which they can replicate, and identifying the clone of recipient cells that has acquired the desired recombinant. A number of possibilities exists for each of these stages, with choices between the different options depending on a variety of factors such as what is known about the cellular expression of the gene being cloned, whether its presence in the cell imparts an easily identifiable phenotypic change in the host cell, why the gene is being cloned (for example, for the further study of the gene itself or for the production of large amounts of the gene product) and the nature of the recipient cell.

The basic strategy for isolating any gene therefore frequently depends on being able to select the clone of interest from within a population of recombinant DNA molecules rather than constructing one specific recombinant. Such a population of recombinant molecules that comprises sufficient members so as to contain nearly all the genes present in a particular organism is usually referred to as a *gene library*, and the major problem faced in cloning any gene will usually be how to select a specific DNA molecule from the gene library. For eukaryotic cells, the collection of recombinant molecules can be either a genomic library, for which quasi-random fragments of genomic DNA are produced by partial digestion with a restriction enzyme and cloned into a plasmid or bacteriophage vector; or a cDNA library, in which the DNA to be cloned is synthesized from mRNA molecules by reverse transcription. These two forms of library differ significantly, since genomic libraries will contain all chromosomal DNA sequences of the organism regardless of whether those sequences are being expressed and whether or not they occur in protein-coding regions (i.e. exons). In contrast, cDNA libraries lack any intron sequences, being derived from mature polyA$^+$ RNA from which introns have been removed by splicing. More importantly though, cDNA libraries only contain copies of genes that were being actively transcribed in the tissue or cells from which the polyA$^+$ RNA was isolated, so that the composition of a cDNA library, unlike a genomic library, will depend on the nature and developmental state of the original cell or tissue. When expression of the cloned gene is of paramount importance the use of a cDNA library is usually obligatory (but see below).

While the choice of the appropriate gene library is of obvious importance, another key choice to be made is that of the vector since in many cases this will dictate the method used to isolate the clone of interest; this in turn is influenced by what we know about the target gene or gene product. In general, most screening techniques involve colony or plaque hybridization with radiolabelled or immunological probes. The power of these methods is such that specific clones can be isolated when the sequence of as little as 10–15 amino acids in the gene product is known.

One of the most versatile vectors for cloning eukaryotic DNA into bacteria in the form of cDNA is a derivative of the bacteriophage λ termed λgt11 (Fig. 2). This vector carries the *E. coli lacz* gene within which is a unique site for the restriction enzyme EcoRI which can be used as a cloning site for cDNA molecules carrying EcoRI linkers. In recombinants derived from λgt11 the *lacz* gene is insertionally inactivated so that the phage fail to produce a functional β-galactosidase. This can be detected by using a simple indicator dye, X-gal, in the agar medium. However, in about one-sixth of the recombinants, the cDNA will be in the correct orientation to maintain the translational reading frame from β-galactosidase to the cDNA. These clones will still have an inactive β-galactosidase, but will now generate fusion proteins in which the N-terminal component is derived from β-galactosidase and the C-terminal component by translation of the cDNA. Lysogens derived from λgt11 recombinants produce significant amounts of fusion protein after induction with IPTG; this protein can be readily detected using antibodies against the cDNA product (Fig. 2), making λgt11 an excellent vector in situations where antibodies against purified antigen are available as well as for nucleotide hybridization screening protocols.[2,3]

In many cases though, alternative approaches have to be used either because antibodies are unavailable or because the tissue in which the target gene is

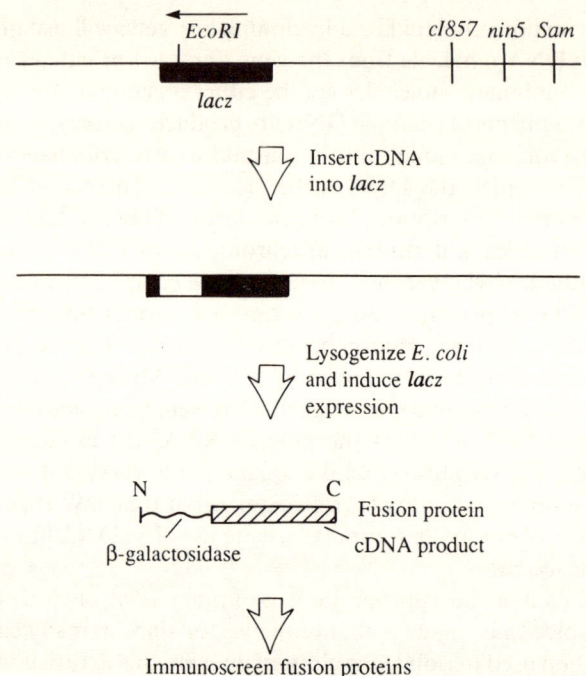

Figure 2   The structure of λgt11 and its use in detecting cloned genes by immunoscreening.

expressed is not known. A good example of the power of selection protocols in this sort of situation is provided by the cloning of the human factor VIII gene which was performed in 1984.[4-7]

2.1. CLONING AND PRODUCTION OF HUMAN FACTOR VIII[4-7]

Human factor VIII is a component of the clotting cascade in which a small stimulus at the beginning of the cascade is amplified through a series of reactions involving about 30 polypeptides leading ultimately to the conversion of soluble fibrinogen to its insoluble form, fibrin, which stabilizes the platelet plug at the site of the wound to arrest bleeding. The absence of any one factor in the cascade blocks the whole process and causes prolonged bleeding. The two most common haemophilic disorders are caused by decreased function of either factor VIII (haemophilia A) which accounts for about 80% of the total cases, or factor IXa (haemophilia B or Christmas disease). Although both these disorders can be controlled by treatment with the appropriate purified plasma concentrate, these products can be contaminated with adventitious agents, most notably HIV-1 and hepatitis viruses. The obvious advantages and benefits that could be derived by producing human factor VIII by genetic engineering, with a potential world market of about £200M in 1990, clearly outweigh any theoretical risks. Nonetheless, merely cloning the factor VIII gene posed a considerable technical challenge.

The site of factor VIII synthesis in the body was not known, nor were antibodies to the purified protein available. It was known that the factor VIII gene was localized on the X chromosome, and a short amino acid sequence from a factor VIII tryptic peptide had been determined. From this peptide sequence it was possible to predict a nucleotide sequence that could encode this region of the factor VIII protein and chemically to synthesize an oligonucleotide with this sequence (see Section 4). By using radiolabelled oligonucleotide as a probe to screen a library of human genomic DNA, derived from a XXXXY cell line, in a phage λ vector by plaque hybridization, clones were isolated that contained part of the factor VIII gene. These clones were then used to isolate other genomic fragments carrying factor VIII sequences and to isolate cDNA fragments that were eventually assembled by subcloning to form a full-length factor VIII cDNA molecule.

The complexity of this process arises in part from the size of the factor VIII gene and its product. Within the X chromosome, the factor VIII gene spans a massive 186 kbp or nearly 0.1% of the total chromosome. The factor VIII protein contains 2351 amino acids including a 19-residue signal sequence and is coded for by a 9 kb mRNA, implying that the majority of the chromosomal sequences in factor VIII are introns. In terms of producing biologically active factor VIII protein, a major problem to be overcome is one that is associated with the synthesis of most eukaryotic proteins, specifically that the active form of the gene product is derived from the primary translation product by extensive post-translational modification. Factor VIII, in common with most serum proteins, is extensively glycosylated, but most importantly, this glycosylation is essential for factor VIII activity. Factor VIII protein synthesized in bacteria or yeast would presumably then be inactive as a clotting agent since bacteria lack glycosylation systems, while those of yeast produce a different glycosylation pattern to that found in human cells.

Synthesis of active factor VIII from the cloned gene therefore requires expression of the cDNA in mammalian cells where the glycosylation patterns are very similar to these of humans. To achieve this, the factor VIII cDNA was subcloned into a plasmid vector so that its expression was under the control of a viral promoter. Factor VIII protein produced in both hamster and monkey kidney cells transfected with this plasmid was active in a sensitive biochemical assay and reduced the clotting time of plasma from haemophiliacs. In both types of cell, although active factor VIII was synthesized, the initial levels of expression were low, with only about 1–5% of the normal plasma concentration of factor VIII in the cell medium. While this is sufficient for initial, pilot studies on the gene and its product, commercial applications for therapeutic compounds require not just basal levels of expression but rather maximization of the expression of the specific gene product.

2.1.1. *Optimizing gene expression*

Optimization of expression can usually increase the original yield several fold by taking into account a number of factors. Foremost amongst these is the strength of the promoter being used to direct transcription, for although heterologous promoter function can frequently be obtained, efficient gene expression using regulatable, strong promoters usually requires the use of a vector-associated homologous promoter element. In *E. coli*, this can be done so that the cloned gene produces its own protein product or, more frequently, a fusion protein.

The three most commonly used expression systems in *E. coli* rely on promoter elements derived from the *lac* and *trp* chromosomal operons and the λp_L promoter. These are naturally efficient, regulatable promoters and all contain base sequences at either the -35 or -10 homologies that approximate to the consensus sequences for these regions.[8] However, by combining part of the *trp* promoter with part of the *lac* promoter, two artificial promoters, *tacI* and *tacII*, have been constructed in which both conserved promoter elements approximate more closely to the consensus.[9] These promoters are inducible by IPTG and are about ten and eight times stronger respectively than the *lac* promoter (Fig. 3). However, the *lac* promoter has an advantage that, in concert with all or part of the *lacz* gene, cloned DNA can be expressed as part of a fusion protein with β-galactosidase.

	−35 Homology		−10 Homology
lac	TTTACA	(18)	TATATT
trp	TTGACA	(17)	TTAACT
λP_L	TTGACA	(17)	GATACT
tacI	TTGACA	(16)	TATAAT
tacII	TTGACA	(17)	TTTAAT
Consensus	TTGACA		TATAAT

Figure 3 Comparison of the sequences of some natural and synthetic bacterial promoter elements.

Amongst the factors that must be considered other than promoter strength are those that affect the translational efficiency, such as codon bias and mRNA secondary structure. A less tangible variable to optimize, though, is the host-cell physiology and the effect of growth conditions and growth rate on gene expression, accumulation, and secretion. These are particularly important parameters for potential commercial applications in which initial small-scale production is scaled-up for large-scale cell growth in fermenters. As yet, there has been no systematic study on the effect of host cell physiology on the production of heterologous protein in either bacteria, yeast or mammalian cells.

2.2. CLONING AND EXPRESSION OF INSULIN[10]

The difficulties inherent in the production of a large protein like factor VIII are not necessarily reduced or ameliorated when the required protein is small like insulin or β-endorphin. In both these cases, cloning and production strategies are constrained by the fact that both of these polypeptides are normally derived from larger precursor molecules by specific polyproteolytic processing. Nonetheless both insulin and β-endorphin can be produced in bacteria using strategies that make use of their limited amino acid composition.

Insulin is synthesized in humans in the β-islet cells of the pancreas as a single polypeptide chain, preproinsulin. This has a signal peptide at the N-terminus that is cleaved as the remainder of the molecule, proinsulin, is transported across intracellular membranes to be stored in vesicles within the cell. Conversion of proinsulin to the active hormone, insulin, occurs within the membrane vesicles by proteolytic cleavage leaving the mature A and B chains from the N- and C-termini of proinsulin respectively, held together by disulphide bridges.

Insulin controls the level of blood glucose, and diseases characterized by lack of insulin can be fatal if untreated. Treatment involves a continuing series of insulin injections to maintain the blood insulin concentration. Insulin for this was usually obtained from pigs or cows, although the use of porcine insulin was sometimes associated with complications as a consequence of slight differences in the amino-acid sequence, making the production of recombinant human insulin an attractive target.

The procedure used to synthesize human insulin in bacteria makes use of three features of insulin: first, it is not modified by glycosylation or phosphorylation; second, because it is small, with 21 amino acids in the A chain and 30 in the B chain, not all of the 20 common amino acids are present in insulin; third, the complete amino-acid sequence of both A and B chains is known. From this sequence it was possible to predict a nucleotide sequence that could encode the A and B chains in bacteria, taking account of codon bias in *E. coli* rather than the natural coding sequence for preproinsulin in humans. Two oligonucleotides were then synthesized chemically (see Section 4), one of 63 nucleotides for the A chain and another of 90 nucleotides for the B chain. These were both extended by three nucleotides at their 5′ termini to add on an ATG codon for methionine and at their 3′ termini to add a translation termination signal so that their overall length was 69 and 96 nucleotides respectively. These synthetic genes for insulin A and B chains were then separately inserted into the *E. coli lacz* gene carried on a plasmid vector such that the reading frame was maintained between β-galactosidase and the insulin chains (Fig. 4). In bacteria, on induction of β-galactosidase expression, these plasmids direct the synthesis of fusion proteins with β-galactosidase sequences at the N-terminus and insulin A or B chain at the C-terminus. The hybrid proteins constituted about 20% of the total cell protein and precipitated as insoluble aggregates that could be redissolved in guanidinium hydrochloride and formic acid.

Release of insulin A and B chains from the fusion proteins exploits the methionine linker between the *lacz* and insulin sequences and the limited amino acid composition of insulin, since neither A nor B chain contains an internal methionine or tryptophan residue. Treatment of the fusion protein with

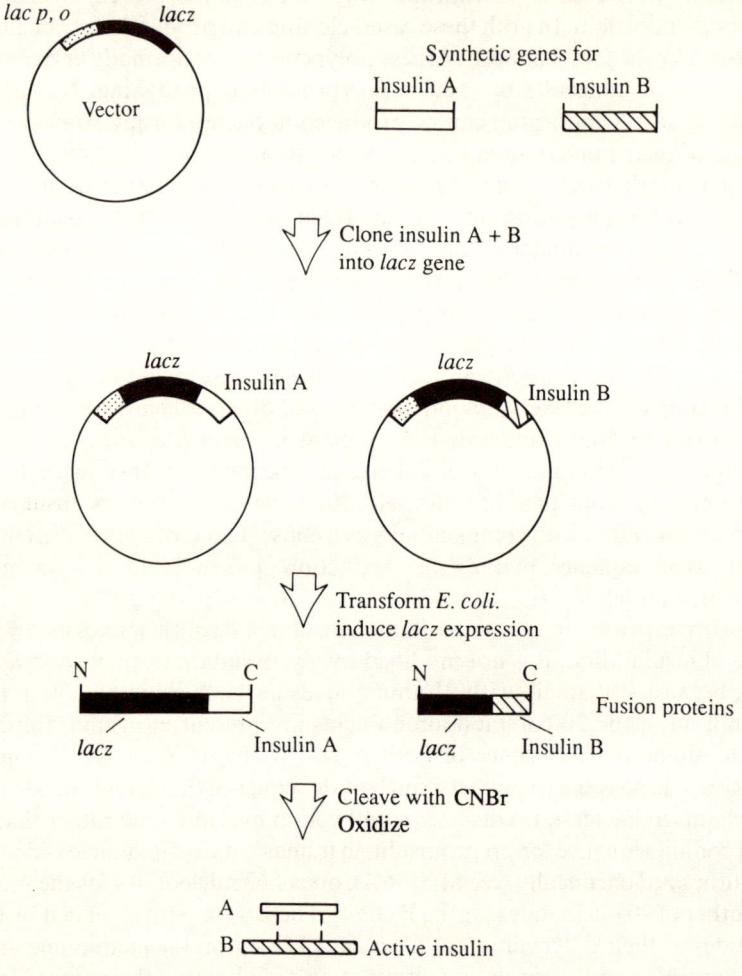

Figure 4 Outline of the method used to produce insulin in *E. coli* using synthetic genes for insulin A and B chains.

cyanogen bromide therefore releases the A and B chains intact; these can be reconstituted into active insulin by disulphide bridge formation using sodium dithionate and sodium sulphite. Insulin produced in this way is biologically active, although the yield of native insulin from disulphide bond formation is relatively poor. More recent efforts to produce recombinant insulin have used DNA specifying the entire proinsulin molecule that, after synthesis, folds correctly and can be converted to active insulin by proteolytic cleavage after purification. An increasing proportion of the world market for insulin (around 2000 kg/year) can be expected to be met by recombinant human insulin produced in this way.

2.3. CLONING AND EXPRESSION OF β-ENDORPHIN[11]

The problems of insulin production are also to some extent encountered in the synthesis of β-endorphin in bacteria, although a different strategy has to be used to produce a protein that contains internal methionine. β-endorphin is a 31 amino-acid opioid peptide synthesized in the pituitary as part of a larger, complex protein that is proteolytically processed to produce at least seven active hormones (Fig. 5). The β-endorphin sequences are at the C-terminus of this precursor protein.

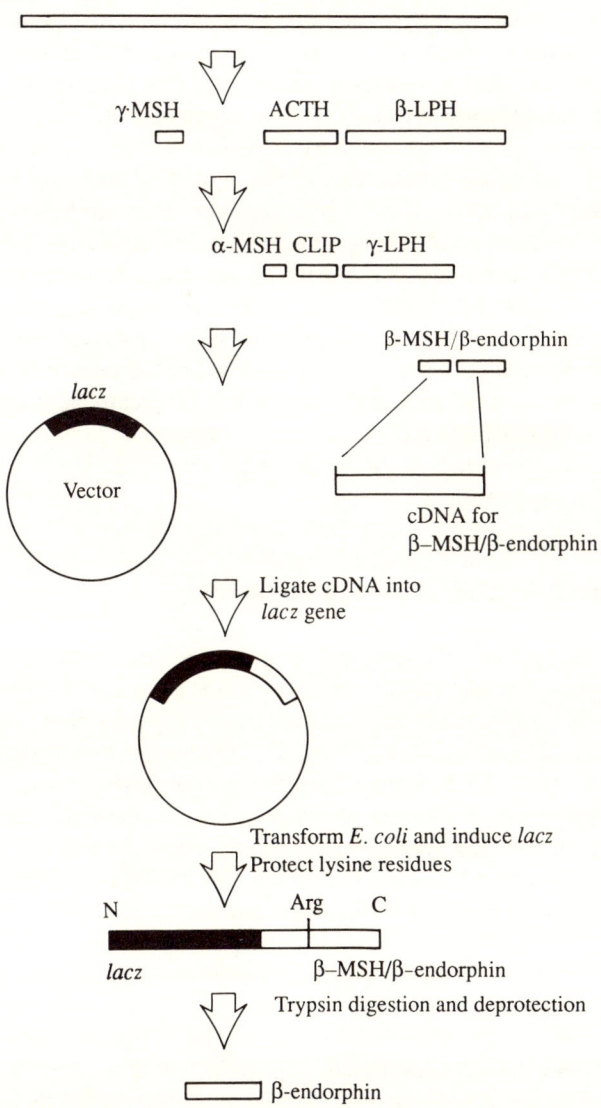

Figure 5 The method used to clone and express β-endorphin in *E. coli*.

To clone the β-endorphin coding sequences in a form that could be used to synthesize potentially active β-endorphin in bacteria required significant technical manipulation of a cloned cDNA for the ACTH/β-LPH precursor protein. A subfragment of this cDNA encoding part of β-MSH was cloned into the *E. coli lacz* gene carried on a plasmid, such that the reading frame between β-galactosidase and the β-MSH/β-endorphin was maintained. After transferring into *E. coli* and inducing with IPTG, this plasmid directs the synthesis of a fusion protein in which the C-terminal component comprises part of β-MSH linked via a connecting dipeptide to β-endorphin. To this point then, the synthesis of β-endorphin is similar to the production of recombinant insulin as a fusion protein, even to the extent that the fusion protein precipitates as an insoluble aggregate.

Unlike insulin though, β-endorphin contains an internal methionine residue which precludes the use of cyanogen bromide to liberate mature β-endorphin. In this case, cleavage of β-endorphin from the fusion protein precursor relies on the fact that β-endorphin is preceded in the β-MSH/β-endorphin molecule by an arginine that can act as a site for proteolytic cleavage by trypsin. Mature β-endorphin contains no other arginine residues, while the five lysine residues in β-endorphin that would also be sites for trypsin hydrolysis can be protected from enzymatic attack by modification with citraconic anhydride. Thus, after dissolving the fusion protein and citraconylation, β-endorphin was released by trypsin digestion and deprotected by removal of the citraconic groups at pH 3. The β-endorphin produced in this way was biologically active as assayed by its ability to bind to opiate receptors in brain membrane preparations and to elicit an opiate-like response in inhibiting the stimulation by prostaglandin E_1 of cAMP in an appropriate cell line.

3. Characterization of cloned DNA

The work and techniques described so far in this chapter have served to establish the basic principles and practices used to isolate genes and to express these cloned genes in heterologous cells to make proteins or peptides of pharmaceutical value or potential. However, in addition to these methods for gene isolation and expression, molecular biologists have also accumulated protocols and skills that can be used to investigate the pharmaceutical or pharmacological potential of previously undefined gene products and to modify in precisely defined ways the properties and activities of known gene products. These techniques and some of their applications are described in this section.

3.1. DETERMINATION OF DNA SEQUENCE

Cloned DNA can be characterized in a number of ways for a variety of purposes. Digesting DNA with a range of restriction enzymes enables the construction of a map of the relative position of the different restriction sites in the cloned DNA, which in turn allows a physical orientation of the DNA and facilitates further subcloning. Hybridization techniques, which rely on the fact that single-stranded nucleic acids can be irreversibly immobilized on a nitrocellulose or nylon filter while retaining the capacity to form double-stranded structures by correct base-pairing with a complementary polynucleotide, are particularly important in this process of characterization. However, the ultimate characterization of any DNA molecule is the determination of the base sequence of that molecule, and it is the development of techniques for rapid accurate DNA sequencing that have been responsible as much as anything for the revolution in biological sciences since the late 1970s.

At that time, two quite different methods for DNA sequencing were described. One of these, developed by Maxam and Gilbert,[12,13] uses double-stranded DNA and has as its central component the cleavage of that DNA using chemical reagents that act specifically at a particular base. Several variations of this method have been described, differing in the particular chemicals used to break the DNA. This method has a significant disadvantage (for the sequencer!) that the chemicals used are all extremely toxic.

In contrast, the method developed by Sanger and his colleagues[14] relies not on DNA degradation but rather on DNA synthesis and the specific inhibition of enzymatic synthesis of DNA by deoxyribonucleoside triphosphate analogues. The key to the Sanger method is the ability to isolate the DNA whose sequence is to be determined in a single-stranded form, and the widespread current use of Sanger's method for sequencing has resulted in part from the development of a series of cloning vectors that produce large amounts of single-stranded DNA. These vectors are based on the *E. coli* filamentous phage M13.[15]

M13 and other filamentous phage such as f1 and fd only infect male bacteria, probably by binding to receptors specified by the F-factor. The phage particle carries a single-stranded circular DNA molecule of about 6400 nucleotides which on infecting a bacterium is converted to a double-stranded replicative form (RF). This RF multiplies in the cell to produce a pool of RF molecules which in turn are used to generate mature single-stranded phage genomes. These are ultimately packaged and released from the infected cell, although unlike most coliphage, release of the phage particles is not associated with host lysis. Infected cells thus become factories for the production of phage particles with single-stranded genomes, leading to extremely high phage titres in the medium.

The life-cycle of M13 has a number of advantages for the molecular biologist. The amount of DNA that can be packaged into a phage particle can be varied enormously; the phage life-cycle includes phases of double-stranded DNA that can be used for cloning and manipulation, and single-stranded DNA that could be used as a template for DNA synthesis and sequencing; and both single- and double-stranded DNA can be used to transfect *E. coli*. However, wild-type M13 itself is a poor cloning vector, lacking good cloning sites and a simple selection or screening system to identify recombinant molecules.

These drawbacks have been remedied by the construction of the M13mp series of vectors which have revolutionized DNA sequencing to the extent that sequencing the entire human genome is now a feasible proposition.[16] M13 has 10 genes that are essential for phage replication, with only a small region of 507 nucleotides that is non-essential. This region was used to introduce two modifications into the M13 genome. The first was to clone into this region a DNA fragment from the *E. coli* genome carrying part of the *lacz* operon, specifically the *laci* gene, the promoter–operator complex and part of the *lacz* gene (*lacz'*) that codes for the β-galactosidase α-peptide. This segment of the *lacz* gene is regulated normally but synthesizes an enzymically inactive peptide that will complement the amino-terminally deleted β-galactosidase encoded by the host chromosomal *lacz*ΔM15 mutation by protein–protein interaction. Functional β-galactosidase activity can be detected as blue plaques by including the dye X-gal in the agar media.

The second modification was to clone into the N-terminal region of the *lacz'* gene a synthetic polylinker that carried a number of unique sites for a variety of different restriction enzymes. This polylinker cloning site maintains the *lacz'* reading frame to synthesize a modified but functional α-peptide that is inactivated on cloning DNA fragments into the polylinker, thereby providing a simple selection procedure for recombinant phage. Since all DNA fragments in the polylinker are flanked by the same *lacz'* sequences a single oligonucleotide primer, universal primer, can be used to prime all DNA synthesis reactions for sequencing using the M13 clone as template. The sequencing reactions themselves rely on two features of DNA polymerase, or more accurately the Klenow fragment of DNA polymerase which lacks the $5'$–$3'$ exonuclease activity: these are the ability to copy a DNA template accurately, and the inability to extend a chain that has incorporated a $2'3'$dideoxynucleotide analogue, since this lacks a $3'$-OH group for extension. By using dideoxy analogues of the four dNTPs, newly synthesized DNA molecules are terminated at specific points in their synthesis dependent on their sequence. By including a radiolabel in the growing DNA chain, resolving the products by acrylamide gel electrophoresis such that molecules differing in size by a single nucleotide can be resolved, and detecting the products by autoradiography, the DNA sequence of the cloned fragments can be read directly from the autoradiograph. More recent improvements to this basic method have included the use of fluorophores and automatic fluorescence detection,[17] and semi-automation of the entire sequencing protocol, such that it is now possible to read about 8000 nucleotides of sequence from a single gel.

3.2. WHAT CAN WE LEARN FROM DNA SEQUENCES?

The importance of the information that can be obtained from DNA sequencing cannot be overestimated. It allows a molecular analysis of the sequences that surround the gene and contribute to the regulation of expression of the gene. This has been particularly important in establishing the rules for promoter function, especially in the design and exploitation of efficient, regulatable highly expressed promoters designed to maximize gene expression. DNA sequencing also identifies the coding region of the gene and predicts the amino-acid sequence of the gene product in the absence of purified protein. This can be important both where the function of the gene product is known in advance and the gene has been cloned for the express purpose of synthesizing that particular gene product, as well as in cases where the cellular function of the protein is known but the enzymic activity of the protein is not. In these situations, the predicted amino-acid sequence of the gene product can be analysed by a number of sophisticated algorithms to predict higher-order structures, potential epitopes within the protein, possible functional domains and perhaps most importantly, by comparing the sequence with databases of other known amino-acid sequences, the potential enzymatic activity of the gene product.

In the case of human factor VIII for example, the amino-acid sequence of the protein predicted from the DNA sequence of the gene had an obvious domain structure with repeated blocks of related sequences. Comparison of the sequence of the factor VIII protein with other sequences revealed that it had significant homology with another clotting protein, factor V, and more surprisingly to another plasma protein, ceruloplasmin, a copper-binding protein whose likely function in the plasma is to act as an oxidizing agent specifically to convert Fe^{2+} to Fe^{3+}. The significance of this latter homology is unclear but suggests that factor VIII may also be a copper-binding protein, although what role this may play in the (as yet unknown) mechanism by which factor VIII catalyses the activation of factor Xa is a matter for speculation.

Comparison of predicted amino-acid sequences with databases of other known protein sequences can also yield insights into the activity of gene products with important biological functions and can help to define functional domains within related proteins. These conserved, functional domains can then be used as targets for direct or indirect therapeutic intervention (see Section 4). An example of this that also illustrates the way in which studies on fundamental biological systems can have important potential pharmaceutical applications comes from studies aimed at understanding the control of cell division in eukaryotes, for which the yeasts *Saccharomyces cerevisiae* and *Schizosaccharomyces pombe* have been used as simple model systems. Work with these organisms has identified two genes that play essential roles in regulating progress through the cell cycle, *CDC*28 in *S. cerevisiae* and *cdc*2 in *S. pombe*. These gene products are functionally homologous and retain their function in the heterologous organism. The molecular mechanism by which they control cell division was immediately suggested when the genes were cloned and sequenced, since their predicted amino-acid sequences contained a region of about 200 residues that was homologous to regions of a number of other proteins, all of which were known to be protein kinases. This family comprises over 70 proteins and includes oncogene products, growth factor receptors, and hormone receptors, and suggests that the control of cell division and cell cycle progress is achieved at least in part by protein phosphorylation.[18-23]

Computer analysis of the predicted *cdc*2 protein kinase sequence also identified potential antigenic epitopes within the protein sequence. Short oligopeptides corresponding to these sequences were synthesized, coupled to carrier proteins, and used to raise antibodies in rabbits. These antibodies had a high affinity for the $cdc2$ gene product even though they were raised to a synthetic oligopeptide. Moreover, these antibodies cross-reacted not only with $S.$ $pombe$ $cdc2$ protein but also with a mammalian protein that was subsequently shown to be mitosis/maturation promoting factor (MPF), which regulates cell cycle progress in vertebrate cells and which also contains the functional conserved catalytic domain characteristic of protein kinases.[24] In this case then, molecular biology has used model systems to identify proteins that regulate cell division and cell proliferation, and which offer themselves as targets for strategies aimed at controlling cell proliferation either by modification of the protein itself (see Section 4) or by the design of specific inhibitors of this particular protein kinase activity (see Section 5). Moreover, the homology between the yeast and vertebrate proteins means that in this case, inhibitors can be tested in the relatively cheap and simple yeast systems before progressing to mammalian cells.

Perhaps the most remarkable application of genetic engineering over the last five years though has been in the design of pharmaceuticals for AIDS therapy, to replace the chemotherapeutic agent AZT which helps in many cases but which can cause problems of toxicity. Infection by HIV-1 has been shown to require the binding of a viral surface glycoprotein, gp120, to a specific antigen on the surface of susceptible cells, the CD4 antigen, which is found principally on a subset of mature T-lymphocytes. Attempts to block this interaction and hence the infective process by raising neutralizing antibodies to gp120 have been largely unsuccessful. The alternative approach of trying to block gp120 binding to cellular CD4 using soluble CD4 in the plasma has been more promising, but any long-term therapeutic value of this is constrained by the short (15 min) half-life of soluble CD4. However, with the sequence and predicted structure of CD4 it was possible both to identify the region of the CD4 molecule that was recognized by gp120 and to construct a DNA molecule to express a hybrid fusion protein, an immunoadhesin, that carried the CD4 gp120-binding region at the N-terminus and part of a human IgG heavy chain at the C-terminus.[25] This protein retains characteristics of CD4 in that it bound gp120 but also had some of the properties of the immunoglobulin component in being able to bind to the F_c receptor (but not C1q) and had a markedly increased plasma half-life of about 48 h. Hybrid molecules such as these immunoadhesins offer enormous potential for the treatment of a variety of infectious diseases, while the scope for their design and production can only increase with further fundamental studies on the molecular mechanisms of viral infection.

4. Modification of cloned DNA

While DNA sequences can be interpreted to yield a wide range of information about particular gene products, features of the gene products such as functional domains and catalytically important residues within the active site remain merely predictions in the absence of confirming biochemical evidence. This is frequently difficult to obtain and, until recently, required inhibitor or chemical modification studies with purified gene products. Now, though, methods are available for the specific alteration of proteins in a directed manner by introducing nucleotide changes into the cloned gene, so-called *in vitro mutagenesis*. Techniques for this range from those that introduce relatively gross changes into the DNA, such as the removal of a complete restriction fragment, to more limited changes that can be produced by chemical modification of bases in single-stranded regions of DNA after restriction enzyme cleavage. However, the most powerful and versatile method for mutagenesis *in vitro* relies on the use of synthetic oligonucleotides to introduce a single, specific nucleotide change at a defined point within a gene, allowing the phenotypic effect of an individual amino-acid substitution to be assessed.

Site-directed mutagenesis is possible because of the concurrent emergence of methods for DNA sequencing and for the rapid, efficient synthesis of long oligonucleotides with a precisely defined base sequence. Until only a few years ago, it was a formidable task to synthesize an oligonucleotide containing as few as 10 residues with significant yield. Now though, a number of different chemistries exist for the rapid, automated synthesis of oligonucleotides up to 100 residues long in significant yield with a cycle time of less than 10 min per residue.

Chemical synthesis of oligonucleotides is based on the ability to form $5'-3'$-phosphodiester bonds while preventing reactions between all of the other potentially reactive groups on the nucleosides. The various chemistries differ in the specific reactions used to synthesize the phosphodiester bonds and the blocking groups used to prevent the unwanted side-reactions. The cycle of synthesis for the phosphoramidite method of synthesis is summarized in Fig. 6. This method is particularly widely used because of the relative stability of the starting materials and the high coupling efficiencies that can be obtained. The synthesis is carried out with the growing nucleotide chain coupled to a solid support such as silica so that excess reagents can be easily removed by filtration.

The starting material for synthesis is the solid support coupled with the derivatized nucleoside that will be the $3'$ nucleotide of the final product coupled to the column via a spacer arm to the $3'$-OH group of deoxyribose. The exocyclic amines of adenine and cytosine are protected by benzoyl groups, the exocyclic amine of guanine by an isobutyryl group while thymine is not modified since it carries no reactive group. The first step in the synthesis is the activation of the nucleoside coupled to the support by removing the dimethoxytrityl $5'$-blocking group with trichloroacetic acid to leave a free $5'$-OH for the addition. The next step, addition, occurs by activating the phosphoramidite with tetrazole to form a highly reactive protonated derivative that readily forms a $5'-3'$ internucleotide link on nucleophilic attack by the $5'$-OH, leaving a dinucleotide coupled by trivalent phosphorus. At this point, unreacted groups are capped by treating with acetic anhydride and diaminopyridine to prevent them taking part in any subsequent rounds of synthesis. Finally the round of synthesis is completed by oxidizing the phosphorus to the more stable pentavalent form in a reaction

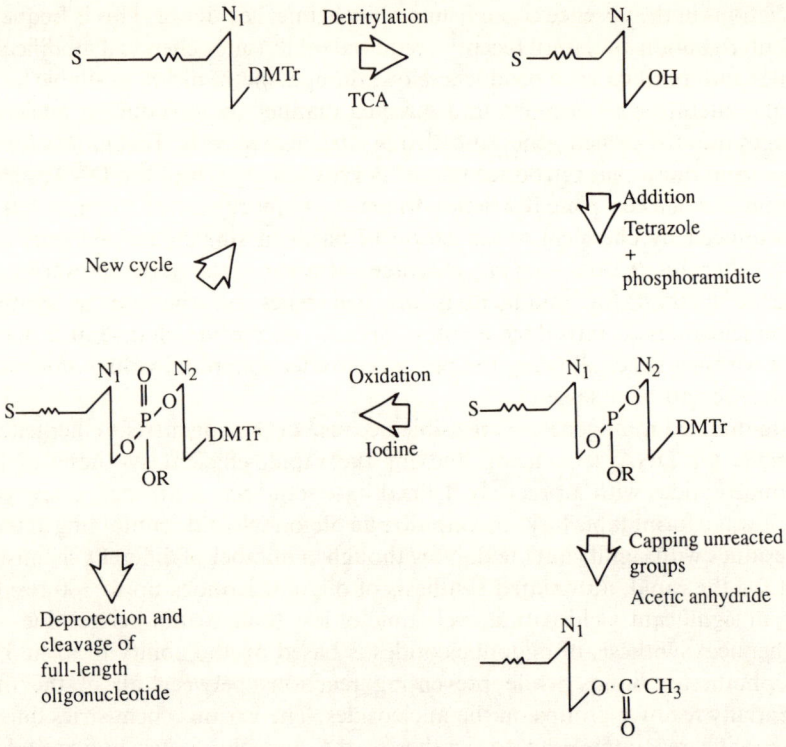

Figure 6 The reaction cycle for oligonucleotide synthesis using phosphoramidite chemistry.

catalysed by iodine. This cycle is then repeated until the required chain elongation is complete, when the nucleotide is removed from the column with ammonia and deprotected by heating in ammonia at 55°C for 12–18 h.

Oligonucleotides themselves have been used to alter patterns of gene expression in several model systems ranging in complexity from bacteria to mammalian cells. In general, this approach uses an oligonucleotide that is an anti-sense molecule to inhibit transcription or translation of the target gene. The application of oligonucleotides in this way is likely to be increased by the development of new chemistries to allow the incorporation of base analogues or modified bases into the growing chain.

One of the most important applications of oligonucleotides is their use, coupled with the availability of single-stranded cloned DNA in M13 vectors, to generate defined point changes in the nucleotide sequence of cloned genes leading to single, specific amino-acid changes in the resulting protein product. Any chosen nucleotide within a long DNA fragment can be changed to any one of the other three bases. When the base sequence of a cloned DNA molecule has been determined, this sequence is used to synthesize an oligonucleotide (usually 17–20 residues) that is complementary to the region to be mutated, but which contains within it the single base change for the required mutation. The mismatched oligonucleotide is annealed to the complementary gene sequence (which occurs despite the mismatch), where it serves as the primer for the synthesis of the remainder of the second strand by DNA polymerase, effectively creating a double-stranded molecule with a single mismatched base pair at the site of mutation (Fig. 7). After introducing the DNA into *E. coli*, the mismatches will be repaired and/or replicated to produce wild-type or fully mutant molecules. In theory, half of the resultant phage should be mutant and half should be wild-type; in practice, considerably fewer mutant phage are found than theory indicates, suggesting that the wild-type sequence is somehow marked for preferential repair.

A number of routes have been developed to increase the frequency with which mutant phage can be recovered. These depend largely on being able to incorporate a selection bias into the second-strand synthesis such that only progeny phage derived from the synthesized mutant strand can survive subsequent rounds of replication in *E. coli*. For example, preparing the original M13 single-stranded template in a strain of bacterium that carries mutations in the *dut* and *ung* genes will result in the incorporation of uracil into the M13 template in place of thymine. After second-strand synthesis *in vitro*, transfection of a wild-type strain of *E. coli* should result in the preferential degradation of all phage strands containing uracil, so that progeny phage will only be formed from the mutant strand. Alternatively, a second oligonucleotide primer can be used in conjunction with the mutant primer to direct the second-strand synthesis. By using a second primer that also carries a mismatch to part of the M13 component of the template, this selection primer can be used to change a restriction enzyme site or an amber mutation in an essential M13 gene. In this last case, transfection of the double-stranded DNA into a strain of *E. coli* that is sup^- allows progeny phage to be formed from the newly-synthesized strand in which the amber codon has been mutated to wild-type, but not from the original template strand which still carries the amber mutation. These protocols all result in high recovery of mutant phage with frequencies in excess of 70%.

While the techniques of DNA cloning, sequencing, and site-directed mutagenesis can be combined to generate specific amino-acid changes in any protein, most studies to date have used these techniques to focus on the question of protein structure and function. That is, for any protein, what effect does changing one amino acid for another have on the structure and folding of the protein, and how does this change in protein structure influence the cellular and enzymatic activity of the protein? In the absence of precise structural information though, this question is usually limited to asking how the activity of a protein is modulated by specific residues within the overall amino acid sequence.

An important example of work of this sort is provided by studies on the oncogene tyrosine kinases that are responsible for the transforming capacity and tumourigenicity of a wide range of retroviruses. Understanding the way that these proteins work is important not only for the insight that it can give into the origins of cancer and the regulation of growth, but also for its potential in

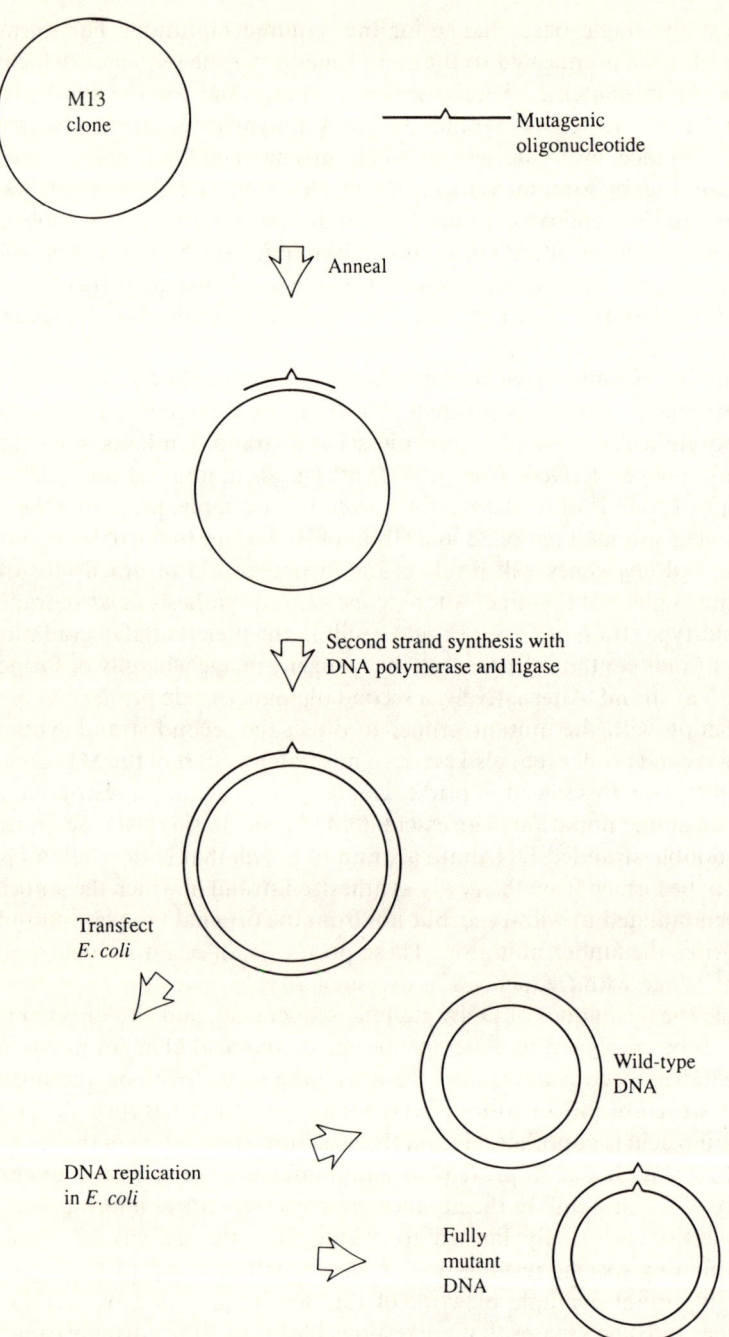

M13 clone

Mutagenic oligonucleotide

Anneal

Second strand synthesis with DNA polymerase and ligase

Transfect E. coli

Wild-type DNA

DNA replication in E. coli

Fully mutant DNA

Figure 7 The use of mismatched oligonucleotides to generate mutations in DNA cloned into M13 single-strand vectors.

identifying possible targets for therapeutic intervention. Mutagenesis studies have been performed on *ras*, *fps*, *mos*, and *src*, the latter three being characterized as tyrosine kinases for which *v-src* and its product $pp60^{v\text{-}src}$ serve as the prototype.

The product of *v-src* contains 526 amino acids and like all other protein kinases is itself a phosphoprotein, the major sites of phosphorylation being serine-17 and tyrosine-416. In addition, lysine-295 plays a key role at the catalytic site by stabilizing the binding of ATP prior to the transfer of the γ-phosphate to the substrate protein. An equivalent essential lysine residue is found in all known protein kinases. Like most eukaryotic protein kinases, the catalytic region of $pp60^{v\text{-}src}$ is located towards the C-terminus of the protein, while the N-terminus contains a membrane binding domain within which the N-terminal glycine is modified by myristylation. Each of these key residues can be mutated individually and their effect on cellular $pp60^{v\text{-}src}$ monitored to determine which components of the protein are needed for its kinase activity, and whether $pp60^{v\text{-}src}$ contains regions other than the kinase domains that are essential for its cellular function. These studies have provided a number of insights into the role of $pp60^{v\text{-}src}$ in oncogenesis and in particular have served to distinguish separate functions needed for transformation and tumourigenicity.[26] In identifying these non-kinase functional domains though, an important question to try to answer is what other protein molecules interact with these domains (if any), and how can this information be applied to prevent transformation and tumour formation by the oncogene protein kinase? Approaches to this are described in the following section.

5. Designs for the future

The application of genetic engineering and recombinant DNA technology to the pharmaceutical industry is in its infancy, with hardly any of the 20 or so biotechnology-derived drugs on the market offering radical advances in the cures for diseases. Rather, to date, nearly all of the proteins that the biotechnology industry can produce and market from heterologous organisms using recombinant DNA techniques, belong to a sub-family of naturally occurring proteins whose therapeutic potential has already been established. Few, if any, novel proteins have yet been produced by rational, directed synthesis. Probably the nearest that we have yet come to such a molecule are two of the approaches targeted at HIV-1 and AIDS therapy, using CD4 immunoadhesins (see Section 4) or tailored pseudo-virus-like particles carrying part of the HIV-1 gp120 surface protein. This situation is unlikely to continue for much longer, given the rate at which the technology of modern molecular research is developing.

Recent advances in gene synthesis and genetic engineering allow us to construct DNA molecules encoding any desired amino acid sequence. To design new proteins, then, we can either modify existing genes or synthesize a completely new DNA sequence, allowing the transition from genetic engineering to protein engineering. The power of protein engineering lies not just in its theoretical implications, because in concert with X-ray crystallography and sophisticated computer algorithms it can be used for the rational design and synthesis of useful, novel proteins for numerous medical applications. It should be possible to modify the optimal reaction conditions, substrate specificity, or substrate-binding affinity of an enzyme; or to design and synthesize entirely novel peptides and proteins that have a precisely defined, predicted activity. This all presupposes concurrent advances not only in the technology of protein and genetic engineering, but in the associated areas of drug delivery and targeting systems and most importantly, in our basic understanding of the way that proteins fold to form higher-order structures.

It is clear that molecular biology has advanced sufficiently since the mid 1970s to allow the bulk production at reasonable efficiency of natural proteins with known therapeutic uses. Further fundamental research will inevitably lead to a better understanding of the way that these proteins can be produced efficiently in large scale using a variety of different specialized host cell lines or strains. What is less immediately obvious is how new proteins with therapeutic value can be designed, synthesized, and exploited; in particular, what criteria or experimental data could be used in the design process and what constraints would this impose on the synthesis? One possible scenario for the design of novel, synthetic peptide pharmaceuticals is described below.

The basis of one possible method for generating novel bioactive peptides lies not in preventing the activity of an enzyme directly, but rather in preventing the protein from forming specific interactions with activating or other ancillary polypeptide factors. This in turn requires that we can identify within a protein those residues or side-chains that are involved in specific protein–protein contacts, and build up from this a three-dimensional picture of the interaction site. An avenue that might lead to such modelling is provided by the ability to construct mutant forms of cloned genes, which are capable of inhibiting the action of the wild-type gene product in the cell, thus causing the cell to be deficient in the function of that gene product. Such mutations are referred to as dominant negative mutations or alleles since their effect is seen even in the presence of the wild-type protein.[27] Mutants with these properties arise because proteins have multiple functional sites or domains that can be altered independently. Thus, dominant negative mutations in general will produce proteins that are capable of carrying out some but not all of the set of functions inherent in the normal wild-type protein. For example, if substrate is limiting, a dominant negative mutant form of the protein might retain the capacity to bind the substrate and effectively compete out the normal protein, but be unable to catalyse the conversion of the substrate to product.

How then might dominant negative mutations be exploited? This can best be illustrated by way of a specific example. Several gene products have been identified in yeast that play a key role in controlling cell growth and division (see Section 3). These proteins are protein kinases, containing the short stretches of amino acids that are conserved in the catalytic domain of all protein kinases. These proteins are also homologous to some of the proteins in maturation promoting factor (MPF) that performs a similar regulatory function in the cell cycle of vertebrate cells. In the yeast *S. cerevisiae*, there is evidence that control of this *CDC28* protein kinase activity is achieved by the assembly–disassembly of a multiprotein complex in which the protein kinase is associated with other proteins of unknown function.[28] While simple chemical inhibitors of protein kinase activity would interfere with this and all other protein kinases in the cell, preventing *CDC28* kinase activity by prohibiting the formation of the required complex would be specific for this particular enzyme and would interfere with cell proliferation without affecting other cellular functions. Moreover, inhibitors of *CDC28* function in yeast might well have a similar inhibitory effect on cell division in vertebrate cells.

To synthesize such inhibitors requires firstly that we can identify those residues within the *CDC28* protein kinase whose side-chains are involved in holding the multiprotein complex together. This could be done by constructing a clone for the conditional expression of the gene, for example by making its expression inducible by galactose, then treating the DNA *in vitro* with chemicals to introduce random mutations into the gene. After transferring the DNA into a normal cell, those clones that have acquired a dominant negative mutation could be identified by their inability to grow when plasmid gene expression was induced by galactose since under these conditions the normal cellular protein encoded by the chromosome would be competed out by the mutant form of the gene product specified by the plasmid. Characterizing a number of dominant negative alleles by DNA sequencing would establish different 'families' of mutation, each defective in a single, different aspect of the overall function of the protein.

To exploit these observations, though, we need to consider what this information really tells us. Each family of dominant negative mutations is defective in a different functional aspect, say for example binding to one component of a polyprotein complex. What this identifies then is the amino acid side-chains that are important in forming and maintaining the complex, which by X-ray crystallography or computer modelling can be assigned to specific coordinates in the three-dimensional protein structure. Thus, if a synthetic peptide could be synthesized in which the key side-chain groups were in the same relative positions held together by short linker peptides, such a synthetic oligopeptide should in theory form a non-functional complex in the cell by replacing the *CDC28* protein kinase in the complex, hence preventing cell division. In this way then, genetic engineering could be intrinsically coupled to protein engineering to generate synthetic, catalytically inactive oligopeptides with potential use as growth inhibitors.

This represents only one way in which current genetic engineering techniques could be coupled with other methodologies to generate novel synthetic peptide and protein pharmaceuticals. While it is always difficult to predict scientific progress, there would seem to be little doubt that early next century, bioengineered synthetic peptides and proteins will be a major therapeutic route for many human pathological conditions. Enormous challenges and rewards are clearly available.

References

1. *Principles of Gene Manipulation* R.W. Old and S.B. Primrose (1985) Blackwell Scientific Publications, Oxford.
2. 'Efficient isolation of genes using antibody probes' R.A. Young and R.W. Davis (1983) *Proc. Natl. Acad. Sci. USA*, **80**, 1194–1198.
3. 'Immunoscreening lambda gt11 recombinant DNA expression libraries' R.A. Young and R.W. Davis (1985). In: *Genetic Engineering*, Vol. 7 (J. Setlow and A. Hollender, eds.) Plenum Press, New York.
4. 'Characterisation of the human factor VIII gene' J. Gitschier, W.I. Wood, T.M. Goralka, K.L. Wion, E.Y. Chen, D.H. Eaton, G.A. Vehar, D.J. Capon and R.M. Lawn (1984) *Nature*, **312**, 326–330.
5. 'Expression of active human factor VIII from recombinant DNA clones' W.I. Wood, D.J. Capon, C.C. Simonsen, D.L. Eaton, J. Gitschier, B. Keyt, P.H. Seeburg, D.H. Smith, P. Hollingshead, K.L. Wion, E. Delwart, E.G.D. Tuddenham, G.A. Vehar and R.M. Lawn (1984) *Nature*, **312**, 330–337.
6. 'Structure of human factor VIII' G.A. Vehar, B. Keyt, D. Eaton, H. Rodriguez, D.P. O'Brien, F. Rotblat, H. Oppermann, R. Keck, W.I. Wood, R.N. Harkins, E.G.D. Tuddenham, R.M. Lawn and D.J. Capon (1984) *Nature*, **312**, 337–342.
7. 'Molecular cloning of a cDNA encoding human antihaemophilic factor' J.J. Toole, J.L. Knopf, J.M. Wozney, L.A. Sultzman, J.L. Buecker, D.D. Pittman, R.J. Kaufman, E. Brown, C. Shoemaker, E.C. Orr, G.W. Amphlett, W.B. Foster, M.L. Coe, G.J. Knutson, D.N. Fass and R.M. Hewick (1984) *Nature*, **312**, 342–347.
8. 'Compilation and analysis of *Escherichia coli* promoter DNA sequences' D.K. Hawley and W.R. McClure (1983) *Nucl. Acids Res.*, **11**, 2237–2255.
9. 'Vectors bearing a hybrid *trp-lac* promoter useful for regulated expression of cloned genes in *Escherichia coli*' E. Amman, J. Brosius and M. Ptashne (1983) *Gene*, **25**, 167–178.
10. 'Expression in *Escherichia coli* of chemically synthesised genes for human insulin' D.V. Goeddel, D.G. Kleid, F. Bolivar, H.L. Heyneker, D.G. Yansura, R. Crea, T. Hirose, A. Kraszewski, K. Itakura and A.D. Riggs (1979) *Proc. Natl. Acad. Sci. USA*, **76**, 106–110.
11. 'Expression of cloned β-endorphin gene sequences by *Escherichia coli*' J. Shine, I. Fettes, N.C.Y. Lan, J.L. Roberts and J.D. Baxter (1980) *Nature*, **285**, 456–461.
12. 'A new method for sequencing DNA' A.M. Maxam and W. Gilbert (1977) *Proc. Natl. Acad. Sci. USA.*, **74**, 560–564.
13. 'Sequencing end-labelled DNA with base-specific chemical cleavages' A.M. Maxam and W. Gilbert (1980) *Methods in Enzymol.*, **65**, 499–560.
14. 'DNA sequencing with chain-terminating inhibitors' F. Sanger, S. Nicklen and A.R. Coulson (1977) *Proc. Natl. Acad. Sci. USA*, **74**, 5463–5467.
15. 'A system for shotgun DNA sequencing' J. Messing, R. Crea and P.H. Seeburg (1981) *Nucl. Acids Res.*, **9**, 309–321.
16. 'A new pair of M13 vectors for selecting either DNA strand of double-digest restriction fragments' J. Messing and J. Vieira (1982) *Gene*, **19**, 269–276.
17. 'Fluorescence detection in automated DNA sequence analysis' L.M. Smith, J.Z. Sanders, R.J. Kaiser, P. Hughes, C. Dodd, C.R. Connell, C. Heiner, S.B.H. Kent and L.E. Hood (1986) *Nature*, **321**, 674–679.
18. 'Gene required in G_1 for commitment to cell cycle and in G_2 for control of mitosis in fission yeast' P.M. Nurse and Y. Bissett (1981) *Nature*, **292**, 558–560.
19. 'The selection of *S. cerevisiae* mutants defective in the START event of cell division' S.I. Reed (1980) *Genetics*, **95**, 561–577.
20. 'Functionally homologous cell cycle control genes in budding and fission yeast' D. Beach, B. Durkacz and P.M. Nurse (1982) *Nature*, **300**, 706–709.

21. 'Primary structure homology between the product of the yeast cell division control gene *CDC28* and vertebrate oncogenes' A. T. Lorincz and S. I. Reed (1984) *Nature*, **307**, 183–185.
22. 'Sequence of the cell division gene *cdc2* from *Schizosaccharomyces pombe*' J. Hindley and G. A. Phear (1984) *Gene*, **31**, 129–134.
23. 'The protein kinase family; conserved features and deduced phylogeny of the catalytic domain' S. K. Hanks, A. M. Quinn and T. Hunter (1988) *Science*, **241**, 42–52.
24. '*cdc2* is a component of the M phase-specific histone H1 kinase; evidence for identity with MPF' D. Arion, L. Meijer, L. Brizuela and D. Beach (1988) *Cell*, **55**, 371–378.
25. 'Designing CD4 immunoadhesins for AIDS therapy' D. J. Capon, S. M. Chamow, J. Mordenti, S. A. Marsters, T. Gregory, H. Mitsuya, R. A. Byrn, C. Lucas, F. M. Wurm, J. E. Groopman, S. Broder and D. H. Smith (1989) *Nature*, **337**, 525–531.
26. 'A mutation at the ATP-binding site of pp60$^{v\text{-}src}$ abolishes kinase activity, transformation and tumorigenicity' M. A. Snyder, J. M. Bishop, J. P. McGrath and A. D. Levinson (1985) *Mol. Cell. Biol.*, **5**, 1772–1779.
27. 'Functional inactivation of genes by dominant negative mutations' I. Herskowitz (1987) *Nature*, **329**, 219–222.
28. 'Control of the yeast cell cycle is associated with assembly/disassembly of the *CDC28* protein kinase complex' C. Wittenburg and S. I. Reed (1988) *Cell*, **54**, 1061–1072.

—— Chapter 7 ——————————————————————

Physiological and pharmacological evaluation of peptide analogues

R. J. Knapp, T. P. Davis, T. F. Burks and H. I. Yamamura

I. Introduction

Peptides are key regulators of a wide variety of physiological processes. Peptide hormones produced by the pituitary regulate many critical functions including growth, reproduction, lactation, water balance, and adrenal function. The secretion of most of these hormones is in turn controlled by neuroendocrine peptides produced by the hypothalamus. Gastrointestinal peptides like cholecystokinin and gastrin influence many digestive processes, while the pancreatic peptides insulin and glucagon are essential for the regulation of carbohydrate metabolism. The recognition that peptides can act as neurotransmitters in the central and peripheral nervous systems further establishes the essential role that this diverse group of chemical substances plays in the control and co-ordination of many physiological processes.

Peptide hormones such as insulin and growth hormone have been used as pharmaceuticals for the treatment of deficiency disorders for many years. Until recently, these peptides had to be isolated from animal tissues. The advent of recombinant DNA technology has permitted the large-scale production of some of these peptides from bacteria transfected with genes encoding these hormones. A second approach now being explored is the chemical synthesis of analogues for endogenous peptides. It is sometimes possible to identify portions of an endogenous peptide which contain the pharmacophore, or bioactive part of the peptide, responsible for its bioactivity. An example is the synthesis of SMS-201-995, an eight-amino-acid peptide, which has the full agonist activity of somatostatin, a 14-amino-acid peptide.[1] The identification of such peptide analogues facilitates the development of purely synthetic peptide pharmaceuticals.

The development of peptide analogues is in many ways similar to the development of more conventional drugs. The basic goals are the same: to produce increasingly potent and selective compounds. Likewise, the methods used for the determination of the biological characteristics of peptides are in many ways similar to those used for other types of compounds. However, the susceptibility of many peptides to proteolytic degradation and their poor absorption from the gut has limited them largely to parenteral administration. It is also clear that most neuropeptide analogues do not readily penetrate into the CNS from the systemic circulation. Thus, the methods used for the screening of peptide analogues must also take into account differences in their stability and ability to reach their sites of action.

The screening of new compounds for their interactions with tissue is an essential part of the drug development process. The analysis of these interactions is ideally a multidisciplinary effort where the bioactivity of new compounds is tested by *in vivo* and *in vitro* bioassays, while receptor interactions are measured directly by radioligand binding methods. Drug development cannot proceed in the absence of effective analytical methods, and the success of a developmental program can hinge on the validity of the analytical methods used for the characterization of the drugs under investigation.

The development of peptide drugs is particularly dependent on bioactivity analysis, because the large size and structural complexity of these molecules greatly impedes the analysis of their conformation. The ability of peptides to assume a large number of different conformational states makes the determination of structure–activity relationships much more difficult than is the case for simpler organic compounds.[2] The conformational flexibility of peptides also facilitates their interaction with different receptors. This makes it difficult to predict the effect of even a minor alteration in a peptide's structure on its bioactivity. Thus, it is essential that the chemists responsible for the synthetic effort receive accurate and timely information on the biological properties of the analogues produced.

1.1. IDENTIFICATION OF APPROPRIATE ASSAYS

The key to successful peptide development is the identification of analytical procedures that measure the desired biological activity of the peptide. Such assays are often based on the activities of prototype drugs in intact animals or isolated animal tissues. Since peptides being developed for drug use are typically analogues of endogenous peptides, the endogenous peptide itself may serve as the prototype drug. The ability of an endogenous peptide to serve as a prototype for the development of an analytical procedure can depend on its stability, since nearly all naturally occurring bioactive peptides are rapidly degraded by tissue. The difficulty of determining the extent of peptide degradation *in vivo* and the general inability to prevent such degradation limits the value of *in vivo* assays for the study of such labile peptides. Better control of peptide degradation can be achieved with *in vitro* bioassays and radioligand binding measurements. Thus, these types of studies are preferable for screening novel peptide analogues with unknown susceptibility to degradation.

2. Analysis of peptide–receptor interactions I: bioassay

2.1. BIOASSAY TECHNIQUES

The objective of any bioassay used for the screening of drugs is to quantitate their relative potency for producing or antagonizing a particular response. This response must in some way be related to the desired *in vivo* effect of the drug. The establishment of this relationship constitutes the validation of the assay. This is not a serious obstacle in some cases, since many peptides regulate clearly defined processes such as the secretion of a hormone which can be studied directly *in vitro*. However, most neuropeptides do not have such a clearly defined function and the validation of any bioassay using animals that attempts to model human CNS activity is difficult.[3]

A second concern is the sensitivity of the assay since it is important to be able to measure the activities of different analogues which are likely to vary considerably in their potency. The ability of the bioassay to measure the maximal response of each of the compounds tested is important for the determination of their relative efficacies in addition to their relative potencies. A sensitive bioassay enables the discrimination of partial agonists from full agonists and can thus facilitate the development of antagonists. Differences in efficacy are of particular importance in the screening of peptides that may have partial agonist activity. Full agonists with low efficacy can be incorrectly identified as partial agonists if the tissue used for their bioassay has a low receptor concentration. Another concern is the variability of the results obtained with different preparations used in the same assay procedure. The statistical quantification of the sensitivity and variability of a bioassay is essential for the meaningful comparison of different drugs.[4–6]

Both *in vivo* and *in vitro* assays can measure the biological activity of a compound. They differ in that the response measured by an *in vivo* assay depends on many factors (e.g. absorption, distribution, and metabolism) beside the interaction of the drug at receptors which in turn may be present in different tissues within the organism. *In vitro* bioassays can measure a more direct response to a drug which is limited to a specific tissue. While the response measured in an isolated tissue preparation is a more direct measure of the actual potency of a drug at the receptor, it cannot be generalized to other tissues and may even vary considerably between tissues. However, the simplicity of measuring a response in an isolated tissue preparation is an advantage in peptide screening.

An important consideration for both *in vitro* and *in vivo* bioassays is that drug effects other than those desired may occur. The test compound may act at multiple receptor types or even on enzymes. The ability to detect multiple actions of a test compound can be an advantage since it allows the prediction of potential side effects. However, the potential interaction of test compounds with different components of the preparation requires that the investigator be able to distinguish between the primary action of the agent and its secondary effects.

2.1.1. in vitro *bioassays*

In vitro bioassays may employ a tissue section or intact organ obtained from an animal, but they can also use primary cell cultures or continuously cultured cells of a clonal cell line. The response measured may be either physiological, such as muscle contraction in a guinea pig ileum preparation, or biochemical, like the secretion of a hormone or the release of a neurotransmitter. Some of the most sensitive bioassays, using cytochemical methods, can measure a response within a single cell.[7] Since the response measured in the tissue preparation may be quite different from the desired pharmacological response in the intact organism, it is essential to establish a correlation between them.

In addition to this consideration is the issue of the identity of the receptor, or receptors, acted upon in an *in vitro* preparation. This depends on the receptors present and the pharmacological selectivity of the agents being tested. The interpretation of results from a bioassay generally rests on the assumption that the activity of the peptide agents being tested occurs through their interaction with a specific receptor in the preparation that is acted on by a prototype drug. It may be necessary to actually demonstrate this relationship to confirm the validity of the assay results. If an antagonist of the prototype drug is available this relationship can be established by the measurement of the antagonist's pA_2 value against both the prototype and the peptide.[8] An *in vitro* preparation is ideal for this since the concentrations of the agonists and antagonist can be precisely controlled but this method can also be applied to *in vivo* assays.[9]

The number of bioassays available for the characterization of different peptides exceeds the number of peptides which have been isolated. Each assay has different characteristics with respect to specificity, sensitivity, and reproducibility. A partial list of established assays for several different types of biologically active peptides is given in Table 1.

2.1.2. in vivo *bioassays*

The bioactivity of most peptides can be studied *in vitro* and it is generally preferable to do so, at least initially, since these assays give a much more direct measurement of the potency of the peptide. Not all peptide bioactivities can be measured *in vitro*, however. Some biological activities depend on an integrated response involving different tissues and possibly different receptors for the same peptide. One example of this is opioid-induced analgesia. Analgesia is a complex result of the co-ordinated activities of many neurons and there is considerable controversy regarding the identity and distribution of the opioid receptors responsible for the mediation of this effect.[10] However, the ability of many of these *in vivo* assays to predict the human analgesic response to opioids is well

Table 1 Bioassays for various biologically active peptides

Peptide	Reference
Adenohypophyseal hormones	
Follicle stimulating hormone	99
Luteinizing hormone and follicle stimulating hormone	100
Prolactin	102
Growth hormone	102
Adrenocorticotropin hormone	103
Neurohypophysial hormones	
Vasopressin and oxytocin	104
Hypothalamic releasing hormones	
Corticotropin releasing hormone	105
Thyrotropin releasing hormone	105
Luteinizing hormone releasing hormone	105
Melanocyte-stimulating hormone inhibiting factor	105
Melanocyte-stimulating hormone releasing factor	105
Growth hormone releasing hormone	105
Somatostatin	105
Opioids	
In vitro bioassays	106
In vivo bioassays (analgesia)	107
Pancreatic hormones	
Insulin	108
Glucagon	109
Gastrointestinal hormones	
Cholecystokinin	110
Gastrin	110
Secretin	110

established.[11] Some of the characteristics exemplified by these assays, including dependence on the site and method of drug administration and assay dependent differences in the observed efficacies of the different families of opioids, are typical of *in vivo* bioassays. Thus, some of these assays will be described here in detail.

The analgesic potencies of opioids differ depending on the assay used. Essential variables include the type (thermal, mechanical, chemical) and intensity of the nocive stimulus, the route of administration (intraperitoneal (i.p.), intrathecal, intracerebroventricular), and the species of animal selected. Tyers[12] has shown that while certain opioids like ketazocine and nalorphine are essentially inactive in thermal tests of analgesia like the hotplate or warm water tail immersion tests, they are potent analgesics in chemical or mechanical tests like acetylcholine-induced writhing and paw pressure. He attributes these observations to differences in the opioid receptor types mediating the response to these stimuli.

A characteristic of analgesia assays is that they do not generally attempt to measure the maximum effects of the analgesic agents. This is due to the ability of the stimulus to injure tissue after prolonged administration. Thus, the application of the stimulus is typically restricted to a time interval short enough to avoid tissue damage. If the animal does not react to the stimulus by the end of this interval, then the response is defined as the 'maximum possible effect'. A result of this approach is that differences in the efficacies of some analgesic drugs cannot be determined by these assays.

2.1.2.1. *Analgesia assays using thermal stimuli*
Thermal analgesic bioassays include the (radiant heat) tailflick test for rats[13] or mice,[14,15] the (warm water) tail withdrawal test for rats,[16] and the hotplate test for mice.[17] A review of these tests and their variations is given by Fennessy and Lee.[18]

The *tailflick* and *warm water tail withdrawal tests* are similar in terms of the type of stimulus (i.e. heat) and the response measured so they tend to produce similar results (morphine $ED_{50} \approx 4.0$–4.6 mg/kg i.p.). There is some question as to whether the tailflick or tail withdrawal is mediated only at the spinal level, or if supraspinal mechanisms play a role. The evidence available suggests that both are involved.[19]

A modification of the warm water tail withdrawal assay, in which a cold ($-10°$) thermal stimulus is used has been recently described.[20] In this assay the tail of a rat is immersed in water–ethylene glycol mixture maintained at $-10°$ by a refrigerated cold water circulating bath. It is possible to demonstrate analgesic activity of κ opioid agonists, including dynorphin A, following intracerebroventricular administration using this assay as well as agonists for the μ and δ opioid receptor types.

The *hotplate test* differs from the other thermal tests in that supraspinal mechanisms are clearly involved since the response measured (either hindpaw lick or escape) requires the co-ordinated movement of the animal's whole body. The sensitivity of the test can be increased by reducing the temperature from $55°$ to $49.5°$ which permits better estimates of the analgesic potency of mixed agonists–antagonists.[21]

2.1.2.2. *Analgesia assays using mechanical stimuli*
The two most common mechanical pain tests in use are the *tail clip method*[22,23] and the *paw pressure test* of Randall and Selitto[24] as modified by Swingle et al.[25]

The *tail clip test* consists of attaching a surgical artery clip to the tail of an animal and measuring the time required for the animal to vocalize, bite at the clip, or otherwise attempt to escape. This test is less reproducible than others since the stimulus intensity varies between different artery clips.

The *paw pressure test* involves the application of a blunt conical point to the volar surface of a rat's paw.[24] The pressure is generally regulated by a mechanical device such as an 'Analgesy-Meter' (Ugo Basile, Milan, Italy). The sensitivity of the test can be increased by the prior inflammation of the paw with a subcutaneous injection of brewer's yeast.[25] The endpoint measured can either be the withdrawal of the paw or a vocalization response by the animal. This test is quite sensitive as the ED_{50} for morphine (subcutaneous) is between 0.9 to 3.0 mg/kg (inflamed paw) and 1.4 to 3.8 mg/kg (uninflamed paw).

2.1.2.3. *Analgesia assays using chemical stimuli*
The most commonly used chemical model of pain is the *writhing* or '*abdominal stretch reflex*' paradigm.[26] The subjects used are generally mice, but rats and other animals will also show similar responses. In this test, the animal is given an intraperitoneal injection of an irritant which causes it to contract its abdominal muscles while arching its back. The endpoints measured can either be the development of the response, or the number of responses which occur over a specific period of time. A variety of irritants have been used including 0.02% 2-phenyl-1,4-benzoquinone,[26] 0.6% acetic acid,[27] or 0.4 mg/kg acetylcholine.[28] The type of irritant used as well as its concentration affects the apparent potency of the drug tested,[28] and must be considered when making comparisons between different studies.

The writhing or 'abdominal stretch' test differs qualitatively from the thermal and mechanical tests as the type of pain produced is different. The intraperitoneal injection of an irritant produces a long-term 'visceral' type of pain which may resemble that produced by disease processes like cancer more than the acute localized pain obtained with the thermal and mechanical stimuli. The stimulus produced by irritants is presumably fairly mild since it is blocked by low doses of analgesics and the animals often engage in normal eating or exploration behaviour between adbominal constrictions. The writhing test is sensitive to most opioid and non-opioid analgesic drugs and their potency in this test correlates well with their potency in human subjects.[28] The test is very sensitive to opiates, as the ED_{50} value for morphine is around 0.45 mg/kg.[18]

2.1.2.4. *Administration of drugs*
Most of the non-peptide opiates are effective analgesics after systemic administration, including subcutaneous (s.c.), intraperitoneal (i.p.), intravenous (i.v.) routes. This is not true for some of the opioid peptides which may be degraded at systemic sites and may be unable to enter the central nervous system (CNS) from the circulation. Peptide opioids can produce an analgesic response after systemic administration as DADLE,[29] DAGO,[30] FK-33,824,[31] and metkephamid[32] (see Table 2 for structures) all produce limited analgesia after systemic administration, though at relatively high doses compared to other routes, but as a general rule peptide opioids are much more potent after intracerebroventricular (i.c.v.) or intrathecal (i.t.) administration. These are not equivalent routes of administration, however. For some test paradigms, κ opioid agonists are potent analgesics in the spinal cord, but are essentially ineffective at supraspinal sites.

The assays just described have been used to screen opioid peptides for analgesic activity and illustrate many of the characteristics of *in vivo* assays. The dependence of the response measured on the site and route of administration is especially important in the assay of peptides and has also been demonstrated for peptide hormones.[33] This problem is especially relevant for the bioassay of neurotransmitter peptides which have a limited ability to enter the CNS after systemic administration. However, this can sometimes be an advantage since it can be exploited to distinguish central from peripheral effects.

Table 2 Opioid peptide structures

Endogenous enkephalins
[Leu^5]enkephalin   Tyr–Gly–Gly–Phe–Leu
[Met^5]enkephalin   Tyr–Gly–Gly–Phe–Met

Linear enkephalin analogues
DADLE   Tyr–D-Ala–Gly–Gly–D-Leu

FK 33–824   Tyr–D-Ala–Gly–ψ-(–C–N–)–Phe–Met(O)-ol, with $\overset{O}{\overset{\|}{C}}$ and N bearing CH_3

DAGO   Tyr–D-Ala–Gly–ψ-(–C–N–)–Phe–NH–$(CH_2)_2$–OH, with $\overset{O}{\overset{\|}{C}}$ and N bearing CH_3

Metkephamid   Tyr–D-Ala–Gly–Phe–ψ-(–C–N–)–Met-NH_2, with $\overset{O}{\overset{\|}{C}}$ and N bearing CH_3

Cyclic enkephalin analogues
[D-Pen^2, L-Cys^5]enkephalinamide   Tyr–D-Pen–Gly–Phe–L-Cys-NH_2
[D-Pen^2, L-Cys^5]enkephalin   Tyr–D-Pen–Gly–Phe–L-Cys
[D-Pen^2, D-Pen^5]enkephalin   Tyr–D-Pen–Gly–Phe–D-Pen

Linear μ opioid agonist
PL-17   Tyr–Pro–ψ–C–N–Phe–D-Pro-NH_2, with $\overset{O}{\overset{\|}{C}}$ and N bearing CH_3

Cyclic μ opioid antagonists
SMS 201, 995   D-Phe–Cys–Phe–D-Trp–Lys–Thr–Cys–Thr-ol
CTP   D-Phe–Cys–Tyr–D-Trp–Lys–Thr–Pen–Thr-NH_2
CTOP   D-Phe–Cys–Tyr–D-Trp–Orn–Thr–Pen–Thr-NH_2

3. Analysis of peptide–receptor interactions II: radioligand binding

3.1. RADIOLIGAND BINDING TECHNIQUES

Radioligand binding can serve as a powerful tool in the development of new peptides. Radioligand binding is used to measure the affinity of the peptide at a specific receptor. By measuring the affinity at two or more receptors it is possible to determine the selectivity of the peptide. While radioligand binding does not provide a measurement of a peptide's potency, which depends on both its affinity and efficacy at a receptor, a great deal of information can be obtained from as little as a milligram of peptide. Discussions of radioligand binding techniques have been recently provided by Burt[34] and by Bennett and Yamamura,[35] while problems specifically related to the measurement of peptide ligand binding are discussed by Hanley.[36]

Radioligand studies can be divided into two types: direct (saturation) and indirect (competitive) binding analyses. Direct binding studies are used to measure the binding of the ligand at all of its sites and requires that the ligand be radiolabelled. Thus, direct binding studies are not used for screening, but are valuable for the characterization of peptides selected for further analysis. Competitive binding studies measure the ability of an unlabelled ligand to compete with a labelled ligand for its binding site. By using a radioligand that is selective for the receptor of interest it is possible to determine the affinity of a peptide for that receptor. It is useful to remember that what is actually measured in the indirect type of assay is the ability of the unlabelled ligand to inhibit the binding of the radioligand. This can result from non-competitive (allosteric) as well as competitive interactions (see below).

3.1.1. Radioligands

The use of radioligand binding for the screening of new peptides depends on the availability of a suitable radioligand. The essential characteristics of a radioligand include:

- its affinity and selectivity for the receptor of interest;
- its specific activity and the radioisotope used for labelling;
- the degree to which it binds non-specifically to tissue.

The affinity of the radioligand will affect the concentration at which it is used in the assay and will determine the methods that can be used for the separation of bound from free ligand. The selectivity of the radioligand determines the extent to which it labels more than one site in a given tissue preparation. The specific activity of the radioligand affects the concentration which must be used in the assay to obtain a sufficient amount of radioactivity for measurement while the isotope used determines the methods that can be used for this measurement. Non-specific binding, which consists of non-receptor-bound radioligand that cannot be displaced by a competitive ligand, affects the amount of specific (receptor) binding that can be measured with the radioligand. Accurate IC_{50} values cannot be measured by inhibition assays in which the specific binding is low. The binding properties of the radioligand with respect to these characteristics should be well established before it is considered for use in a screening assay.

The affinity of a radioligand for its receptor site, as measured by its equilibrium dissociation constant or K_d, should be $<10\,\text{nM}$ and preferably $<2\,\text{nM}$. Radioligands with K_d values of $>10\,\text{nM}$ are likely to dissociate from their receptor sites during any wash procedure used to reduce non-specific binding. To selectively label a receptor, a radioligand should have about 100-fold greater affinity for that receptor over any other binding site. It is possible to restrict the binding of a less selective ligand to the desired receptor by either the selection or treatment of the tissue preparation used (see below).

Nearly all radioligands are labelled with either ^3H or ^{125}I. Ligands labelled with ^{125}I have much higher specific activities (1000–2200 Ci/mmole) than those labelled with ^3H (20–80 Ci/mmole). Some peptide hormones can be directly labelled with ^{125}I by oxidative methods using either chloramine-T[37] or lactoperoxidase.[38] However, direct labelling of small peptides or peptides with methionine or cystine amino acids often destroys their ability to bind to receptors. Peptides can sometimes be iodinated by the reaction of a primary amino group (either at the N-terminal or an ε-amino group) with ^{125}I labelled Bolton–Hunter reagent (N-succinimidyl-3-(4-hydroxyphenyl) propionate). Since it is preferable to use the lowest concentration of radioligand possible in competitive binding, the high specific activity of ^{125}I-labelled radioligands is an advantage. Other advantages of the use of ^{125}I labelled ligands are the greater efficiencies of γ-counters ($\approx 70\%$) compared to the liquid scintillation counters used to measure ^3H ($\approx 50\%$), the lower cost of the radioligand, and the elimination of the need to add expensive liquid scintillation cocktail to the samples. The low specific activity of ^3H labelled ligands is compensated for by their much longer half-life (12 years compared to 60 days for ^{125}I labelled ligands) and relative safety. Tritium labelling is also much less likely to affect the interaction of the ligand with its receptor, and some peptides cannot be iodinated without a loss of binding activity.

All ligands, labelled or otherwise, bind non-specifically to tissue. Non-specific binding is operationally defined as that portion of total binding which cannot be blocked by a concentration of a competitive ligand sufficient to occupy all of the receptor sites. The degree to which a ligand non-specifically binds to tissue depends on its ionic character, hydrophobicity, and other physical properties which are intrinsic to the particular compound. The percentage of total binding that is non-specific increases with increasing ligand concentration. Radioligands whose non-specific binding is $>50\%$ of their total binding at their K_d concentration are generally not suitable for inhibition studies. Non-specific binding affects the precision of the IC_{50} values measured in inhibition assays. If non-specific binding is high such that the combined errors in the measurement of total binding (absence of inhibitor) and non-specific binding are a substantial fraction of the difference between them (i.e. specific binding) then the error in the inhibition measurements (presence of inhibitor) will also be large.

3.1.2. Receptor preparation

The use of radioligand binding for screening peptide analogues depends on the availability of a tissue preparation with a sufficient receptor concentration (>10 fmol/mg protein) to allow accurate measurements of binding. The obvious choice for neurotransmitter peptides is CNS tissue, usually brain although it is sometimes advantageous to use spinal cord instead. Examples of peripheral tissues used for binding would include pancreas tissue for cholecystokinin, pituitary tissue for hypothalamic releasing hormones like LHRH, and adipose tissue for insulin binding studies. Sometimes it is not possible to find a tissue with a sufficiently high concentration of receptors for a particular ligand. $[^3H][Nle^4,$

D-Phe^7]α-MSH is an example of such a ligand. Receptors for α-MSH appear to be widely distributed throughout the periphery, but there are no tissues with very high concentrations of these receptors.^{39} We were able to characterize the binding of the α-MSH analogue despite this obstacle by using membranes prepared from murine B16 melanoma cells (unpublished data).

The selectivity of the radioligand can sometimes be enhanced by the choice of a tissue which lacks the other sites recognized by the radioligand. Another approach is to block the competing sites either by using a reversibly binding ligand which is selective for these sites or by irreversibly alkylating them. The complete occupation of a secondary site requires a high concentration (100 times its K_d or more) of a reversible ligand which must be sufficiently selective to not bind the primary site of the radioligand. The expense of this approach can easily become prohibitive for use in screening procedures.

Alkylation can be accomplished either by using a selective ligand having a chemically reactive functional group able to form a covalent bond to the receptor or by the use of the selective protection method. An example of a selective receptor alkylating reagent is β-funaltrexamine which selectively alkylates μ opioid receptors.^{40} A less selective reagent that alkylates all opioid receptor types, but not other kinds of receptors, is β-chlornaltrexamine.^{41} Unfortunately there are few such reagents available for other receptor types, which limits the applicability of this approach.

The selective protection method depends on the ability of a selective, but reversibly binding ligand, to protect its preferred site against inactivation by a nonselective alkylating reagent. Phenoxybenzamine^{42} and N-ethylmaleimide^{43} have been used as alkylating agents for this purpose. Selective protection is performed by incubating tissue with a large excess of the selective ligand followed by the addition of the alkylating reagent. The tissue must then be thoroughly washed to remove the ligand and allow binding to the protected site. It is typically observed that there is a substantial loss of the sites protected even when high concentrations of protecting ligands are used. The success of this approach depends heavily on the selectivity of the protecting ligand.

The relative concentration of receptor to ligand is an important consideration in any radioligand binding assay. If the receptor concentration present in a tissue is low then it will be necessary to use a relatively high tissue concentration in the assay which will result in high non-specific binding. However, a high concentration of receptors in the assay incubates can result in free (unbound) ligand depletion which can introduce serious errors in the analysis of the binding data.^{44} The models used for the analysis of binding data depend on the assumption that the free ligand (labelled or unlabelled) is $>90\%$ of the total ligand added. Depletion of the radioligand in a competitive binding assay can produce a Hill slope value of <1 and may result in either an over- or underestimate of the IC_{50} value, while depletion of the unlabelled ligand will always result in an overestimate of the IC_{50} value.

3.1.3. *Assay conditions*

Many peptides are vulnerable to proteolytic enzyme attack, and the assay conditions used must serve to prevent their degradation (Table 3). This problem is increased for screening assays since large numbers of different analogues must be tested which can vary in their susceptibility to different enzymes. Furthermore, most of the proteolytic enzymes studied have the ability to hydrolyze a variety of biologically active peptides.[45] Thus, it is useful to add a variety of protease inhibitors to the assay medium. Protease inhibitors can be roughly divided into three groups. They include peptides (e.g. bestatin), metal chelators (e.g. EDTA and 1,10-phenanthroline) and irreversible enzyme inhibitors (e.g. phenylmethylsulfonyl fluoride). A list of protease inhibitors which have proved useful is given in Table 3. However, care must be taken to be certain that the proper inhibitors are selected for a given peptide.

Many different assay media have been used in radioligand binding so that it is impossible to specify which is best. In general one should use the medium that was determined to be optimal for the radioligand when it was characterized. If the radioligand selected is not a peptide, then it may be necessary to modify the procedure used for its characterization by adding protease inhibitors and an antiadsorbant to prevent the loss of the unlabelled peptides tested to test tube and pipette surfaces. The most useful antiadsorbant is bovine serum albumin (BSA). While there have been reports that BSA may inhibit the binding of some ligands,[46] our experience has been that reductions of total binding are usually due to a reduction of non-specific binding with specific binding actually being increased. An alternative approach is to siliconize the surfaces of everything used to contain the peptide, including incubation tubes and pipette tips.

If it is necessary to modify the assay medium used for the characterization of a radioligand, then the affinity of the radioligand should be measured by saturation analysis in the new medium to determine if it is different under the new conditions. This is important if K_i values are to be calculated from the IC_{50} values measured during the screening assays (see Section 3.1.6).

3.1.4. *Assay temperature and incubation time*

Peptide degradation can be reduced by decreasing the incubation temperature at a cost of having to increase the incubation time. Most peptides can be safely incubated at 25°C, though some assays are carried out at 0–4°C to reduce degradation. It is important to determine the incubation time required for the radioligand to reach steady state binding at the concentration to be used in the assay. This will depend on the concentration of radioligand used, and there is a trade-off between the desirability of using a low concentration of radioligand (see below) and the shorter incubation time allowed by the use of higher concentrations. If the specific binding of the radioligand declines after reaching a maximum value, then it is likely that it is being degraded and the assay medium or conditions should be changed to prevent this.

Table 3 Proteolytic enzymes and inhibitors

Enzyme	EC number	Sites of action†	Inhibitors
Aminopeptidase M	EC 3.4.11.2	N-terminal L-amino acids excluding X–Pro bonds	puromycin (10^{-6} M) bestatin (10^{-5} M) 1,10-Phenanthroline
Leucine aminopeptidase	EC 3.4.11.1	N-terminal amino acids; does not cleave Lys or Arg	EDTA bestatin
Pyroglutamate aminopeptidase	EC 3.4.19.3	N-terminal pyroGlu	iodoacetamide
Carboxypeptidase A	EC 3.4.17.1	C-terminal cleavage of L-amino acids with free α-carbonyl	EDTA
Carboxypeptidase B	EC 3.4.17.2	C-terminal cleavage of L-Lys and L-Arg β-endorphin: X–Lys^{28}–Lys^{29} X–Tyr^{27}–Lys^{28} X–Ile^{23}–Lys^{24} X–Phe^{18}–Lys^{19} X–Glu^8–Lys^9	L-Arg (5×10^{-4} M) EDTA
Chymotrypsin A	EC 3.4.21.1	cleaves aromatic carboxyl bonds (e.g. Tyr, Phe, Trp) but also Met, Asp, Leu, Glu β-endorphin: Tyr^1–Gly^2–X Phe^4–Met^5 Met^5–Thr^6	aprotinin L-Trp (2×10^{-2} M) chymostatin

Table 3 (cont'd)

Enzyme	EC number	Sites of action†	Inhibitors
		$Phe^{18}–Lys^{19}$ $Tyr^{27}–Lys^{28}$ neurotensin: $\quad Tyr^3–Gln^4$ $\quad Tyr^{11}–Ile^{12}$	
α- and β-Trypsin	EC 3.4.21.4	hydrolysis at Arg–X and Lys–X β-endorphin: $\quad Lys^9–Ser^{10}$ $\quad Lys^{19}–Asn^{20}$ $\quad Lys^{24}–Asn^{25}$ $\quad Lys^{28}–Lys^{29}$ $\quad Lys^{29}–Gly^{30}$ neurotensin: $\quad Lys^6–Pro^7$ $\quad Arg^8–Arg^9$	aprotinin leupeptin soybean trypsin inhibitor
Proline endopeptidase	EC 3.4.21.26	hydrolysis at Pro–X β-endorphin: $\quad Pro^{13}–Leu^{14}$ neurotensin: $\quad Pro^7–Arg^8$ $\quad Pro^{10}–Tyr^{11}$	p-chloromercuric benzoate (10^{-5} M)
Cathepsin C	EC 3.4.14.1	N-terminal cleavage of dipeptides; blocked by N-terminal Lys, Arg, or Pro	iodoacetate formaldehyde
Glycyl-Glycine dipeptidase	EC 3.4.13.11	cleavage of Gly–Gly dipeptides β-endorphin: $\quad Gly^2–Gly^3$	EDTA (10^{-3} M)

Table 3 (cont'd)

Enzyme	EC number	Sites of action†	Inhibitors
Peptidyl dipeptidase A (angiotensin converting enzyme)	EC 3.4.15.1	sequential hydrolysis of dipeptide fragments from C-terminal	captopril (10^{-5} M)
Endopeptidase-24.15	EC 3.4.24.15	cleaves peptides (<22 residues) with non-polar amino acids in P_1, P_2, and P_3^1 positions. β-endorphin: Leu^{14}–Val^{15} neurotensin: Arg^8–Arg^9	N–[1–(R,S)–carboxy-2-phenethyl]-Ala–Ala–Phe–Pab
Endopeptidase-24.11 (enkephalinase)	EC 3.4.24.11	hydrolysis of X–Y bonds where Y is a hydrophobic amino acid (e.g. Phe, Leu, Ile, Val, Tyr, Trp) β-endorphin: Gly^3–Phe^4 Pro^{13}–Leu^{14} Ile^{22}–Ile^{23} Ala^{26}–Tyr^{27} neurotensin: Tyr^{11}–Ile^{12}	thiorphan phosphoramidon 1,10-phenanthroline

† The general description of enzyme activity is followed by examples of specific bonds hydrolyzed in either β-endorphin or neurotensin which are provided for illustration.

3.1.5. *Separation of bound radioligand*

It is seldom practical to screen large numbers of peptides using centrifugation methods, because of the considerable amount of time required to prepare the samples for radioactivity measurement. This makes vacuum filtration the most reasonable separation method. Low-affinity radioligands with K_d values > 10 nM should be avoided since significant radioligand dissociation is likely to occur during filtration wash steps. Glass fibre filters (e.g. Whatman GF/B) allow rapid filtration, while still permitting quantitative retention of the membrane particles produced from tissue by homogenization. Glass fibre filters must be treated prior to filtration to reduce non-specific radioligand binding. Filter binding of some ligands can be greatly reduced by treating the filters with 0.1% polyethyleneimine in distilled water, but we have observed that treating glass fibre filters for 1 h with a solution of 1.0 mg/ml BSA is better for certain peptide radioligands such as $[^{125}I]CCK_8$. It is important to actually measure the binding of radioligand to the treated filters in the absence of tissue under the filtration conditions used in the assay to ensure that the treatment is effective.

3.1.6. *Analysis of binding data*

The analysis of radioligand binding data is based on the use of mathematical relationships derived from models of receptor–ligand interactions which are dependent on a number of basic assumptions.[47–49] These assumptions include:

- that all ligand–receptor interactions (labelled *and* unlabelled) have reached equilibrium;
- that only a small fraction of the total concentrations of the ligands added are bound;
- that the ligands are homogeneous (i.e. chemically pure) and are not degraded during the incubation.

Experimental conditions which allow deviations from these requirements can produce inaccurate binding parameters. In the case of an inhibition experiment it is also necessary for the inhibitor to actually compete for the same site labelled by the radioligand. Inhibition data which are best fitted to a one-site model but give a pseudo-Hill slope of less than unity may be evidence for a negative allosteric interaction.[50]

While IC_{50} values can be obtained from competitive inhibition data by the use of a simple log-logit transformation, the availability of computer programs (i.e. LIGAND,[51] LUNDON,[52] and others) based on non-linear regression analysis methods for one or more sites provides a better alternative. Besides the relative ease with which data can be analysed by these programs, they provide sensitive tests for the identification of multiple site interactions.

Tests for multiple site interactions require that the data be fitted to different equations that model ligand binding to one or more sites. Differences in the 'goodness of fit' of the data between the different models are represented by differences in the residual sums of squares (i.e. the SSE) of the deviations of the points from the fitted curve. The F ratio test is commonly employed by the non-linear regression programs to determine the statistical significance of an apparent improvement in fit between different models. The value of F can be calculated using the formula:

$$F = \frac{(\text{SSE}_1 - \text{SSE}_2 / (df_1 - df_2)}{\text{SSE}_2 / df_2}$$

where SSE_1 and SSE_2 are the sum of the squares for the deviations of the data from the two models and df refers to the degrees of freedom which is the difference between the number of data points and the number of parameters (e.g. K_d, B_{max}) fitted by the model. These values should be supplied by the program used, which often performs the required calculations of significance automatically. The significance of the difference in SSE values for two models can also be determined using a table of critical values of the F distribution for different degrees of freedom.

The IC_{50} value measured for a competitor is only an approximation of its actual affinity for the receptor labelled by the radioligand. The relationship between the dissociation constant (K_i) of the competitive inhibitor and its IC_{50} value is given by the Cheng–Prusoff[53] formula:

$$K_i = \frac{\text{IC}_{-50}}{1 + ([L] / K_d)}$$

where $[L]$ is the free molar concentration of the radioligand and K_d is its dissociation constant. It is important to recognize that this formula only holds for competitive interactions at one site. From this formula it can be seen that the denominator will approach unity, and the IC_{50} will approach K_i, when the concentration of radioligand becomes small relative to its K_d. The better estimate of K_i is one reason for using a low radioligand concentration. More important, however, is that by using a concentration of radioligand that is below its K_d, the binding is to a great extent restricted to the desired receptor. A practical reason is that less inhibitor is required to obtain an IC_{50} value.

3.1.7. Determination of assay conditions for inhibition studies

The basic elements of any screening assay that measures the competitive inhibition of radioligand binding are the same regardless of the ligand or tissue preparation used. It will be assumed for the purposes of this discussion that the analyst has chosen an appropriate radioligand for the receptor of interest that has been reasonably well characterized with regard to its affinity and stability under a defined set of conditions.

The first requirement is to have a tissue preparation for these studies. The tissue selected should be that with the highest concentration of the desired receptor available in reasonable quantity. The receptor density should be of the order of 100 fmole/mg protein. Approximately 1 g of tissue would be required to prepare 100 samples at this receptor density, which is sufficient to analyse four peptides at 10 different concentrations. If this amount of tissue can be obtained from just a few animals then it is reasonable to use fresh tissue for the screening assay. This is advantageous since fresh tissue provides the maximum amount of specific binding. If such large numbers of animals are required that the preparation of tissue becomes burdensome, then it may be preferable to prepare a large batch of tissue and freeze aliquots. The tissue should be divided into aliquots sufficient to do an entire day's assay and rapidly frozen with either dry ice or liquid nitrogen. Most tissue preparations can be stored at $-70°$ for up to a month without substantial deterioration.

The second requirement is for an unlabelled ligand that can be used for the determination of non-specific binding. The ideal ligand for this purpose would be an antagonist specific for the receptor. Unfortunately there are few antagonists for peptide receptors, but it is still useful to use a second ligand that differs from the labelled ligand. The objective is to have two ligands whose preferred receptor selectivities overlap, but whose secondary sites do not. Thus, the specific binding defined by the total binding of the radioligand less its binding in the presence of a large concentration of the unlabelled ligand should be limited to only the receptor of interest. The concentration of unlabelled ligand used should be between 100 to 1000 times its K_i for the receptor. It is useful to actually measure the required concentration by producing an inhibition curve of the unlabelled ligand against the radioligand. The inhibition curve should reach a plateau and remain constant over a tenfold concentration range.

Once the three requirements of a radioligand, tissue preparation, and unlabelled ligand for defining non-specific binding have been met, it is possible to determine the optimal concentrations of the radioligand and tissue preparation to be used for the assay. This can be accomplished using a 4×4 matrix in which the specific binding of four different radioligand concentrations ($2 \times$, $1 \times$, $0.5 \times$, and $0.1 \times K_d$) is measured at each of four different tissue concentrations (2%, 1%, 0.5%, and 0.25% w/v). The optimal radioligand and tissue concentrations will be the lowest of each that produces between 500 and 1000 c.p.m. of specific binding where the total binding is less than 10% of the added counts. The value of 500 c.p.m. represents the lowest practical amount of specific binding and it would certainly be better if more counts could be obtained under the same restrictions. This will depend on the affinity of the radioligand and its specific activity.

3.1.8. *Detection of unlabelled peptide degradation*

The degradation of peptides during assay is one of the most difficult problems to deal with. A simple approach to determine if an IC_{50} value obtained for a peptide is inaccurate due to degradation is to perform a preincubation experiment. The peptide should be added at a concentration of 10 times its measured IC_{50} to a microcentrifuge tube containing the same amount of tissue used in the assay. A second tube containing tissue which has been boiled for 5 min can be used as a control for loss due to non-specific binding. The tissue is separated by centrifugation after incubation under the same conditions used in the assay and aliquots of the supernatant equal to a tenth of the total volume are removed for a single concentration determination of competitive inhibition. If the inhibition measured is between 60–40%, then degradation of the peptide is probably not important. If the inhibition is much less than this then it is necessary to use different protease inhibitors or otherwise modify the assay conditions (see above) to correct the problem. This kind of experiment is not difficult to carry out and is a valuable test to show that the IC_{50} values measured are accurate.

4. The development of peptides for the μ and δ opioid receptors

4.1. OPIOID RECEPTORS

Three important breakthroughs have been responsible for the rapid expansion of opioid research. The demonstration of opiate binding by Pert and Snyder,[54] Simon et al.,[55] and Terenius[56] based on the use of stereoselectivity described by Goldstein[57] enabled the direct characterization of opiate–receptor interactions. This was followed by the isolation of endogenous peptide ligands (i.e. the enkephalins) for these receptors.[58] Shortly thereafter Martin et al.[59] and Gilbert and Martin[60] identified three different opioid receptor types (μ, κ, and σ). The identification of the δ opioid receptor followed a year later.[61]

These three breakthroughs have stimulated the development of a multitude of synthetic opioid peptides which have significant therapeutic potential.[62] We have participated in the development of several opioid peptides where the approach of conformational restraint was used to increase their selectivities for different opioid receptor types. Two examples, the development of δ opioid receptor selective enkephalin analogues and the development of somatostatin analogues which act as selective μ opioid antagonists will be described to illustrate the role of bioassay and radioligand binding methods in peptide development.

4.2. CYCLIC ENKEPHALIN ANALOGUES

Solution and solid-phase conformational studies suggest that the enkephalins can adopt folded low-energy conformations in which the C-terminal amino acid lies in close proximity to the amino acid in the 2 position. Many enkephalin analogues having a covalent linkage between the amino acids in the 2 and 5 positions have been produced to determine if this folded conformational state resembles that of the receptor-bound enkephalins.[63] The cyclic enkephalin analogues to be described represent an effort to produce selective opioid ligands whose rigidity facilitates the study of their conformational properties. The development of these enkephalin analogues depended on the use of radioligand inhibition measurements and isolated tissue bioassays. These bioassay preparations have been extensively used in the analysis of opioid activity[64] and will be described in detail.

4.2.1. Guinea pig ileum bioassay for opioids

The observation that opiates act to inhibit the contraction of the electrically stimulated guinea pig ileum longitudinal muscle–myenteric plexus preparation (GPI) established this as one of the most widely used preparations for the evaluation of opioid drugs.[65] Contractions of the electrically stimulated mouse vas deferens (MVD) are also inhibited by opioids.[66]

Small intestine contains two major muscle coats, longitudinal and circular muscles, and two major intramural plexuses of nerves, the myenteric and the submucous plexus. In GPI, the longitudinal muscle is innervated almost exclusively by postganglionic parasympathetic nerve fibres which release acetylcholine. The circular muscle is innervated by a variety of peptide-containing nerve fibres, in addition to cholinergic fibres. When the longitudinal muscle is teased away in strips from the intestine, the myenteric plexus largely adheres to the muscle, and the cholinergic fibres are preserved. Transmural electrical stimulation of strips of longitudinal muscle–myenteric plexus causes a release of acetylcholine, which acts at the muscarinic cholinergic receptors of the smooth muscle to induce contractions. The amplitude of these contractions is proportional to the amount of acetylcholine released from the nerves. Opioid agonists act presynaptically on the cholinergic fibres to inhibit release of acetylcholine[65] and this inhibition is proportional to the concentration of the opioid agonist.[67] Measurement of the amplitude of electrically-induced contractions of the GPI thus provides a convenient system for the assessment of opioid agonist activity. An advantage of the GPI is that it contains functional μ and κ, but not δ, opioid receptors.[68] Thus, κ receptor activation by δ agonists can be measured in the presence of μ receptor antagonists (see below) using this preparation.

The GPI bioassay used for the analysis of the cyclic enkephalins employed the guinea pig longitudinal muscle–myenteric plexus preparation as described by Kosterlitz et al.[69] Segments of ileum were removed from adult male Hartley guinea pigs and strips of the longitudinal muscle containing myenteric plexus were prepared. These strips were attached to isometric force transducers in a 20 ml organ bath containing Krebs–bicarbonate buffer that was continuously saturated with a gas mixture of 95% oxygen and 5% carbon dioxide at 37°. The strips were stretched to a resting tension of 1.0 g after a 15 min equilibration period without tension. Contraction was induced by transmurally stimulating the tissue between two platinum electrodes with 0.4 ms pulses at 0.1 Hz using a supramaximal voltage. The twitch response was recorded using a Soltec multichannel recorder. The concentrations of peptides used were tested in random order and a period of 15 min separated each testing interval during which the tissue preparation was washed several times to prevent the development of acute tolerance. The response produced by the peptide on the twitch tension was measured after an exposure of 3 min. The change measured in the twitch height was calculated as a percent of the control response determined just prior to the addition of the peptide. Dose–response curves and statistical calculations were using procedures described by Tallarida and Murray.[70]

4.2.2. Mouse vas deferens bioassay for opioids

Electrical stimulation of the MVD induces the release of norepinephrine which acts on α-adrenergic receptors to contract the smooth muscle of the MVD. The release of this transmitter is presynaptically inhibited by opioid agonists in a concentration-dependent manner.[71] In contrast to the GPI, the MVD is unusually sensitive to enkephalins and δ opioid receptor agonists and served as a 'prototype' tissue preparation for the identification of the δ opioid receptor.[61] However, this tissue also contains both μ and κ in addition to δ opioid receptors.[61,72,73]

The basic procedure used for the MVD assay was similar to that of Hughes et al.[71] and used the same isometric signal transducer apparatus used in the GPI assay. Paired vasa deferentia were removed from adult male ICR mice and attached to the signal transducer as a single unit. The same Krebs–bicarbonate buffer was used except for the exclusion of magnesium. Following a 15 min equilibration period without tension, the tissue was placed under 0.5 g of resting tension. Contractions were induced using 2 ms pulses of transmural stimulation at 0.1 Hz and supramaximal voltage. Drug application and data analysis were as described for the GPI assay.

4.2.3. Development of cyclic enkephalin analogues

The first group of enkephalin analogues had D-penicillamine ($\beta\beta$-dimethyl cystine) substitutions in the 2 position and either D- or L-Cys substitutions in the 5 position of the enkephalin sequence. The presence of the thio groups on these amino acids provided a way to covalently join the 2 and 5 positions to produce a conformationally constrained cyclic peptide. All of these peptides had amide blocked C-terminals. The use of conformational restraints in peptide development has recently been reviewed.[74,75] These peptides were evaluated for their pharmacological selectivity using the GPI and MVD preparations and for their receptor binding selectivity by competitive radioligand binding methods.

The Leu-enkephalin analogue [D-Ala^2, D-Leu^5]enkephalin (DADLE) was included among the peptides tested and found to have IC_{50} values of 24.3 and 0.27 nM in the GPI and MVD, respectively, giving DADLE a selectivity ratio of 90. In comparison, Kosterlitz et al.[76] obtained IC_{50} values of 47.8 and 0.54 nM in the GPI and MVD which give a selectivity ratio of 88. Of the cyclic enkephalinamide analogues tested in this study, [D-Pen^2, L-Cys^5] enkephalinamide was the most selective, being 32-fold more potent in the MVD assay compared to the GPI assay.

In a second study[77] the same cyclic peptides were prepared without the C-terminal amide group. This relatively minor change in structure resulted in a major change in the relative potencies of these analogues. Thus, the [D-Pen^2, L-Cys^5]enkephalin peptide was 666-fold more potent in the MVD than GPI in contrast to the 32-fold difference observed for its amide analogue. The change in relative potencies is due to a substantial (11-fold) increase of potency in the MVD assay with a smaller (twofold) decrease of potency in the GPI.

In addition to bioactivity, the relative *affinities* of the analogues were measured by radioligand binding to brain homogenates using $[^3H]$naloxone to label μ opioid receptor sites and $[^3H]$DADLE to label δ receptor sites. In the first study of the enkephalinamide analogues, the ratio of the IC_{50} values for $[D\text{-}Pen^2, L\text{-}Cys^5]$enkephalinamide determined by competitive inhibition of $[^3H]$naloxone and $[^3H]$DADLE binding was 22, consistent with the potency ratio of 32. However, in the second study the affinity ratio determined for $[D\text{-}Pen^2, L\text{-}Cys^5]$enkephalin by competitive inhibition was 15.2, suggesting that there was no change in selectivity. A probable reason for this discrepancy is the low binding selectivity of the two radioligands. A careful study of the relative binding affinities of different opioids for the μ, δ, and κ receptor sites by James and Goldstein[78] showed that naloxone has only 18-fold less affinity for δ than μ opioid receptors while DADLE has only 2.1-fold less affinity for μ compared with δ opioid receptors. The latter observation for DADLE contrasts sharply with the apparent selectivity of this peptide measured in the two bioassays and serves as a classic example of the difference between potency, as measured by bioassay, and affinity, as measured by radioligand binding. The two techniques complement each other and one may actually serve to make up for deficiencies in the other.

At first it was thought that a double substitution of Pen in both 2 and 5 positions would not permit ring closure for steric reasons and all of the early cyclic analogues produced by Hruby and colleagues had a Cys substitution in either the 2 or 5 positions. The observation that ring closure could be obtained with bis-penicillamine substituted enkephalin analogues led to the synthesis of $[D\text{-}Pen^2, L\text{-}Pen^5]$enkephalin and $[D\text{-}Pen^2, D\text{-}Pen^5]$enkephalin.[79,80] These two analogues (DPLPE and DPDPE) are highly selective for the δ opioid receptor. DPDPE, the more selective of the two, has IC_{50} values of 6930 nM in the GPI and 2.19 nM in the MVD assays. Thus, the potency ratio for this analogue is 3164. Again, the $[^3H]$naloxone : $[^3H]$DADLE affinity ratio (175) substantially underestimated the potency ratio. However, different results were obtained in a later study by Akiyama et al.[81] in which $[^3H]$DPDPE binding to NG 108–15 neuroblastoma × glioma hybrid cells, which contain only the δ type of opioid receptors, and $[^3H]$DPDPE binding to brain membranes was used to measure affinity at the δ receptor. In this study IC_{50} values for DPDPE measured at the NG 108–15 δ opioid receptor and at brain μ receptors labelled by $[^3H]$naloxone gave an affinity ratio of 3463 which is quite close to the 3164 ratio obtained from the GPI and MVD bioassays. The affinity ratio obtained when both radioligands were used to label μ and δ receptors in brain membranes was 1234. Both ratios, 3463 and 1234, are closer to the bioassay results than the 175 value measured when $[^3H]$DADLE was used to label δ receptors, suggesting again that $[^3H]$DADLE may be labelling sites other than the δ receptor. These results further emphasize the importance of radioligand selectivity for the measurement of binding to a particular receptor when a tissue (e.g. brain) having a heterogeneous population of receptors is used.

4.3. DEVELOPMENT OF μ OPIOID ANTAGONISTS

The development of μ opioid antagonist peptides differs from the development of DPDPE just described. In contrast to the development of DPDPE, which is derived from the basic structure of the enkephalins, the μ opioid antagonists were developed from an analogue of somatostatin (SMS 201–995) that acts as a potent agonist at somatostatin receptors.[82] The structure of SMS 201–995 is:

$$\text{D-Phe–}\overline{\text{Cys–Phe–D-Trp–Lys–Thr–Cys}}\text{–Thr(ol)}$$

In addition to its somatostatin activity, however, this cyclic peptide acts as a weak opioid antagonist with some selectivity for the μ receptor.[83] Thus, the task of developing a selective μ opioid receptor antagonist from this analogue required that analogues be screened for their binding affinity at three different sites: the somatostatin, δ opioid, and μ opioid receptors. In addition it was necessary to screen for bioactivity (agonist and antagonist) at the μ and δ opioid receptors.

Radioligand binding studies offered a more direct and, relative to the use of three different bioassays, a much faster means of determining the receptor selectivities of the somastostatin analogues. Thus, radioligand binding studies were used as the primary screen in the development process. Somatostatin-like binding activity was measured using ^{125}I-[des-Ala^1, Gly^2-(desamino-Cys^3, Tyr^{11})-3,14-dicarba]somatostatin (CGP 23,996), a somatostatin analogue designed to allow direct iodination using chloramine-T.[84] Non-specific binding at the somatostatin receptor was defined using unlabelled somatostatin. Binding affinity at the μ opioid receptor was estimated using [^{3}H]naloxone as the tracer and naltrexone to define non-specific binding, while affinity at the δ receptor was estimated using [^{3}H]DADLE with [Met^5]enkephalin used to define non-specific binding. All of the competitive inhibition studies used membranes prepared from whole rat brains suspended in 50 mM Tris buffer containing BSA and bacitracin.

Eight different analogues of the basic SMS 201–995 structure were prepared and screened by radioligand binding. Of these, the analogue with the greatest selectivity for the μ opioid receptor, referred to for convenience as CTP, has the structure:[85]

$$\text{D-Phe–}\overline{\text{Cys–Tyr–D-Trp–Lys–Thr–Pen}}\text{–Thr-NH}_2$$

CTP has the highest affinity for the site labelled by [^{3}H]naloxone with an IC$_{50}$ value of 3.5 nM. The IC$_{50}$ values against [^{3}H]DADLE (950 nM) and [^{125}I]CGP 23,996 (690 nM) were both much higher. The IC$_{50}$ value just given for the competitive inhibition of [^{3}H]DADLE was obtained from a one-site model and is actually a poor estimate of the δ receptor affinity of CTP. The Hill coefficient measured for the one-site model was 0.33 suggesting that this model was inappropriate. Reanalysis of the data showed that a significant improvement in fit was obtained using a two-site model with IC$_{50}$ values of 19 nM at the high-affinity site and 24 000 nM at the low-affinity site. The relative concentrations of the high- and low-affinity sites were 43 and 56%, respectively. This observation is consistent with the low selectivity previously described for $[^3H]DADLE$ and demonstrates the value of computer analysis using multiple-site models.

The same GPI and MVD preparations used for the bioassay of the enkephalin analogues were used to study the bioactivity of CTP.[68] Intrinsic agonist activity was measured directly while antagonist activity was determined by pA_2 measurements using PL-17 as the agonist in the GPI and MVD. The initial addition of CTP to the GPI preparation produced a small (20–30%) inhibition of the twitch response which was lost after about 2 min and was not observed in subsequent additions of CTP to the same preparation. A similar response was observed in the MVD and was not a result of opioid activity as it could not be blocked by naloxone. This agonist activity appeared to be related to action at somatostatin receptors since it was also produced by somatostatin and the induction of acute somatostatin tolerance by pretreatment of the preparations with this peptide abolished it. CTP also showed weak δ opioid agonist activity in the MVD at concentrations over $3.0\,\mu M$. In contrast to the somatostatin-like activity, which was only observed at CTP concentrations between 10 and 100 nM, the opioid agonist activity produced by high concentrations of CTP in the MVD could be blocked by either naloxone or the selective δ opioid antagonist ICI 174,864.

The *in vitro* antagonist activity of CTP was measured by Schild analysis using three different concentrations of CTP to shift the concentration–response curves of PL-17 in the GPI and MVD. The pA_2 value for CTP in the GPI was 7.1 ± 0.17 with a slope value of -0.97. Since the slope was not significantly different from unity, the pA_2 value can be transformed to give a valid estimate of the K_d of CTP in the GPI which is 79 nM. This value is 22-fold greater than the IC_{50} value measured against naloxone binding to brain membranes, which is itself an overestimate of the true K_i. Two possible explanations can be advanced:

- The trivial explanation that the differences are due to the differences in the conditions used (e.g. different buffers, temperatures, etc.). However, antagonist binding is typically less influenced by such differences.
- That the receptors bound by CTP in the GPI and rat brain membrane preparations are different.

The data available do not confirm either possibility, but the potential problem of receptor differences should always be considered when measurements are made in different tissues (see discussion). The pA_2 value measured using PL-17 as the agonist in the MVD was 6.9 ± 0.16 which is similar to that measured in the GPI. This observation supports other studies which have shown that the MVD contains functional μ opioid receptors. CTP failed to antagonize the agonist activities of either DPDPE or the κ selective opioid U 50488 in the MVD consistent with the conclusion that these agonists do not act on μ receptors in this preparation.

The antagonist properties of CTP were also investigated *in vivo* using two of the analgesia bioassays (hotplate and abdominal stretch) previously described.

CTP and the other opioids used were administered by either i.c.v. (hotplate) or i.t. (abdominal stretch) injection.

CTP antagonized the analgesic effects of both μ receptor selective (PL-17, morphine, and DAGO) and δ receptor selective (DPDPE) agonists, but not the κ receptor selective agonist U 50488, in both assays.[86] However, a marked difference in CTP antagonist activity was observed between the most selective μ agonist (PL-17) and DPDPE with CTP being about 30-fold more effective against the μ agonist in the hotplate test. Schild analysis of CTP antagonism suggested that while the interaction of the antagonist with the μ agonists was competitive (slope values not significantly different from unity), the interaction with DPDPE was not (slope $= -0.55$). Because of its great selectivity in *in vitro* bioassays and ligand binding studies, CTP probably does not interact significantly with δ opioid receptors. In fact, the kinetics of CTP interactions with DPDPE suggest non-competitive effects of CTP at the δ receptor.[86] Non-competitive (allosteric) interactions between μ and δ opioid receptors have been postulated.[87,88] Thus, it is difficult to characterize the interaction of CTP with DPDPE, but CTP clearly does act as a competitive antagonist at the μ opioid receptor.

It is evident from the preceding discussion that the bioassay of an antagonist is a more formidable task than that of an agonist. However, the CTP bioassays provided information about the somatostatin-like activity and interaction with DPDPE that would not have been expected from the results of the radioligand binding studies. These observations have resulted in the development of additional CTP analogues with minimal somatostatin-like activity and increased μ opioid receptor selectivity.[89,90]

5. Discussion

The desirability of screening peptides by both bioassay and radioligand binding techniques has been emphasized here. These two techniques give fundamentally different kinds of information. Bioassay measures mainly potency, which is dependent on affinity, efficacy, and tissue specific factors such as receptor reserve, while radioligand binding measures only binding affinity and cannot provide information about efficacy. The basic information that is required of any new peptide is a knowledge of its efficacy at the specific receptors on which it acts. A peptide may have full agonist activity (positive efficacy) characterized by its induction of the maximum response the tissue is capable of producing or it may have antagonist activity (zero efficacy) characterized by its ability to block the response of an agonist without producing an effect of its own. The efficacy of a peptide may also lie between these extremes resulting in partial agonist activity characterized by an inability to produce more than a submaximal response. While efficacy related differences in radioligand binding have been observed for some well-characterized receptors,[91] differences in efficacy are best measured by bioassay. The measurement of a response does not, however, identify the receptor or receptors responsible for its mediation. The identification of the receptor(s) mediating the activity of a drug depends on the availability of selective antagonists which are often unavailable for receptors acted on by peptides. Furthermore, the use of an isolated tissue preparation for the bioassay may prevent the observation of interactions with secondary sites if these are not present or are not efficiently coupled to the response measured. The identification of the site or sites recognized by a drug is best determined by radioligand binding. The advantages of radioligand binding for the identification of receptors recognized by a ligand include the ability to selectively label specific receptors in a variety of tissues containing multiple receptor types. Agonists and antagonists can be screened in the same radioligand binding assay. This is not true for bioassays, as different procedures are required for the measurement of agonist or antagonist activity.

It has been suggested that agonist binding and signal transduction (receptor activation) depend on different conformational features of the same peptide.[92] This suggests that modification of different structural features of a peptide may independently alter either its affinity for a specific receptor or its efficacy. Thus, information on both affinity and efficacy are required to characterize the effect of a structural change.

The validity of comparing the results obtained from a bioassay to those obtained by radioligand binding depends on the assumption that both are measuring effects at the same receptor. Problems resulting from the misidentification of the site(s) labelled by DADLE were discussed in the sections on the development of cyclic enkephalins and somatostatin analogues. Similar problems can occur in bioassays since the tissues used seldom, if ever, contain homogeneous populations of receptors. This problem is increased by the frequent necessity of using different tissues for radioligand binding and bioassay studies. This practice increases the possibility that different receptor subtypes are being acted on.

There is substantial evidence for peptide receptor heterogeneity in CNS and peripheral tissues. Multiple receptor subtypes have been observed for neurokinin,[93] cholecystokinin,[94] and opioid[95] receptors, to name only a few. While the development of highly selective agonists and antagonists for many peptide receptors offers some relief from this problem, the use of animal tissue whose total receptor population is not defined will always carry with it some risk of ligand interaction at an unknown receptor. Two current developments in biochemical pharmacology which may provide a solution to this problem are the characterization of receptor-activated second-messenger systems in clonal cell lines and the transfection of cells with genes encoding specific receptors.

With the exception of ligand-gated ion channels, the response measured in even the simplest bioassays (e.g. secretion of a substance) is the result of a culmination of events initiated by the activation of a second messenger system. The best studied are linked to either adenylate cyclase or polyphosphoinositide systems. The inhibition of adenylate cyclase activity by μ and δ opioid agonists in many tissues is well established[96] and a number of peptides appear to stimulate inositide turnover.[97] The activation of these second-messenger systems represents a clearly defined response to agonist binding to its receptor and can be measured using cultured cells. By selecting a clonal cell line with a defined receptor for a specific peptide one could measure bioactivity and ligand binding with confidence that the same receptor was being studied in each. At present, the search for such cells involves a lengthy process of screening different cell lines and characterizing their receptors with no assurance that the receptor population will actually be homogeneous. The identification of genes encoding specific receptors offers an alternative solution.

The development of recombinant gene technology permits the isolation, identification, and expression of genes encoding specific proteins.[98] This technology has been applied to the identification of genes encoding several different receptors which have in turn been expressed in cells lacking these receptors. Thus, it is possible to transform cells to produce a specific receptor and then mass-produce the cells in culture. The receptors expressed in the cell lines that have been produced so far appear to couple to the second-messenger systems already present and can be used to measure second-messenger responses. By selecting an appropriate promoter for the receptor gene, it is also possible to obtain a very high receptor density which would facilitate the use of the cells in radioligand binding.

The development of stable cell lines expressing specific human receptor types coupled to defined second-messenger systems would provide the ultimate analytical tools for peptide development. While such tools would greatly reduce our dependence on animal-based systems for the analysis of drug activity, they would not eliminate the need for *in vivo* models. The final test of any drug is its ability to act in the intact organism, and this depends on biopharmaceutic properties which cannot be fully modelled in isolated systems. These properties are especially critical to the successful use of peptide drugs. Thus, peptide chemists will continue to depend on the skill of pharmacologists to determine the bioactivity of their products.

Acknowledgements

Supported in part by USPHS grants.

References

1. 'SMS 201-995: A very potent and selective octapeptide analogue of somatostatin with prolonged action' W. Bauer, U. Briner, W. Doepfner, R. Haller, R. Huguenin, P. Marbach, T.J. Petcher and J. Pless (1982) *Life Sci.*, **31**, 1133–1140.
2. 'A perspective on the application of physical methods to peptide conformational-biological activity studies' V.J. Hruby (1985). In: *The Peptides*, Vol.7 (S. Udenfriend and J. Meienhofer, eds.) pp. 1–14. Academic Press, New York.
3. 'Contributions of industrial research to basic neuropsychopharmacology: Preclin-

ical screening and discovery' A. Weissman and B. K. Koe (1987). In: *Psychopharmacology: The Third Generation of Progress* (H. Y. Meltzer, ed.) pp. 1649–1657. Raven Press, New York.

4. *Biostatistics* A. Goldstein (1969) Macmillan, New York.
5. *Statistical Method in Biological Assay* (D. J. Finney (1952) Hafner, New York.
6. *The Dose–Response Relation in Pharmacology* R. J. Tallarida and L. S. Jacob (1979) Springer-Verlag, New York.
7. *The Cytochemical Bioassay of Polypeptide Hormones* J. Chayen (1980) Springer-Verlag, Berlin.
8. 'pA_2 and receptor differentiation: A statistical analysis of competitive antagonism' R. J. Tallarida, A. Cowan and M. W. Adler (1979) *Life Sci.*, **25**, 637–654.
9. 'Determination of pharmacological constants: Use of narcotic antagonists to characterize analgesic receptors' A. E. Takemori (1974). In: *Narcotic Antagonists* (M. C. Braude, L. S. Harris, E. L. May, J. P. Smith and J. E. Villarreal, eds) pp. 335–344. *Advances in Biochemical Psychopharmacology*, Vol. 8, Raven Press, New York.
10. 'Mediation of analgesia by multiple opioid receptors' R. J. Knapp, F. Porreca, T. F. Burks and H. I. Yamamura (1989). In: *Drug Treatment of Cancer Pain in a Drug-Oriented Society* (C. S. Hill and W. S. Fields, eds) pp. 247–289. Raven Press, New York.
11. 'Predictive value of analgesic assays in mice and rats' R. I. Taber (1974) *Adv. Biochem. Psychopharmacol.*, **8**, 191–211.
12. 'A classification of opiate receptors that mediate antinociception in animals' M. B. Tyers (1980) *Br. J. Pharmacol.*, **69**, 503–512.
13. 'A method for determining loss of pain sensation' F. E. D'Amour and D. L. Smith (1941) *J. Pharmacol. Exp. Ther.*, **72**, 74–79.
14. 'The type of analgesic–receptor interaction involved in certain analgesic assays' G. Hayashi and A. E. Takemori (1971) *Eur. J. Pharmacol.*, **16**, 63–66.
15. 'The increased efficacy of narcotic antagonists induced by various narcotic analgesics' F. C. Tulunay and A. E. Takemori (1974) *J. Pharmacol. Exp. Ther.*, **190**, 395–400.
16. 'The inhibitory effect of fentanyl and other morphine-like analgesics on the warm water induced tail withdrawal reflex in rats' P. A. J. Janssen, C. J. E. Niemegeers, and J. G. H. Dony (1963) *Arzneim. Forsch.*, **13**, 502–507.
17. 'Synthetic analgesics. II. Dithienylbutenyl and dithienylbutylamines' N. B. Eddy and D. Leimbach (1953) *J. Pharmacol. Exp. Ther.*, **107**, 385–393.
18. 'The assessment of and the problems involved in the experimental evaluation of narcotic analgesics' M. R. Fennessy and J. R. Lee (1975). In: *Methods in Narcotics Research* (S. Ehrenpreis and A. Neidle, eds.) pp. 73–99. Marcel Dekker, New York.
19. 'The tail-flick test' W. L. Dewey and L. S. Harris (1975). In: *Methods in Narcotics Research* (S. Ehrenpreis and A. Neidle, eds.) pp. 101–109. Marcel Dekker, New York.
20. 'Antinociceptive action of intracerebroventricularly administered dynorphin and other opioid peptides in the rat' P. J. Tiseo, E. B. Geller and M. W. Adler (1988) *J. Pharmacol. Exp. Ther.*, **246**, 449–453.
21. 'Quantification of the analgesic activity of narcotic antagonists by a modified hot-plate procedure' J. P. O'Callaghan and S. G. Holtzman (1975) *J. Pharmacol. Exp. Ther.*, **192**, 497–505.
22. 'Analgesic properties of 4-ethoxycarbonyl-1-(2-hydroxy-3-phenoxypropyl) 4-phenylpiperidine (B.D.H.200) and some related compounds' C. Bianchi and A. David (1960) *J. Pharm. Pharmacol.*, **12**, 449–459.
23. 'Experimental observations on Haffner's method for testing analgesic drugs' C. Bianchi and J. Franceschini (1954) *Br. J. Pharmacol.*, **9**, 280–284.
24. 'A method for measurement of analgesic activity on inflamed tissue' L. O. Randall and J. J. Selitto (1957) *Arch. Int. Pharmacodyn. Ther.*, **111**, 409–419.

25. 'Quantal responses in the Randall–Selitto assay' K.F. Swingle, T.J. Grant, and D.C. Kvam (1971) *Proc. Soc. Exp. Biol. Med.*, **137**, 536–538.

26. 'A method for evaluating both non-narcotic and narcotic analgesics' E. Siegmund, R. Cadmus, and G. Lu (1957) *Proc. Soc. Expl Biol. Med.*, **95**, 729–731.

27. 'Acetic acid for analgesic screening' R. Koster, M. Anderson, and E.J. de Beer (1959) *Fed. Proc.*, **18**, 412.

28. 'The abdominal constriction response and its suppression by analgesic drugs in the mouse' H.O.J. Collier, L.C. Dinneen, C.A. Johnson, and C. Schneider (1968) *Br. J. Pharmacol.*, **32**, 295–310.

29. 'Enhanced analgesic activity of D-Ala^2 enkephalinamids following D-isomer substitutions at position five' J.D. Belluzzi, L. Stein, W. Dvonch, S. Dheer, M.I. Gluckman and W.H. McGregor (1978) *Life Sci.*, **23**, 99–104.

30. 'Analogues of β-LPH$_{61-64}$ possessing selective agonist activity at mu-opiate receptors' B.K. Handa, A.C. Lane, J.A.H. Lord, B.Z. Morgan, M.J. Rance and C.F.C. Smith (1981) *Eur. J. Pharmacol.*, **70**, 531–540.

31. 'A synthetic enkephalin analogue with prolonged parenteral and oral analgesic activity' D. Röemer, H.H. Buescher, R.C. Hill, J. Pless, W. Bauer, F. Cardinaux, A. Closse, D. Hauser and R. Huguenin (1977) *Nature*, **268**, 547–549.

32. 'Metkephamid, a systematically active analog of methionine enkephalin with potent opioid delta-receptor activity' R.C.A. Frederickson, E.L. Smithwick, R. Shuman and K.G. Bemis (1981) *Science*, **211**, 603–605.

33. 'The importance of biopharmaceutic considerations in biological assays of LHRH and its analogs' D.W. Hahn, A. Phillips-Probst and J.L. McGuire (1981) *Endocrin. Res. Commun.*, **8**, 273–283.

34. 'Receptor binding methodology and analysis' D.R. Burt (1986). In: *Receptor Binding in Drug Research* (R.A. O'Brien, ed.), pp. 3–29. Marcel Dekker, New York.

35. 'Neurotransmitter, hormone, or drug receptor binding' J.P. Bennett and H.I. Yamamura (1985). In: *Neurotransmitter Receptor Binding*, 2nd edn (H.I. Yamamura, S.J. Enna and M.J. Kuhar, eds.) pp. 61–89. Raven Press, New York.

36. 'Peptide binding assays' M. Hanley (1985). In: *Neurotransmitter Receptor Binding*, 2nd edn (H.I. Yamamura, S.J. Enna and M.J. Kuhar, eds.) pp. 91–102. Raven Press, New York.

37. 'The preparation of ^{131}I-labeled human growth hormone of high specific radioactivity' F.C. Greenwood, W.M. Hunter and J.S. Glover (1963) *Biochemical J.*, **89**, 114–123.

38. 'Solid state lactoperoxidase: A highly stable enzyme for simple, gentle iodination of proteins' G.S. David (1972) *Biochem. Biophys. Res. Commun.*, **48**, 464–471.

39. 'Specific receptors for α-melanocyte-stimulating hormone are widely distributed in tissues of rodents' J.B. Tatro and S. Reichlin (1987) *Endocrinology*, **121**, 1900–1907.

40. 'A novel opioid receptor site directed alkylating agent with irreversible narcotic antagonistic and reversible agonist activities' P.S. Portoghese, D.L. Larson, L.M. Sayre, D.S. Fries and A.E. Takemori (1980) *J. Med. Chem.*, **23**, 233–234.

41. 'Synthesis and pharmacologic characterization of an alkylating analogue (chlornaltrexamine) of naltrexone with ultralong-lasting narcotic antagonist properties' P.S. Portoghese, D.L. Larson, J.B. Jiang, T.P. Caruso and A.E. Takemori (1979) *J. Med. Chem.*, **22**, 168–173.

42. 'Specific protection of the binding sites of D-Ala^2-D-Leu^5-enkephalin (δ-receptors) and dihydromorphine (μ-receptors)' L.E. Robson and H.W. Kosterlitz (1979) *Proc. R. Soc. Lond. Biol. Sci.*, **205**, 425–432.

43. 'Selective protection of stereospecific enkephalin and opiate binding against inactivation by N-ethylmaleimide: Evidence for two classes of opiate receptors' J.R. Smith and E.J. Simon (1980) *Proc. Natl. Acad. Sci. USA*, **77**, 281–284.

44. 'Ligand dissociation constants from competition binding assays: Errors associated with ligand depletion' A. Goldstein and R.W. Barrett (1987) *Mol. Pharmacol.*, **31**, 603–609.

45. 'Are there neuropeptide-specific peptidases?' A.J. Turner, R. Matsas and A.J. Kenny (1985) *Biochem. Pharmacol.*, **34**, 1347–1356.

46. 'Characterization of central cholecystokinin receptors using a radioiodinated octapeptide probe' L. P. Wennogle, D.J. Steel and B. Petrack (1985) *Life Sci.*, **36**, 1485–1492.

47. 'Mathematics of hormone–receptor interaction. I. Basic principles' D. Rodbard (1973). In: *Receptors for Reproductive Hormones* (B. W. O'Malley and A. R. Means, eds) pp. 289–326. Plenum Press, New York.

48. 'Quantitative analysis of drug–receptor interactions: I. Determination of kinetic and equilibrium properties' G. A. Weiland and P. B. Molinoff (1981) *Life Sci.*, **29**, 313–330.

49. J. P. Bennett and H. I. Yamamura (1985). In: *Neurotransmitter Receptor Binding*, 2nd edn (H. I. Yamamura, S. J. Enna and M. J. Kuhar, eds.) pp. 61–89. Raven Press, New York.

50. 'Estimation of the affinities of allosteric ligands using radioligand binding and pharmacological null methods' F. J. Ehlert (1988) *Mol. Pharmacol.*, **33**, 187–194.

51. Biosoft, P.O. Box 580 Milltown, New Jersey 08850, USA.

52. Lundon Software, Inc., P.O. Box 21820, Cleveland, Ohio 44121, USA.

53. 'Relationship between the inhibition constant (Ki) and the concentration of inhibitor which causes 50 percent inhibition (IC_{50}) of an enzymatic reaction' Y.-C. Cheng and W. H. Prusoff (1973) *Biochem. Pharmacol.*, **22**, 3099–3102.

54. 'Opiate receptor: Demonstration in nervous tissue' C. B. Pert and S. H. Snyder (1973) *Science*, **179**, 1011–1014.

55. 'Stereospecific binding of the potent narcotic analgesic [^{3}H]etorphine to rat brain homogenate' E. J. Simon, J. M. Hiller and I. Edelman (1973) *Proc. Natl. Acad. Sci. USA*, **70**, 1947–1949.

56. 'Characteristics of the "receptor" for narcotic analgesics in synaptic plasma membrane fraction from rat brain' L. Terenius (1973) *Acta Pharmacol. Toxicol. (Copenhagen)*, **33**, 377–384.

57. 'Stereospecific and nonspecific interactions of the morphine congener levorphanol in subcellular fractions of mouse brain' A. Goldstein, L. I. Lowney and B. K. Pal (1971) *Proc. Natl. Acad. Sci. USA*, **68**, 1742–1747.

58. 'Identification of two related pentapeptides from the brain with potent agonist activity' J. Hughes, T. W. Smith, H. W. Kosterlitz, L. A. Fothergill, B. A. Morgan and H. R. Morris (1975) *Nature*, **258**, 577–580.

59. 'The effects of morphine- and nalorphine-like drugs in the nondependent and morphine-dependent chronic spinal dog' W. R. Martin, C. G. Eades, J. A. Thompson, R. E. Huppler and P. E. Gilbert (1976) *J. Pharmacol. Exp. Ther.*, **197**, 517–532.

60. 'The effects of morphine- and nalorphine-like drugs in the nondependent, morphine-dependent, and cyclazocine-dependent chronic spinal dog' P. E. Gilbert and W. R. Martin (1976) *J. Pharmacol. Exp. Ther.*, **198**, 66–82.

61. 'Endogenous opioid peptides: Multiple agonists and receptors' J. A. H. Lord, A. A. Waterfield, J. Hughes and H. W. Kosterlitz (1977) *Nature*, **267**, 495–499.

62. 'Multiple opioid receptors and novel ligands' R. J. Knapp, K. N. Hawkins, G. K. Lui, J. E. Shook, J. S. Heyman, F. Porreca, V. J. Hruby, T. F. Burks and H. I. Yamamura (1990). In: *Adv. Pain Res. Ther.* (C. Benedetti, C. R. Chapman and G. Giron, eds) pp. 45–85 Raven Press, New York.

63. 'Research topics in the medicinal chemistry and molecular pharmacology of opioid peptides–present and future' R. S. Rapaka (1986) *Life Sci.*, **39**, 1825–1843.

64. 'Methods used for the study of opioid receptors' F. M. Leslie (1987) *Pharmacol. Rev.*, **39**, 197–249.

65. 'The action of morphine and related substances on contraction and on acetylcholine output of coaxially stimulated guinea-pig ileum' W. D. M. Paton (1957) *Br. J. Pharmac. Chemother.*, **12**, 119–127.

66. 'A new example of a morphine-sensitive neuro-effector junction: Adrenergic transmission in the mouse vas deferens' G. Henderson, J. Hughes and H. W. Kosterlitz (1972) *Br. J. Pharmacol.*, **46**, 764–766.

67. 'In vitro models in the study of structure–activity relationships of narcotic analgesics' H. W. Kosterlitz and A. A. Waterfield (1975) *Ann. Rev. Pharmacol. Toxicol.*, **15**, 29–47.

68. 'Pharmacologic evaluation of a cyclic somatostatin analog with antagonist activity at *mu* opioid receptors *in vitro*' J. E. Shook, J. T. Pelton, W. S. Wire, L. D. Hirning, V. J. Hruby and T. F. Burks (1987) *J. Pharmacol. Exp. Ther.*, **240**, 772–777.

69. 'The effects of adrenaline, noradrenaline and isoprenaline on inhibitory α- and β-adrenoceptors in the longitudinal muscle of the guinea-pig ileum' H. W. Kosterlitz, R. J. Lydon and A. J. Watt (1970) *Br. J. Pharmacol.*, **39**, 398–413.

70. *Manual of Pharmacological Calculations with Computer Programs* R. J. Tallarida and R. B. Murray (1981) Springer-Verlag, New York.

71. 'Effect of morphine on adrenergic transmission in the mouse vas deferens. Assessment of agonist and antagonist potencies of narcotic analgesics' J. Hughes, H. W. Kosterlitz and F. M. Leslie (1975) *Br. J. Pharmacol.*, **53**, 371–381.

72. 'Assessment in the guinea-pig ileum and mouse vas deferens of benzomorphans which have strong antinociceptive activity but do not substitute for morphine in the dependent monkey' M. Hutchinson, H. W. Kosterlitz, F. M. Leslie, A. A. Waterfield and L. Terenius (1975) *Br. J. Pharmacol.*, **55**, 541–546.

73. 'Comparison of dynorphin-selective kappa receptors in mouse vas deferens and guinea pig ileum' B. M. Cox and C. Chavkin (1983) *Mol. Pharmacol.*, **23**, 36–43.

74. 'Conformational restrictions of biologically active peptides via amino acid side chain groups' V. J. Hruby (1982) *Life Sci.*, **31**, 189–199.

75. 'Design of conformationally constrained cyclic peptides with high delta and mu opioid receptor specificities' V. J. Hruby (1986). In: *Opioid peptides: Medicinal Chemistry. NIDA Research Monograph* **69** (R. S. Rapaka, G. Barnett and R. L. Hawks, eds) pp. 128–147. US Government Printing Office, Washington D.C.

76. 'Effects of changes in the structure of enkephalins and of narcotic analgesic drugs on their interactions with μ and δ-receptors' H. W. Kosterlitz, J. A. H. Lord, S. J. Paterson and A. A. Waterfield (1980) *Br. J. Pharmacol.*, **68**, 333–342.

77. 'Conformationally constrained cyclic enkephalin analogs with pronounced delta opioid receptor agonist selectivity' H. I. Mosberg, R. Hurst, V. J. Hruby, J. J. Galligan, T. F. Burks, K. Gee and H. I. Yamamura (1983) *Life Sci.*, **32**, 2565–2569.

78. 'Site-directed alkylation of multiple opioid receptors I. Binding selectivity' I. F. James and A. Goldstein (1984) *Mol. Pharmacol.*, **35**, 337–342.

79. 'Cyclic penicillamine containing enkephalin analogs display profound delta receptor selectivities' H. I. Mosberg, R. Hurst, V. J. Hruby, K. Gee, K. Akiyama, H. I. Yamamura, J. J. Galligan and T. F. Burks (1983) *Life Sci.*, **33** (suppl. 1), 447–450.

80. 'Bis-penicillamine enkephalins possess highly improved specificity toward δ opioid receptors' H. I. Mosberg, R. Hurst, V. J. Hruby, K. Gee, H. I. Yamamura, J. J. Galligan and T. F. Burks (1983) *Proc. Natl. Acad. Sci. USA*, **80**, 5871–5874.

81. 'Characterization of $[^3H][2$-D-penicillamine, 5-D-penicillamine]-enkephalin binding to δ opiate receptors in the rat brain and neuroblastoma-glioma hybrid cell line (NG 108–15)' K. Akiyama, K. W. Gee, H. I. Mosberg, V. J. Hruby and H. I. Yamamura (1985) *Proc. Natl. Acad. Sci. USA*, **82**, 2543–2547.

82. 'SMS 201–995: A very potent and selective octapeptide analogue of somatostatin with prolonged action' W. Bauer, U. Briner, W. Doepfner, R. Haller, R. Huguenin, P. Marbach, T. J. Petcher and J. Pless (1982) *Life Sci.*, **31**, 1133–1140.

83. 'Opiate antagonist properties of an octapeptide somatostatin analogue' R. Maurer, B. H. Gaehwiler, H. H. Büescher, R. C. Hill and D. Röemer (1982) *Proc. Natl. Acad. Sci. USA*, **79**, 4815–4817.

84. 'Somatostatin receptor binding in rat cerebral cortex: Characterization using a non-

reducible somatostatin analogue' A. Czernik and B. Petrack (1983) *J. Biol. Chem.*, **258**, 5525–5530.

85. 'Conformationally restricted analogs of somatostatin with high μ-opiate receptor specificity' J. T. Pelton, K. Gulya, V. J. Hruby, S. P. Duckles and H. I. Yamamura (1985) *Proc. Natl. Acad. Sci. USA*, **82**, 236–239.

86. '*Mu* opioid antagonist properties of a cyclic somatostatin octapeptide *in vivo*: Identification of *mu* receptor-related functions' J. E. Shook, J. T. Pelton, P. K. Lemcke, F. Porreca, V. J. Hruby and T. F. Burks (1987) *J. Pharmacol. Exp. Ther.*, **242**, 1–7.

87. 'Morphine allosterically modulates the binding of [^{3}H]leucine enkephalin to a particulate fraction of rat brain' R. B. Rothman and T. C. Westfall (1982) *Mol. Pharmacol.*, **21**, 538–547.

88. 'Allosteric coupling between morphine and enkephalin receptors *in vitro*' R. B. Rothman and T. C. Westfall (1982) *Mol. Pharmacol.*, **21**, 548–557.

89. 'Design and synthesis of somatostatin analogues with specific topographical properties results in highly potent and specific μ opioid receptor antagonists with greatly reduced binding at somatostatin receptors' W. Kazmierski, W. S. Wire, G. K. Lui, R. Knapp, J. E. Shook, T. F. Burks, H. I. Yamamura and V. J. Hruby (1988) *J. Med. Chem.*, **31**, 2170–2177.

90. '[^{3}H]CTOP, a potent and highly selective peptide for mu opioid receptors in rat brain' K. N. Hawkins, R. J. Knapp, G. K. Lui, K. Gulya, W. Kazmierski, Y.-P. Wan, J. T. Pelton, V. J. Hruby and H. I. Yamamura (1989) *J. Pharmacol. Exp. Ther.*, **248**, 73–80.

91. M. Hanley (1985). In: *Neurotransmitter Receptor Binding*, 2nd edn. (H. I. Yamamura, S. J. Enna and M. J. Kuhar, eds.) pp. 91–102. Raven Press, New York.

92. 'Binding and information transfer in conformationally restricted peptides' V. J. Hruby and M. E. Hadley (1985). In: *Design and Synthesis of Organic Molecules Based on Molecular Recognition* (G. van Binst, ed.) pp. 269–289. Springer-Verlag, Berlin.

93. 'The tachykinins: a family of peptides with a brood of "receptors"' S. H. Buck and E. Burcher (1986) *Trends Pharmacol. Sci.*, **7**, 65–69.

94. 'The cholecystokinin receptor' S. A. Rosenzweig and J. D. Jamieson (1986). In: *The Receptors*, Vol. 4 (P. M. Conn, ed.) pp. 213–251. Academic Press, New York.

95. 'Opioid receptors: Multiplicity and sequelae of ligand receptor interactions' K.-J. Chang (1984). In: *The Receptors*, Vol. 1 (P. M. Conn, ed.) pp. 1–81. Academic Press, New York.

96. 'The molecular basis of opioid receptor function' W. F. Simonds (1988) *Endocrine Rev.*, **9**, 200–212.

97. 'Inositol trisphosphate, a novel second messenger in cellular signal transduction' M. J. Berridge and R. F. Irvine (1984) *Nature*, **312**, 315–321.

98. 'Heterologous expression of excitability proteins: Route to more specific drugs?' H. A. Lester, *Science*, **241**, 1057–1063.

99. 'Bioassays of follicle stimulating hormone' C. Wang (1988) *Endocrine Rev.*, **9**, 374–377.

100. 'Gonadotrophins' W. R. Butt (1979). In: *Hormones in Blood*, Vol. 1, 3rd edn (C. H. Gray and V. H. T. James, eds) pp. 412–471. Academic Press, New York.

101. 'Prolactin' S. Franks (1979). In: *Hormones in Blood*, Vol. 1, 3rd edn (C. H. Gray and V. H. T. James, eds) pp. 280–331. Academic Press, New York.

102. 'Growth hormone measurement-bioassay' A. E. Wilhelmi (1973). In: *Methods in Investigative and Diagnostic Endocrinology* (S. A. Benson and R. S. Yalow, eds.) pp. 296–302. North Holland, Amsterdam.

103. 'Adrenocorticotrophin and lipotrophin' L. H. Rees and P. J. Lowry (1979). In: *Hormones in Blood*, Vol. 1, 3rd edn (C. H. Gray and V. H. T. James, eds) pp. 141–144. Academic Press, New York.

104. 'Bioassay and radioimmunoassay of oxytocin and vasopressin' T. Chard and M. L.

Forsling (1976). In: *Hormones in Human Blood. Detection and Assay* (H. N. Antoniades, ed.) pp. 488–516. Harvard University Press, Cambridge.

105. 'Hypothalamic releasing and inhibiting hormones and factors' A. Arimura and A. V. Schally (1979). In: *Hormones in Blood*, Vol. 1, 3rd edn (C. H. Gray and V. H. T. James, eds) pp. 1–53. Academic Press, New York.
106. '*In vitro* models in the study of structure–activity relationships of narcotic analgesics' H. W. Kosterlitz and A. A. Waterfield (1975) *Ann. Rev. Pharmacol. Toxicol.*, **15**, 29–47.
107. 'Pharmacology of opioids' W. R. Martin (1983) *Pharmacol. Rev.*, **35**, 283–323.
108. 'Insulin bioassay: Isolated rat epididymal adipose tissue' P. M. Beigelman (1976). In: *Hormones in Human Blood: Detection and Assay* (H. N. Antoniades, ed.) pp. 254–257. Harvard University Press, Cambridge.
109. 'Bioassay and immunoassay of plasma glucagon and glucagon-like substances of gastrointestinal origin' A. S. Luyckx and P. J. Lefebure (1976). In: *Hormones in Human Blood: Detection and Assay* (H. N. Antoniades, ed.) pp. 293–324. Harvard University Press, Cambridge.
110. 'Gastro-intestinal hormones II. Gastrin, cholecystokinin, and secretin' G. J. Dockray (1979). In: *Hormones in Blood*, Vol. 2, 3rd edn (C. H. Gray and V. H. T. James, eds) pp. 357–399. Academic Press, New York.

— *Chapter 8* —

Expectations for peptide pharmaceuticals: a computational perspective

A. M. Brass, D. J. Ward and J. Li

I. Introduction

The aim of this chapter is to show how advances in different aspects of computer software and hardware are expected to complement experimental techniques in the design and conformational characterization of novel peptide drugs. The sections of the chapter reflect the types of analyses described in this book. Section 2 describes improved ways of sampling phase space by slight modification of existing molecular dynamics programs. Section 3 describes hardware developments within computing, where parallelism is encouraging new ways of looking at simulation techniques.

Sections 4 and 5 look at different aspects of drug–receptor interactions. The techniques of free-energy perturbation allow us to computationally test the effect of residue changes on some property of the molecule, for example binding. In Section 5, techniques are explained which give greater reality to simulations of molecular interaction; these are improved consideration of coulombic effects, and quantum mechanics inclusion in molecular dynamics simulation.

1.1. COMPUTER MODELLING OF BIOMOLECULES

Progress in the computer modelling of biomolecules is currently being made in three main areas; firstly, new algorithms are being developed which perform the simulations more efficiently. Secondly, novel and more powerful computers are being developed, allowing larger systems to be simulated or existing systems to be simulated for longer. Thirdly, novel techniques are becoming available, allowing new information about molecular systems to be obtained. These areas are linked.

Where molecular modelling simulations are very demanding of computer time, one approach is to use more efficient algorithms, which can provide more information about the system for the same computing power. Two such algorithms are described below; the first, the *hybrid algorithm*, allows biomolecular simulations to explore phase space more efficiently. Another, the P^3M *algorithm*, gives a significant improvement in simulations run for systems with long-range coulombic interactions.

Developments in computer technology are constantly increasing both the speed and the cost effectiveness of computer simulations. Of particular interest is the development of parallel architecture computers which promise to significantly increase the power and flexibility of molecular modelling programs. As supercomputers get faster, larger molecules can be simulated for longer times, and as new and relatively cheap mini-supercomputers become available then it is possible for an individual research group to have a workstation as powerful as most mainframes, dedicated to running molecular modelling simulations. Improvements in computer technology and architecture will directly lead to new information about biological systems.

New techniques are also being developed, which expand the amount of information which can be obtained from molecular modelling simulations. Two such techniques which have come to prominence in the last few years are *free-energy calculations* and *quantum simulations*. Free-energy simulations allow for the calculation of quantities such as entropy, solvation energy, and Gibbs free energy, which are of great importance in understanding many biological processes but which are not accessible from standard MD simulations. Biological molecules are in some respects close to the border of the domain of applicability of classical mechanics, and algorithms are now available which allow for certain parts of proteins to be treated quantum mechanically.

2. The speeding of molecular dynamics structural refinement

The chapters on NMR and X-ray conformational analysis showed a number of similarities and differences in the information which can be derived from such studies. From NMR, we can check on chemical structure, and obtain backbone dihedral angles from coupling constants, and interatomic distances. With interatomic distances, constrained molecular dynamics simulations can be performed. Depending on the size and complexity of the molecule being investigated, the choice of starting conformation could be crucial. The crystal structure, if available, is probably the best choice of starting conformation, certainly for proteins.

The NMR studies reported reveal *solution* behaviour, and therefore comparisons with X-ray structures may provide fundamental insight into conformational behaviour in different environments. The results mentioned in Chapter 2 indicate differences from the crystal structures of proteins mainly in the conformation of side-chain groups at the surface. With X-ray crystal data, structural refinement is usually performed using energy minimization, which is restricted to a minimum local to the starting point. As discussed in Chapter 3, the use of molecular dynamics is preferable to energy minimization, because it can help the search escape from local minima. For both NMR and X-ray analysis, the following algorithm is expected to find application.

2.1. HYBRID ALGORITHM

Many quantities we wish to measure in MD simulations can be expressed as configurational averages (see Chapter 4, section 2.2.6), and any technique which improves the speed of convergence of these averages will lead to more efficient simulation code. The convergence of the configurational averages depends on the efficiency with which phase space is sampled. In a Monte Carlo simulation[1] the system is started at some point in configuration space, and a trial state is constructed by randomly perturbing the initial state. The energy of the trial state is calculated (H_1) and compared with the initial energy (H_0). The quantity $\exp[-(H_1-H_0)/kT]$ is calculated and compared with a random number (χ) chosen from the interval $[0,1]$. If χ is less than $\exp[-(H_1-H_0)/kT]$ then the new state is accepted and used as the new starting point, otherwise it is rejected and a new trial state is constructed. Configurational averages of thermodynamic quantities can be measured over the set of all accepted states. In order for the acceptance rate to be reasonably high (around 50%) the trial state needs to be close to the initial state. If the various states are thought of as following a path through configurational space, then the step length of this path is very small. It takes a great many steps to sample configurational space adequately, i.e. many tens of thousands of Monte Carlo updates must be calculated before the average values calculated over the Monte Carlo run approximate the true configurational averages.

For typical values of the timestep, the configurational distance between different MD simulation steps is larger than the distance between typical Monte Carlo steps. This is because MD simulations use information about the local derivative of the iso-energy surface (the force) to choose the next configuration. However, MD simulations explore phase space in a deterministic rather than in a random fashion. Although this means that MD simulations can take bigger steps through configuration space, it also means that certain regions of configuration space may not be properly explored (particularly if there are harmonic modes in the system). Although the chaotic nature of MD trajectories (see Chapter 4) will ultimately ensure that all of configuration space is properly explored, it could take a very long MD run to ensure its complete sampling.

Ideally, a simulation technique would combine the large step size of molecular dynamics with the random directions of Monte Carlo simulations.[2] One algorithm designed to combine the best properties of stochastic (Monte Carlo) and deterministic (MD) algorithms is the *hybrid* algorithm[3] which has been developed by high-energy physicists to model the behaviour of quarks in a nucleon. The algorithm combines several simulation techniques (Monte Carlo, MD, and Langevin) such that the resultant 'hybrid' is typically an order of magnitude more efficient than any of the component algorithms. The implementation of the algorithm is straightforward and only needs the addition of a few extra lines of code to an existing MD simulation. The algorithm consists of the following stages:

Step 1. Generate a set of random velocities chosen from a Gaussian distribution with a width appropriate to the temperature at which the simulation is to be run – initialize the particle velocities with these numbers (this is the Langevin step). Calculate H_0, the total energy of the simulation.

Step 2. Apply an updating operator $T(\delta)$ to the configuration. $T(\delta)$ must be reversible in the sense

$$T^{-1}(\delta) = T(-\delta)$$

and area-preserving (see Chapter 4)

$$(d\mathbf{x}, d\mathbf{v}) = (d\mathbf{x}', d\mathbf{v}')$$

Step 3. Calculate H_1, the new energy of the system. Calculate $\exp[-(H_1-H_0)/kT]$ and a random number χ chosen from the interval [0,1]. If χ is less than $\exp[-(H_1-H_0)/kT]$ then accept the new configuration, otherwise reject it and construct a new trial state.

Step 4. Steps 1–3 are then repeated until the configurational averages have equilibrated.

The power of the *hybrid* algorithm comes from using a sequence of MD steps as the updating operator T. Using the notation introduced in Chapter 4, section 2.2.2, T is expressed as:

$$T(\delta) = T_v(\delta/2)T_x(\delta)T_v(\delta) \ldots T_x(\delta)T_v(\delta/2)$$

The initial and final half-steps in velocity are to symmetrize $T(\delta)$ so as to ensure that $T^{-1}(\delta) = T(\delta)$. The number of $T_x(\delta)T_v(\delta)$ steps included in $T(\delta)$ is a variable to be chosen by the programmer. The size of the MD timestep used should be chosen such that the acceptance rate of the algorithm is approximately 80%. This means that the size of the timestep used in the hybrid algorithm is typically twice the size of the timestep used in normal MD simulations, and the final accept/reject step ensures that energy is conserved.

It is possible to show that the above algorithm satisfies detailed balance and will therefore tend to the correct probability distribution. Recently, it has been demonstrated that a variant of the hybrid algorithm can be designed which leads to a removal of the finite-step-size error to any desired order.[4] This algorithm may well have applications in the calculation of perturbative free energies (see below).

We have tested the applicability of this algorithm for refining the X-ray structure of bovine pancreatic trypsin inhibitor (PTI). The X-ray structure of this protein lacks the hydrogens. After these have been added, it is necessary to equilibrate the structure at a biological temperature (300 K). This was done in two different ways – firstly via a traditional MD run with temperature rescaling switched on, and secondly via a hybrid simulation. The systems were said to be equilibrated when the size of the fluctuations in the potential energy had reached a constant value. The graphs showing the size of potential energy fluctuations as a function of the number of iterations for both MD and hybrid simulations are shown in Fig. 1. From this graph it can be seen that the hybrid simulation equilibrated at least an order of magnitude faster than did the MD simulation (it was also found that the hybrid simulation of PTI found a lower energy final state than did the MD simulation).

From these preliminary results it appears that the hybrid simulation does indeed search configuration space more efficiently than an MD simulation and that in applications where it is necessary to calculate configurational averages, the hybrid algorithm should be used in preference to traditional MD.

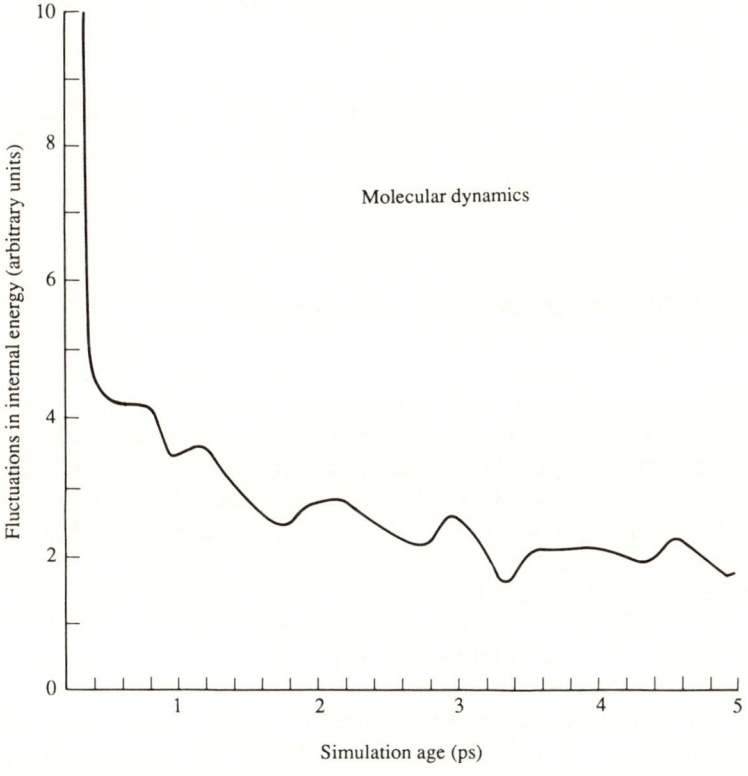

(a)   Simulation age (ps)

Figure 1 The size of the fluctuations in the internal energy of PTI measured as a function of the number of simulation timesteps. It can be seen that the size of the internal energy fluctuations reaches the equilibrium value much faster for (b) the hybrid simulations than for (a) the traditional MD simulations.

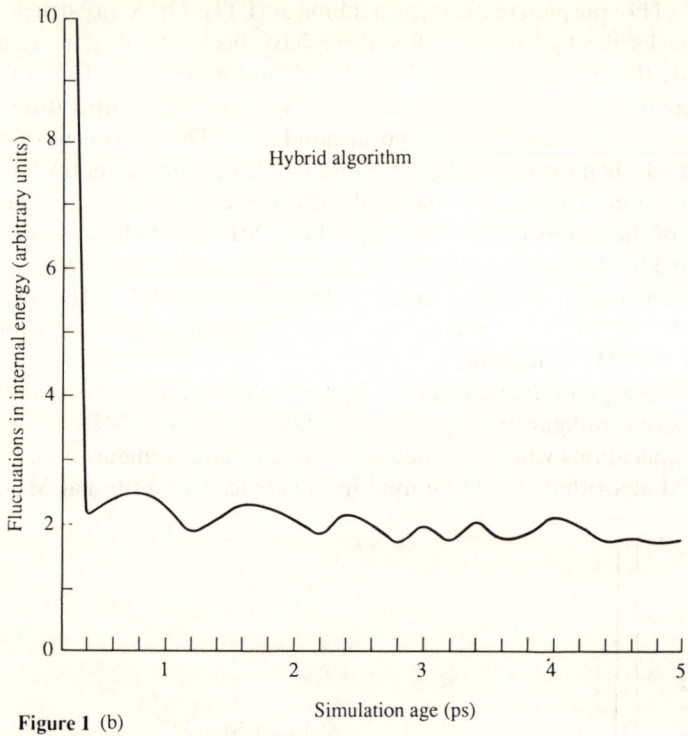

Hybrid algorithm

Figure 1 (b)

Simulation age (ps)

3. The influences of hardware development

The continuing developments in computer architecture and performance are influencing biomolecular modelling in two main ways. Firstly, dedicated graphics workstations, which can be as powerful as minicomputers, have become available at prices which are affordable to individual research groups. These machines (made by Silicon Graphics, Ardent, Convex, Alliant, FPS, etc.) provide several megaflops of computing power, and sometimes excellent graphics capabilities, which can be dedicated to molecular modelling. In practice, this means that it is feasible to run simulations of small, solvated peptides on machines in research laboratories, with real-time graphical display of the results. Even though these machines are not as fast as the true 'supercomputers', for example the CRAY X-MP, the fact that all available machine cycles can be dedicated to the molecular modelling application means that it is possible to run simulations that previously would certainly have required a supercomputer.

Another way in which improvements in computer architecture are influencing molecular modelling is through the increasing speed of the true supercomputers, with the demands of molecular modellers pushing them to their limits. Every increase in supercomputing power increases the range of systems and processes on which the techniques of molecular modelling can be applied. At present, most commercially available supercomputers (e.g. the CRAY series of machines, Cyber-205, Amdahl VP series, etc.) are vector architecture machines.[5] Their main advantage is the ease of programming and the amount of code that has been optimized to run on them.

The main disadvantage of the vector computer design is the lack of potential for radically increasing the speed of calculation. Vector processor computers typically have one processor, and even though the CRAY X-MP can have several processors, it is very difficult for the average user to get access to all of the processors to use within one program. The only way to get an order of magnitude increase in speed in such machines is to redesign the central processing unit radically, and recent history suggests that when such a redesign becomes available it will be very expensive.

Although computers have become considerably more powerful, most have not facilitated, or encouraged, changes in algorithms and algorithm development. An alternative approach is offered by parallel architecture computers.

3.1. PARALLELISM IN MOLECULAR MODELLING

Rather than relying on one expensive high-performance processor, parallel computers use several hundred (or even thousands) of simpler processors to obtain high performance. There are two main designs of parallel computer currently available: Single Instruction Multiple Data (SIMD), and Multiple Instruction Multiple Data (MIMD). In SIMD computers, for example the International Computers Ltd (ICL) Distributed Array Processor[5] (DAP), several thousand processors perform exactly the same instruction but on different data sets. For example, consider a matrix addition operation:

```
DO 1 I = 1,64
DO 1 J = 1,64
    A(I,J) = B(I,J) + C(I,J)
1 CONTINUE
```

On a serial computer this operation would take 4096 separate addition operations. The DAP has 4096 processors arranged as as 64×64 square grid, each processor having communication links with its four nearest neighbours. The elements of the matrices $A(,)$, $B(,)$ and $C(,)$ would be mapped directly onto the processors so that, for example, the data for $A(2,2)$ would be stored on the processor in position (2,2) on the grid. The ICL DAP is an example of a distributed memory computer, that is, rather than the data in the matrices being stored in one central store it is distributed among all the processors. The matrix addition can now be done using a single addition for every processor. For example, the processor in position (2,2) stores the values of A(2,2), B(2,2), and C(2,2) and can perform the calculation $A(2,2) = B(2,2) + C(2,2)$. In an SIMD machine all processors must perform the same operations simultaneously. Although such machines have proved to be very useful in many branches of physics and image processing, their architecture is too rigid for many applications in molecular modelling, where MIMD architectures hold much more promise.

MIMD machines, for example the Edinburgh Concurrent Supercomputer,[6,7] can perform different calculations simultaneously. An MIMD machine would typically be made up of several hundred (or thousands) of processors, each capable of communicating with other processors via a communication topology defined by the user, and each capable of performing different computational tasks. This flexibility greatly increases both the range of problems that can be implemented on these computers as well as the difficulty of writing efficient code for them.

In order to benefit fully from the parallel architecture computers, it is necessary to completely redesign the molecular modelling algorithms. Although in the past it was necessary to rewrite the application code for each different type of parallel computer, recent software developments have made the design of *portable* parallel programs a real possibility. These software tools (e.g. CS-tools) enable the logical topology of the application program to be defined by procedure calls within a conventional high-level FORTAN or C program. These procedure calls distance the programmer from the actual topology of the parallel computer being used, and the programmer need only design an 'ideal' topology for the application program. CS-tools then implement this topology on the parallel computer (e.g. CS-tools running on a MEIKO Computing Surface). The stumbling block for most people is that writing code for parallel machines is *different* from writing code for serial machines.

As an example of designing a program to run on a parallel computer, consider the case of a molecular dynamics simulation of a cluster of N identical particles, where the particle–particle interaction is long-ranged. We will design the algorithm to run on a MEIKO Computing Surface which consists of a number of interconnected INMOS transputers. The transputer is a fast RISC processor (1.5 Mflops) which has four bidirectional communication links which can be connected to other transputers. The connectivity of the transputers can be chosen by the user. For the above example we will connect the transputers in a ring configuration (see Fig. 2). The major part of any MD program is the calculation of the interparticle forces, and in traditional FORTRAN the N-body force calculation would be written as:

```
DO 1 I=1,N
    DO 2 J=1,I-1
        Compute Fij (force on particle I due to particle J)
        Total force on I = Total force on I + Fij
        Total force on J = Total force on J - Fij
2   CONTINUE
1 CONTINUE
```

The inner loop is repeated N times. Naively we could map each repetition of the inner loop onto a different processor, and processor M would compute:

DO 2 J = 1,M–1
    Compute Fij (force on particle I due to particle J)
    Total force on I = Total force on I + Fij
    Total force on J = Total force on J – Fij
2 CONTINUE

The problem with this approach is that the load is not equally balanced among the processors, and processor $M = 2$ will finish long before processor $M = N$. Load balancing is a major consideration in parallel programs, and care should be taken to ensure that although different processors are doing different jobs, some are not waiting for other processors to finish. In the worst case, the program can deadlock: this means that a processor is waiting for an input from another processor which will never arrive.

A more efficient approach to coding the N-body algorithm on a ring of transputers is shown in Fig. 3. In this algorithm, each processor is given the

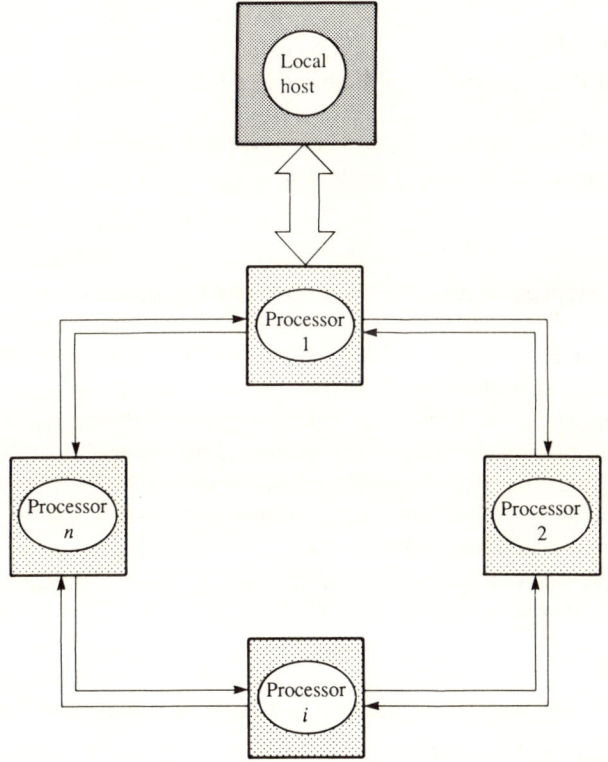

Figure 2   Configuration of n transputers as a ring. The outer ring is used for distributing tasks to each processor, and the inner ring is for transportation of results from processors.

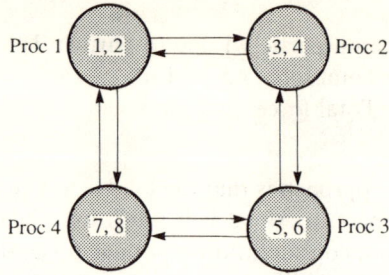

Figure 3 Illustration of an approach to parallel distributive processing. An eight-particle system on a four-transputer ring.

coordinates of a few particles, and the first stage is to calculate the interactions. The next stage is to 'rotate' the ring once, passing the coordinates of the molecules from each processor to its neighbour. The interaction between the 'resident' and 'visiting' particles on each processor can then be calculated, and by rotating the ring several times, depending on how many processors are available, it is possible to calculate all the N-body interactions. Rotating the 'ring' back the other way allows the total force on each particle to be calculated. The above algorithm ensures that all the processors will be active all the time and that the amount of information that need be passed between transputers is minimal. Using the above algorithm it is possible to write code on a MEIKO Computing Surface with 16 transputers which runs more than 30 times faster than in serial mode on a VAX 11/750.[8]

4. Drug–receptor interactions I: free-energy calculations

Developments in gene technology, discussed in Chapter 6, allow the base arrangement in a gene to be modified in a precise manner to give a chosen change in a protein sequence. This means that structure–activity relations are now available for proteins, and therefore receptors, whereas previously SAR was only available for ligands. Recent developments have suggested ways in which not only conformational aspects of mutations can be modelled, but also their effects on the biological actions of the mutant.

In principle, the configurational free energy, F, of a system can be calculated directly from Z, the partition function:[9]

$$F = -kT \ln Z$$

Using elementary statistical physics it is possible to re-express F as:

$$\begin{aligned} F &= kT \ln\langle \exp[+U/kT]\rangle \\ &= kT \ln[1 + \langle U\rangle/kT + \ldots] \\ &= \langle U\rangle - \langle U\rangle^2/(2!kT) + \ldots \end{aligned}$$

where U is the configurational energy of the system.

Although this expression could, in principle, be calculated from the fluctuations in the configurational energy of the simulation, the rate of convergence of the averages of the higher powers of $\langle U \rangle$ would be far too slow to be calculable in any reasonable MD simulation.[10]

Although the absolute free energy of a system is difficult to calculate, the powerful mathematical device of analytic continuation does allow the calculation of the difference in free energy between two states. Consider two states with partition functions Z_0 and Z_1. The difference in free energy between them, ΔF, is defined by:

$$\Delta F = -kT \ln(Z_1/Z_0)$$

Define a coupling parameter, λ, such that as λ varies from 0 to 1, $U(\lambda)$ varies smoothly from U_0 to U_1. This defines an analytic continuation:

$$F(\lambda) = -kT \ln Z(\lambda)$$

Once we have this analytic continuation of F then there are several well-established techniques, e.g. energy perturbation,[11] thermodynamic integration,[12] or potential of mean force calculations[13] which allow differences in free energy to be calculated. For these techniques to work, it is not necessary that the path from state 0 to state 1 is physical. For example, consider the calculation of the difference in free energy between a protein and a mutated version of the same protein in which residue R_0 is changed to residue R_1. If we label the original protein as state $\lambda = 0$ and the mutated protein as state $\lambda = 1$, then an intermediate state has a residue which is a mixture of R_0 and R_1. There is no problem in performing calculating configurational averages of such a protein, even though it does not exist in nature.

Several free-energy calculations can be combined, for example to give information on relative binding affinities for substrates to enzymes in water[14] (see Fig. 4). Using the result $\Delta G_1 - \Delta G_2 = \Delta G_3 - \Delta G_4$, measurements of ΔG_1, ΔG_2, and ΔG_3 give the relative affinity binding, ΔG_4.

The free-energy calculations described above, although providing much useful insight in biological processes, are not foolproof.[15] Free-energy calculations suffer from the multiple minima problem just as much as MD calculations; if the states 1 and 0 are not true ground states then any free-energy measurements are meaningless. Similarly, the values of free energy calculated from MD simulations appear to be very sensitive to round-off errors, the form of the potentials used, cut-off radii, the way in which MD is implemented, etc.

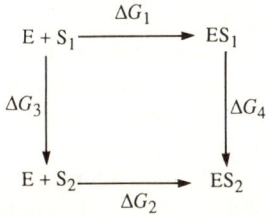

Figure 4 A schematic representation of the thermodynamic cycle used to calculate the relative binding affinity $(\Delta G_1 - \Delta G_2)$.

5. Drug–receptor interactions II: simulation procedures

Although it has been possible for some time to graphically model drug–receptor interaction, this procedure is very subjective, and usually assumes a sterically-allowed 'fit', rather than considering post-binding conformational change. Interaction potentials have also been used, and yet the types of force field used, normally representing the environment as vacuum, suggests important interactions are not being considered adequately, or are being ignored.

In the modelling of a drug–receptor interaction, the most significant interactions are at the receptor/binding site, especially for enzyme–substrate interactions when bonds are being broken. Some reports use an all-atom representation at the binding site, and united-atom for the rest of the molecule. Others have incorporated quantum effects into the simulation, and aspects of this technique are discussed in Section 5.2. First, we consider a way in which coulombic effects can be more adequately treated.

5.1. P³M ALGORITHM

Proteins are charged molecules, so a significant part of their interaction energy arises from the coulombic interaction. In order to model solvated peptides it is necessary to include many hundreds of water molecules in the simulation, and water molecules carry partial charges. Therefore, many of the protein and peptide simulations which we would wish to perform involve the calculation of the coulombic interaction between many hundreds or thousands of molecules. The coulombic interaction is notoriously difficult to handle in computer simulations because this interaction is so long-ranged that every particle will interact electrostatically with every other particle in the system. In small systems with simple charged species this interaction has been calculated using the Ewald sum technique which splits the coulombic calculation into two parts, one in real space and one in reciprocal space. Although this technique works well for a few hundred charged species, the number of arithmetic operations needed for its calculation scales as N^2 where N is the number of charged species in the simulation. The computational cost becomes too high when N gets beyond a few hundred (as it does in the case of most proteins).

The approach taken in most biological simulations is to smoothly let the coulombic interaction go to zero as some predefined cut-off radius.[16] There are two disadvantages to the approach. The first is that it is unphysical and goes against the philosophy of MD calculations (which is that we want to include all the interactions as accurately as we can in order to build the most realistic model of the biomolecule in the computer) and it is difficult to know what errors are introduced into the simulation by the smoothing process. The second problem is that the cut-off radius of the coulombic interaction is still large, and therefore, coulombic interactions are still expensive to calculate.

One way to solve this problem is to use the particle–particle, particle–mesh

(P^3M) algorithm.[17] This algorithm can be viewed as an extreme form of the Ewald summation technique in which *all* the calculation is done in reciprocal space. This algorithm splits the particle–particle force into two parts. F_s, a short-range force calculated using a cut-off radius, and F_r, a long-range smoothly varying force which is calculated using a mesh approximation. The new part of this algorithm is the calculation of F_r on a mesh. A cubic mesh is placed over the simulation cell. The charges on the particles in the simulation are then mapped onto this mesh. Poisson's equations can be solved on the mesh using a Green's function technique. A difference equation can be used to calculate the electric fields at every point on the mesh. The forces on the original particles can be extrapolated from the values of the electric potential on the mesh points. Mathematically this process can be summarized as follows:

Step 1. Charge assignment

$$\rho(\mathbf{x_p}) = q\Sigma W(\mathbf{x_i} - \mathbf{x_p})$$

where $\mathbf{x_p}$ labels a mesh point, $\mathbf{x_i}$ is the position of particle i, W is the charge assignment function and $\rho(\mathbf{x_p})$ is the charge density at mesh point $\mathbf{x_p}$.

Step 2. Solve for the potential on the mesh

$$\varphi(\mathbf{x_p}) = \Sigma G(\mathbf{x_p} - \mathbf{x_p},)\rho(\mathbf{x_p})$$

where G is an optimized Green's function and $\varphi(\mathbf{x_p})$ is the electric potential at $\mathbf{x_p}$.

Step 3. Calculate the electric field

$$E(\mathbf{x_p}) = -D\varphi(\mathbf{x_p})$$

where E is the electric field and D is a finite difference operator.

Step 4. Interpolate from the Mesh Defined Forces to forces on the particles

$$F(\mathbf{x_i}) = q\Sigma W(\mathbf{x_i} - \mathbf{x_p})E(\mathbf{x_p}).$$

This technique has two advantages over the Ewald summation technique. The first is that the operation count for this algorithm goes as $N \ln N$ not N^2, as the convolution in Step 2 can be calculated via a fast Fourier transform technique.[18] The second is that Steps 1–4 can be handled analytically, that is, it is possible to find the functional form for the error introduced by the P^3M approximation and via functional differentiation to choose an optimal Green's function G, such that this error is minimized. Although the P^3M algorithm is more complex to code than a simple coulombic cut-off, it repays the added effort in that it provides both a more efficient and more accurate method of calculating coulombic interactions in MD simulations.

5.2. QUANTUM SIMULATIONS

Experimentally, it is known that there are several phenomena in biomolecular dynamics for which quantum effects are important, for example, the binding reactions of haem proteins[19] and electron transfer reactions.[20] Similarly, zero-point motion effects will contribute to the overall thermodynamics.[9] Quantum simulations of ferrocytochrome c have been carried out.[21] These simulations used the relationship between the Feynman path integral formulation of quantum statistical mechanics and the classical statistical mechanics of polyatomic molecules.[22] This relationship has been extensively used for simulations of simple quantum systems such as electrons in simple fluids.[23] Although such simulations are computationally expensive, it may sometimes be appropriate to treat certain critical residues quantum mechanically in a binding site.

The quantum mechanical partition function for a single particle can be written as:

$$Q(1/kT) = \text{Tr}\{\exp[-H/kT]\}$$

where Tr signifies that the trace of the matrix should be calculated. Using a high-temperature expansion, $Q(1/kT)$ can be rewritten as:

$$Q(1/kT) = \lim_{p \to \infty} Q_p(1/kT)$$

where $Q_p(1/kT)$ is the partition function of the effective classical Hamiltonian, H_{eff}:

$$H_{eff} = \Sigma p_i^2/2m' + \Phi_p(x_1, \ldots, x_p, 1/kT)$$
$$\Phi_p(x_1, \ldots, x_p, 1/kT) = (1/2)k_p\Sigma(x_i - x_{i+1})^2 + (1/p)\Sigma V(x_i)$$
$$k_p = mpk^2T^2/h^2$$

which is the Hamiltonian of p particles (particle i at position x_i) connected sequentially by harmonic springs with spring constant k_p. m' is an arbitrary mass chosen to determine the timestep of the MD sampling. For quantum simulations of proteins a value of $p = 10$ was found to give a good approximation of the quantum limit.[21]

Acknowledgements

Andy Brass was supported by EEC Biotechnology Action Programme (BAP) Contract 0149–UK. David Ward and Jin Li were supported in part by SERC grants GR/D 98754 and GR/E 38177, respectively.

References

1. 'Equation of state calculation by fast computing machine' N. Metropolis, A. W.

Rosenbluth, M. W. Rosenbluth, A. H. Teller and E. Teller (1955) *J. Chem. Phys.*, **21**, 1087–1092.

2. 'Stochastic quantization vs. the microcanonical ensemble: getting the best of both worlds' S. Duane (1985) *Nucl. Phys.*, B, **257**, 652–662.

3. 'Hybrid Monte Carlo' S. Duane, A. D. Kennedy, B. J. Pendleton and D. Roweth (1987) *Phys. Lett.*, B, **195**, 216–222.

4. 'Higher-Order hybrid Monte Carlo algorithms' M. Creutz and A. Gocksch (1989) *Phys. Rev. Lett.*, **63**, 9–12.

5. *Parallel Computers*, 2nd edn, R. W. Hockney and C. Jessope (1988) Adam Hilger, London.

6. 'Physics on parallel computers, part 1: The new technology' K. C. Bowler and R. D. Kenway (1987) *Contemp. Phys.*, **28** (6), 573–598.

7. 'Physics on parallel computers, part 2: Applications' K. C. Bowler and R. D. Kenway (1988) *Contemp. Phys.*, **29** (1), 33–55.

8. 'A study of parallel molecular dynamics algorithms for *N*-body simulations on a transputer system' J. Li, A. Brass, D. J. Ward and B. Robson (1990) *Parallel Computing*, **14**, 211–222.

9. *A Course in Statistical Mechanics* H. L. Freidman (1985) Prentice-Hall, Englewood Cliffs.

10. 'Free energy via molecular simulation: A primer' D. L. Beveridge and F. M. DiCapua (1989). In: *Computer Simulation of Biomolecular Systems: Theoretical and Experimental Applications* (W. F. van Gunsteren and P. K. Weiner, eds) pp. 1–26. ESCOM, Leiden.

11. 'Monte Carlo simulation of differences in free energies of hydration' W. L. Jorgensen and C. Ravimohan (1985) *J. Chem. Phys.*, **83**, 3050–3054.

12. 'Free energy of hydrophobic hydration: A molecular dynamics study of noble gases in water' T. P. Straatsma, H. J. C. Berendsen and J. P. M. Postma (1986) *J. Chem. Phys.*, **85**, 6720–6727.

13. 'A Monte Carlo simulation of the hydrophobic interaction' C. S. Pangali, M. Rao and B. J. Berne (1979) *J. Chem. Phys.*, **71**, 2975–2981.

14. 'Free energy calculation by computer simulation' P. A. Bash, U. C. Singh, R. Langridge and P. A. Kollman (1987) *Science*, **236**, 564–569.

15. 'Free energy perturbation calculations: Problems and pitfalls along the gilded road' D. A. Pearlman and P. A. Kollman (1989). In: *Computer Simulation of Biomolecular Systems: Theoretical and Experimental Applications* (W. F. van Gunsteren and P. K. Weiner, eds) pp. 101–119. ESCOM, Leiden.

16. 'Structural and energetic effects of truncating long ranged interactions in ionic and polar fluids' C. L. Brooks, B. M. Pettitt and M. Karplus (1985) *J. Chem. Phys.*, **83**, 5897–5908.

17. *Computer Simulations using Particles* R. W. Hockney and J. W. Eastwood (1981) McGraw-Hill, New York.

18. *The Fast Fourier Transform* O. E. Brigham (1974) Prentice-Hall, Englewood Cliffs.

19. 'Rate theories and puzzles of hemeprotein kinetics' H. Frauenfelder and P. G. Wolynes (1985) *Science*, **229**, 337–345.

20. *Tunneling in Biological Systems* (B. Chance, ed.) (1979) Academic Press, New York.

21. 'Quantum simulation of ferrocytochrome *c*' C. Zheng, C. F. Wonf, J. A. McCammon and P. G. Wolynes (1988) *Nature*, **334**, 726–728.

22. *Statistical Mechanics* R. P. Feynman (1972) Benjamin, London.

23. 'Excess electrons in simple fluids. II. Numerical results for hard sphere solvent' A. L. Nicholls, D. Chandler, Y. Singh and D. M. Richardson (1984) *J. Chem. Phys.*, **81**, 5109–5116.

Index

active form (conformer), 3, 9, 19, 83–5
  prediction
    for TRH, 106–10
    for enkephalin, 110–13
    for oxytocin, 114–23
agonist, 10, 212, 234–6
  development of, 160–2
  residue substitutions, 111
AIDS, 1, 2
  AZT, 200
  CD4 and HIV-1, 200, 205
  Retrovir, 2
amide bonds, *also see* peptide bonds
  mimics, in design, 154
amino acids
  impurities, 139
  list (Table 1), 4
  L-, D-, 4, 5, 158–9
  in peptide design, 139
    secondary structure bias, 143–4
    topographical bias, 144–9
  residue, 3
  structure, 3
  also see side-chain; backbone
analgesia, 213–16
analogues
  see dermorphin

  see dynorphin analogues
  see enkephalin analogues
  see luteinizing hormone releasing
    hormone
  see oxytocin
  see TRH analogues
antagonist, 10, 212, 234–6
  see luteinising hormone releasing
    hormone
  residue substitutions, 119–23
  see somatostatin
attractor, 11
  also see phase space
avian pancreatic polypeptide (APP), 7

backbone, 5–7, 18–19
  cyclizations
    backbone to backbone, 153–4
    side-chain to backbone, 153
  determination of conformation by
    NMR, 35–41
bacteriophage λ, 189–90
  λgt11, 189–90
beta-endorphin
  cloning and expression, 194–6
bio-active conformation, *see* active form
bioassay, 212–16, 234–6 (Table 1, 214)

analgesia, 215–16
frog skin bioassay, 175–7
guinea pig ileum, 165–72, 229–31, 233
in vitro, 213
in vivo, 213–16
lizard skin bioassay, 175–7
mouse vas deferens, 165–72, 230–1, 233
multiple bioassay system, 157–8, 160
receptor selectivity, 160
techniques, 212–13
biological activity
of LHRH, 173
of opioids, 110, 162
of oxytocin, 114
of TRH, 106–7
Boltzmann's constant, 91
bond length, 4
examples, 7
bound conformer, 49, 85
also see active form
boundary conditions, 90
periodic boundary, 90
Bragg equation, 50, 51
Miller indices, 51
Brookhaven database, 49
BUBU, see enkephalin analogues

Cambridge Structural (Crystallographic)
Database (CSD), 4, 5, 49, 70, 73
CDC28, see protein kinase
cDNA library, 189
CD4, see AIDS
cell cultures, see cell lines
cell lines, 235–6
and second messengers, 236
neuroblastoma-glioma, 231
central dogma (of molecular biology),
185
central nervous system (CNS), 212
CNS-selective drugs, 106
tests for drug potency, 107
chaos, 84, 99, 100
continuous system, 100
criteria for onset, 100–105
Green's method, 105
Lyapunov exponent, 100–103
method of overlapping resonances,
104
power spectra, 103–4

discrete system, 99
universality, 100
chemical formula, 4, 68
chemical modification, 13, 142–3
also see amino acids; backbone; side-
chain conformation
chemical synthesis, see synthetic methods
chimaera, 186
chirality, 4, 52
chromatography
and purity of peptides, 156–7
cis, 7, 144
cloned DNA characterization, 196–200
COLOC, see NMR
configuration, 4
also see backbone; proline; side-chain
conformation
conformation, 4
backbone, 37–41
bound, 49
conformational constraint, 9–13,
19–20, 162–77
conformational restraint, 9–10
environment, 20
in solution, 18–20
low-energy, 48, 49, 107
preferred, 8, 108
selection, 9
side-chain, 41–2
conformation-activity, 13, 85
conformational classes
for TRH conformers, 109
constraints, 20
simulation of H-bonds, 116–20, 123–7
use of constrained ligands, 162–77
also see disulphide
COSY, see NMR
coulombic effects, 254–5
coupling constants, see NMR
calculations on enkephalin, 110
crystal, 50, 56
Bragg equation, 50
Miller indices, 51
packing, 64
refinement, 51, 59–61
space groups, 50
symmetry elements, 50
temperature factor, 51
unit cell, 50

crystal structures (Table 3), 69
crystallization, 52–3
  heavy atom derivatives, 51, 53
  methods, 53
  also see crystal; X-ray crystallography
crystallography, *see* X-ray
    crystallography
cyclosporin, 70, 71

DADLE, *see* enkephalin analogues
DAGO, *see* enkephalin analogues
deamino-oxytocin, *see* oxytocin
delta (δ) receptor, 13
  agonists, 166–71, 228–31
dermorphin
  analogues (μ-agonists), 165–6
dihedral angles, *see* torsion angles
distance geometry, 37–9
  DISGEO, 38
disulphide, 65, 114, 120, 123, 166–71,
    175–6
  constrained by penicillamine, 149–52
  cyclizations
    changing ring size, 150
    side-chain to side-chain, 149
  helicity, 120
  receptor interaction, 149
DNA cloning, *see* gene
DNA-protein interaction, 49
DNA sequence modification, 196–200
dominant negative mutations, 206–7
DPDPE, *see* enkephalin analogues
DPLPE, *see* enkephalin analogues
drug companies, 1, 2
drug delivery, 2, 211
drug design
  experimental aspects, 12–14
  theoretical aspects, 10–12
  also see peptide design; rational design
drug discovery, 8–10
  multidisciplinary, 136–8, 210–11
drug-receptor interaction, 68, 70, 108–9,
    233–6
  enzyme-substrate interaction, 9–10
  free energy calculations, 252–3
  schematic, 106
  signal transduction 3
  simulation procedures, 254–6
  also see radioligand binding

drugs
  classification
    generic, 1
    orphan, 1
DSBULET, *see* enkephalin analogues
DSLET, *see* enkephalin analogues
DTLET, *see* enkephalin analogues
DTLETBU, *see* enkephalin analogues
dynorphin, 112–13, 171–2, 215–16
dynorphin analogues
  U50,488, 233–4
  binding to μ-receptors, 171–2

E. Coli lacz gene, 185–6, 189–90, 192–8
  detection using X-gal, 189–90
  expression of *B*-galactose, 189–90
E.COSY, *see* NMR
ED_{50}, 215–16
efficacy, 212, 234–5
β-endorphin *see* beta-endorphin
energy calculation, 84, 86
energy minimization, 12, 73, 86–7
  on oxytocin, 117
enkephalin, 64–6, 110–13
  analogues, 110–13
  biological activity, 110
  low energy conformer, 112
  residue substitution, 111–13
  structures, 217
  also see opioids
enkephalin analogues
  BUBU, 169–70
  DADLE, 170, 216–17, 230–5
  DAGO, 166, 216–17, 234
  DPDPE, 13, 150, 152, 166–71, 217,
    230–4
  DPLPE, 167–9, 231
  DSBULET, 169–70
  DSLET, 166, 168
  DTBULET, 169–70
  DTLET, 168–70
  DTLETBU, 169–70
  FK-33,824, 216–17
  metakephamid, 216
  also see enkephalin
enzyme inhibitors, 222–4
equilibration
  MD, 91
  equilibrium properties, 92

FK-33,824, *see* enkephalin analogues
F-ratio, *see* radioligand binding
free energy calculations, 10, 252–3
functional groups, 9
    pharmacophore, 9

gauche, 41, 145–9, 165
    also see side-chain conformation
gene
    cloning, 186–96
    isolation and expression, 188–90
        of β-endorphin, 194–6
        of human factor VIII, 190–2
        of insulin, 192–4
    promotors, 192
gene library, 188–9
genetic engineering, 13
global minimum, 68, 91, 127
    also see multiple minima
glycosylation
    in factor VIII production, 191
growth hormone, 2, 13, 210
guinea pig ileum assay, *see* bioassay

Hamiltonian, 11, 90, 91, 99, 104, 256
    and chaos, 99, 104
hardware developments, *see* molecular
        modelling
HIV-1, *see* AIDS
HLA2, 71
homology
    -based modelling, 73
    sequence analysis, 73
    topological equivalence, 73
human factor VIII, 190–2, 199
    glycosylation, 191
    promotor, 192
    role in disorders, 190
hybrid algorithm, 245–58
hydrogen bonds
    simulations on oxytocin, 116–20,
        124–7

IC$_{50}$, *see* radioligand binding
insulin, 2, 13, 61–3, 65–8, 71
    A- and B-chains, 193–4
    cloning and expression, 192–4
    published structure, 67 (Table 2)
    also see colour plates

inter-proton distances, 36–7
isomer, 4

kappa (κ)-receptor, 216, 228–30
    dynorphin analogues, 171–2
    also see dynorphin
Karplus equation, 31, 35, 41
K_d, *see* radioligand binding
K_i, *see* radioligand binding

lacz gene, *see* E. Coli *lacz* gene
leader substance, 8–9
    development of a, 159, 210
leu-enkephalin, *see* enkephalin;
        enkephalin analogues; opioids
LHRH, *see* luteinizing hormone releasing
        hormone
ligand
    ligand-receptor interaction, 9
    also see drug-receptor interaction
luteinizing hormone releasing hormone,
        2, 173–5
    agonists, 173–4
    antagonists, 174–5
lysozyme, 7, 8, 53

M13, 197–8, 202–4
    also see vectors; bacteriophage λ
manifold, 11, 96
    attractor, 11
    iso-energy level, 11
    also see chaos
MEIKO computing surface, 251–2
    also see transputers
α-melanotropin, 150–1, 175–7, 220
    frog skin assay, 175–7
    lizard skin assay, 175–7
metabolic stability, 161–2
met-enkephalin, *see* enkephalin;
        enkephalin analogues; opioids
metkephamid, 216, 217
microcanonical ensemble, *see* molecular
        dynamics
MIMD, *see* Multiple Instructions
        Multiple Data
minimum energy conformer, 11–12
    local minima, 12
    also see global minimum

model
  development of, 159
  of peptide-receptor interaction, 106
modelling procedures, 105–29
  cooling, 12, 123–7
  fitting, 117
  heating, 12, 123–7
  mapping, 17
  residue substitutions, *see* agonists;
    antagonists
  starting conformations, *see* starting
    conformers
  template forcing, 128
molecular dynamics, 12, 19, 22, 30, 73,
    84, 87, 245–8, 253–4
  accuracy, 93
  algorithm, 87
    leapfrog, 89
  chaos, 94–6
  ensembles, 90
  initial conditions, 89
  integrating the equations of motion, 88
  microcanonical ensemble, 91
  restraint molecular dynamics, 39–41
  simulated annealing, 61
  simulations on oxytocin, 123–7
  structural refinement, 244–5
    hybrid algorithm, 245–8
  timestep, 90
molecular graphics, 10, 61, 73
  FRODO, 61
  HYDRA, 61
  QUANTA, 73
molecular mechanics, 10, 86, *also see*
    energy minimization
  SIMPLEX, 86
  GLOBEX, 86
molecular modelling, 10–12, 243–4
  companies, 10
  hardware developments, 248–52
    parallelism, 249–52
    transputers, *see* transputers
  homology-based, 73
  peptide modelling, 105
  derivation of protocols, 105–27
molecular representation, 86
  all-atom, 86
  graphics, 62, 63
  united-atom, 86

Monte Carlo simulations, 245
morphine, *see* opiates
mouse *vas deferens* assay, *see* bioassay
α-MSH, *see* α-melanotropin
mu (μ) receptor, 13, 228–30
  antagonists, 146–8, 162–6
  also see dynorphin analogues;
    somatostatin
Multiple Instructions Multiple Data,
    249–52
multiple minima, 11
mutants, 70, 206–7, *also see* dominant
    negative mutations

neural networks, 10
  also see transputers
neurotransmitter (NT), 1, 137
  NT-neuropeptide coexistence, 106
Newton's equations of motion *see*
    molecular dynamics
NMR, 8, 12, 13, 47, 244
  COLOC, 33
  COSY, 24–5, 31
  coupling constants
    heteronuclear, 31, 33
    homonuclear, 31–2
    J coupling constants, 22, 35
  DISCO, 31
  E.COSY, 2, 31, 32
  generation of 3-D structures, 35–42
  interproton distances, 36–7, 57
  NOE, 19, 22, 25, 26, 30, 33, 36–8, 42, 61
    determination of, 33–5
  NOESY, 25–7, 30, 33–6
  ROE, 26, 33, 34, 36
  ROESY, 25, 30, 33–6
  ROTO, 26–8, 30
  TOCSY, 23–5, 27, 30
nuclear magnetic resonance spectrocopy,
    see NMR
nuclear overhauser enhancement (NOE),
    see NMR
NOESY, *see* NMR

oligonucleotide, 200–5
opiates
  morphine, 13, 215–16
  naloxone, 16, 167, 171–2
opioid analogues

see dermorphin
see dynorphin analogues
see enkephalin analogues
opioids
    biological actions, 162
    see beta-endorphin
    see dermorphin
    see dynorphins
    see enkephalins
    receptors, 13, 162, 228–30
    structures, (Table 2), 217
    also see mu, delta and kappa receptor
oxytocin, 83, 114–27, 152
    analogues
        deamino-oxytocin, 5, (stereo-plot),
            114, 118
        Pen¹-oxytocin, 114
    biological actions, 114
    receptors, 13
    sequence, 114
    simulations
        energy minimization, 114–23
        molecular dynamics, 123–7
    structural formula, 115

pancreatic trypsin inhibitor (PTI), 246–8
parallelism, *see* molecular modelling
penicillamine, *see* disulphide;
            somatostatin; oxytocin; DPDPE;
            α-MSH
peptide bond, 3, 7
    cis/trans, 7
    fixing in simulations, 86
    mimics, 154
    also see amide bonds
peptide design, 10–14, 47–50, 70–3, 83–5,
            135–8, 210–12
peptide hormones
    background 135–8
    functions, 210
peptide minetics, 154–5
    amide bond replacement, 154
peptide precursors, 3, 21
peptide structure, 3–8, 18–20
    primary, 4
    secondary, 4 *also see* secondary
        structure
    tertiary, 4
peptide synthesis, *see* synthetic methods

peptides and proteins
    similarities and differences, 3–8, 20–2
periodic boundary conditions, 90
phage, *see* bacteriophage
pharmocophore, 9, 108–10, 210
    scaffold, 9
    template, 9
phase space, 11–12, 89, 96
    circles, 97
    cylinders, 98
    development of descriptions, 96
    phase portrait, 96
    qualitative theory of dynamics, 96
    sphere, 99
    torus, 11, 97–9
phi, 5–7, 18–19, 33, 143
plasmid, *see* vector
polypeptide, 3, 48
potency, 160
    prediction, 107
potential energy surface, 118
    also see phase space
preferred conformations, 8
primary structure, 4
proline
    endo-configuration, 108
    exo-configuration, 108
prolonged bioactivity, 161–2
promotors
    optimizing gene expressions, 192
protein engineering, 13, 69–73, 205–7
protein folding, 3, 49
protein kinase, 199–200, 206–7
    CDC28, 199–200
protein structure, 3–8, 19–20
proteolytic enzymes and inhibitors,
            (Table 3) 222–4
protocols, *see* modelling procedures
psi, 5–7, 18–19, 33, 144
PTI *see* pancreatic trypsin inhibitor
P³M algorithm, 254–5
purification of peptides
    chromatography, 156
    in synthesis, 155–7
pyroglutamate formation, 7

quantitative structure-activity
            relationships (QSAR), 10, 84
quantum mechanics, 10, 11, 256

radioligand binding, 217–28
  analysis of binding data, 225–6
  conditions, 221, 226–7
  F-ratio, 226
  IC_{50}, 219–20, 225–8, 230–2
  K_d, 218, 220, 226
  K_i, 221, 226–7
  receptor preparation, 219–20
Ramachandran map (φ/ψ plot), 6, 108
  also see modelling procedures
rational design, 9, 83, 127–9, 159
receptor
  environment, 4
  selectivity, 160 also see opioids,
    oxytocin
recombinant DNA, 186–8
  expression of receptors, 235–6
residue substitutions
  agonist, 111
  antagonist, 119–23
resolution, 50–2
restraints, see conformation; constraints;
    molecular dynamics
restriction enzymes, 196
R-factors, see crystal
Ro 09–0198, 38
ROE, see NMR
ROESY, see NMR
root mean square (RMS) deviation
  oxytoxin, 124–7
ROTO, see NMR

second messengers, 235–6
secondary structure, 4, 19, 48, 84, 143
  alpha helix, 6, 19, 84
  beta-turn, 64, 119 173–4
  beta-sheet, 19, 65, 84
  gamma-turn, 119
  mimics, 154
  prediction, 6, 84
  random coil, 20
sequence-structure relationships (SSR),
    85
side-chain conformation, 119
  chemical synthesis
    constraints, 144–9
    side-chain to backbone, 153
    side-chain to side-chain, 149–52
  conformational optimization see
    modelling procedures

determination by NMR, 41–2
  stereochemistry, 145–9
  topography and ligand specificity, 145–
    52
  also see gauche, trans
signal transduction, 3
SIMD, see Single Instruction Multiple
    Data
simulated annealing, see molecular
    dynamics
Single Instruction Multiple Data
    (SIMD), 249
single isomorphous replacement (SIR),
    57
site-directed mutagenesis, 2, 188, 201–5
SMS–201–995, see somatostatin
solid phase synthesis, 140–2
  also see synthetic methods
solution conformers, 8
solvent, 21
  C_6D_6, 21
  $CDCl_3$, 21
  DMFA, 64
  DMSO, 21, 114
  SDS, 21
solvent modelling
  periodic boundary conditions, 90
  reaction field, 108
somatostatin, 150
  analogues
    CGP 23,996, 232
    CTAP, 163–5
    CTOP, 163–4
    CTP, 163–5, 232–4
    SMS–201–995, 163, 210, 232
    structure, 214
  development of μ-receptor antagonists,
    162–6
starting conformers
  oxytocin, 115
  TRH, 108
stereochemistry
  side-chain, 145–9
structure-activity relationships (SAR), 10,
    84, 157–9
  for TRH, 108
structure refinement, 59, 60
  constrained molecular dynamics, 39–41
  hybrid, 245–8
  least squares, 59, 61

structure solution, 56, 57
   difference map, 57
   Patterson method, 56, 57
synthesis, *see* synthetic methods
synthetic methods
   practical methodology, 155–9
   purity, *see* purification of peptides
   solid phase, 140–2
   solution methods, 139–40

target functions, 11–12
tertiary structure, 4
   calculation by NMR, 35–42
thyrotropin releasing hormone (TRH),
      83, 106–10
   analogues, *see* TRH analogues
   low energy conformers, 107
tissue plasminogen activator (TPA), 13
TOCSY, *see* NMR
topography, *see* side-chain conformation
topology, *see* chaos
torsion angles (Dihedrals), 5–6
   for oxytocin, 15–19, 121–4
   for TRH, 107–8
torus *see* phase space
Trans, 7, 41, 86, 144–9
   also see backbone; side-chain
      conformation
transcription, 185

translation, 185
transputers, 250–2
   MEIKO computing surface, 251–2
   ring topology, 250–2
TRH, *see* thyrotropin releasing hormone
TRH analogues
   CG3509, 107
   CG3703, 107
   (3MeHis)TRH, 106–10
   MK771, 107
   RX74355, 107
   RX77368, 107
Trp repressor, 72
turn, *see* secondary structure

vectors, 186
   M13, 197–8, 202–4
   also see bacteriophage λ
v-src, 205

X-ray crystallography, 12, 13, 47, 244–45
   Bragg equation, 50–51
   data collection, 55–6
   Laue method, 55, 56
   Patterson function, 57
   refinement, 51–2
   resolution, 51–2
   structure solution, 56–61
   also see crystal; crystallization